Bioreactors

Bioreactors

Sustainable Design and Industrial Applications in Mitigation of GHG Emissions

Edited by

LAKHVEER SINGH

Department of Environmental Science, SRM University, Amaravati, India

ABU YOUSUF

Department of Chemical Engineering and Polymer Science, Shahjalal University of Science and Technology, Sylhet, Bangladesh

DURGA MADHAB MAHAPATRA

TERI-Deakin NanoBiotechnology Centre, The Energy and Resources Institute, Gurugram, India

Department of Biological and Ecological Engineering, Oregon State University, Corvallis, OR, United States

ELSEVIER

Elsevier
Radarweg 29, PO Box 211, 1000 AE Amsterdam, Netherlands
The Boulevard, Langford Lane, Kidlington, Oxford OX5 1GB, United Kingdom
50 Hampshire Street, 5th Floor, Cambridge, MA 02139, United States

British Library Cataloguing-in-Publication Data
A catalogue record for this book is available from the British Library

Library of Congress Cataloging-in-Publication Data
A catalog record for this book is available from the Library of Congress

ISBN: 978-0-12-821264-6

For Information on all Elsevier publications
visit our website at https://www.elsevier.com/books-and-journals

Publisher: Brian Romer
Acquisitions Editor: Peter Adamson
Editorial Project Manager: Akeh Andrae
Production Project Manager: Kamesh Ramajogi
Cover Designer: Greg Harris

Typeset by MPS Limited, Chennai, India

Contents

List of contributors

Spyridon Achinas
Faculty of Science and Engineering, University of Groningen, Groningen, The Netherlands

Vasileios Achinas
Institute for Life Science and Technology, Hanze University of Applied Sciences, Groningen, The Netherlands

Mohamad Faizal Ahmad
Centre for Environmental Sustainability and Water Security (IPASA), Research Institute of Sustainable Environment (RISE), School of Civil Engineering, Faculty of Engineering, Universiti Teknologi Malaysia (UTM), Skudai, Johor, Malaysia; School of Civil Engineering, Faculty of Engineering, Universiti Teknologi Malaysia, Skudai, Malaysia

Md. Saiful Alam
Department of Petroleum & Mining Engineering, Shahjalal University of Science and Technology, Sylhet, Bangladesh

Nithiya Arumugam
Department of Engineering & Technology, Razak Faculty of Technology and Informatics, Universiti Teknologi Malaysia, Kuala Lumpur, Malaysia

S. Arun
Department of Biosciences and Bioengineering, Indian Institute of Technology Guwahati, Guwahati, India

Saira Asif
Department of Botany, PMAS Arid Agriculture University, Rawalpindi, Pakistan

Awais Bokhari
Department of Chemical Engineering, Biomass Processing Cluster, HICOE-Centre for Biofuel and Biochemical Research, Institute of Self-Sustainable Building, Universiti Teknologi PETRONAS, Seri Iskandar, Perak, Malaysia; Chemical Engineering Department, Biomass Conversion Research Center (BCRC), COMSATS University Islamabad (CUI), Lahore, Pakistan

Rashmi Chandra
Tecnologico de Monterrey, School of Engineering and Science, Toluca, Mexico

Shreeshivadasan Chelliapan
Department of Engineering & Technology, Razak Faculty of Technology and Informatics, Universiti Teknologi Malaysia, Kuala Lumpur, Malaysia

Lai Fatt Chuah
Malaysia Marine Department Northern Region, Gelugor, Penang, Malaysia

Mohd. Fadhil Md. Din
Centre for Environmental Sustainability and Water Security (IPASA), Research Institute of Sustainable Environment (RISE), School of Civil Engineering, Faculty of Engineering, Universiti Teknologi Malaysia (UTM), Skudai, Johor, Malaysia; School of Civil Engineering, Faculty of Engineering, Universiti Teknologi Malaysia, Skudai, Malaysia

Shirin Shafiei Ebrahimi
School of Education, Universiti Teknologi Malaysia, Johor Bahru, Malaysia

Gerrit Jan Willem Euverink
Faculty of Science and Engineering, University of Groningen, Groningen, The Netherlands

Che Ku Mohammad Faizal
Faculty of Engineering Technology, Universiti Malaysia Pahang, Pahang, Malaysia

Pooja Ghosh
Centre for Rural Development and Technology, Indian Institute of Technology, New Delhi, India

S. Hariharan
Department of Biotechnology, Sri Venkateswara College of Engineering, Sriperumbudur, India

M. Amirul Islam
Faculty of Chemical and Natural Resource Engineering, Universiti Malaysia Pahang, Pahang, Malaysia

P. Jaibiba
Department of Biotechnology, Sri Venkateswara College of Engineering, Sriperumbudur, India

Hesam Kamyab
Department of Engineering & Technology, Razak Faculty of Technology and Informatics, Universiti Teknologi Malaysia, Kuala Lumpur, Malaysia

Ahasanul Karim
Faculty of Engineering Technology, Universiti Malaysia Pahang, Pahang, Malaysia

Shefa Ul Karim
Department of Geological Sciences, Chiang Mai University, Chiang Mai, Thailand

Md. Maksudur Rahman Khan
Faculty of Chemical and Natural Resource Engineering, Universiti Malaysia Pahang, Pahang, Malaysia

Santhana Krishnan
Centre for Environmental Sustainability and Water Security (IPASA), Research Institute of Sustainable Environment (RISE), School of Civil Engineering, Faculty of Engineering, Universiti Teknologi Malaysia (UTM), Skudai, Johor, Malaysia; School of Civil Engineering, Faculty of Engineering, Universiti Teknologi Malaysia, Skudai, Malaysia

Manoj Kumar
Department of Biosciences and Bioengineering, Indian Institute of Technology Guwahati, Guwahati, India

Corey A. Laamanen
Bharti School of Engineering, Laurentian University, Sudbury, ON, Canada

D.M. Mahapatra
Biological & Ecological Engineering, Oregon State University, Corvallis, OR, United States

Leow Zi Yan Michelle
Department of Chemical Engineering, Biomass Processing Cluster, HICOE-Centre for Biofuel and Biochemical Research, Institute of Self-Sustainable Building, Universiti Teknologi PETRONAS, Seri Iskandar, Perak, Malaysia

S. Naga Vignesh
Department of Biotechnology, Sri Venkateswara College of Engineering, Sriperumbudur, India

S.N. Naik
Centre for Rural Development and Technology, Indian Institute of Technology Delhi, New Delhi, India

Mohd Nasrullah
Faculty of Civil Engineering Technology, Universiti Malaysia Pahang, Kuantan, Malaysia

Kannan Pakshirajan
Department of Biosciences and Bioengineering, Indian Institute of Technology Guwahati, Guwahati, India

Surjith Ramasamy
Department of Biosciences and Bioengineering, Indian Institute of Technology Guwahati, Guwahati, India

Shahabaldin Rezania
Centre for Environmental Sustainability and Water Security (IPASA), Research Institute of Sustainable Environment (RISE), School of Civil Engineering, Faculty of Engineering, Universiti Teknologi Malaysia (UTM), Skudai, Johor, Malaysia; School of Civil Engineering, Faculty of Engineering, Universiti Teknologi Malaysia, Skudai, Malaysia

Arunaditya Sahay
Birla Institute of Management Technology, Greater Noida, India

N.K. Sahoo
Centre for Rural Development and Technology, Indian Institute of Technology Delhi, New Delhi, India

Shivali Sahota
Centre for Rural Development and Technology, Indian Institute of Technology, New Delhi, India

S.N. Sahu
Centre for Rural Development and Technology, Indian Institute of Technology Delhi, New Delhi, India

Carlos Eduardo Gámez Sánchez
Tecnologico de Monterrey, School of Engineering and Science, Toluca, Mexico

John A. Scott
Bharti School of Engineering, Laurentian University, Sudbury, ON, Canada

Subhanjan Sengupta
Birla Institute of Management Technology, Greater Noida, India

Goldy Shah
Centre for Rural Development and Technology, Indian Institute of Technology, New Delhi, India

Lakhveer Singh
Faculty of Civil and Environmental Engineering, University Malaysia Pahang, Kuantan, Malaysia

Arindam Sinharoy
Department of Biosciences and Bioengineering, Indian Institute of Technology Guwahati, Guwahati, India

Shazwin Mat Taib
Center for Coastal and Ocean Engineering, Research Institute for Sustainable Environment, Universiti Teknologi Malaysia, Kuala Lumpur, Malaysia

Md. Sifat Tanveer
Department of Petroleum & Mining Engineering, Shahjalal University of Science and Technology, Sylhet, Bangladesh

Huong Trinh
Chemical Engineering Department, Universiti Teknologi PETRONAS, Ipoh, Malaysia

Janet Alejandra Gutiérrez Uribe
Tecnologico de Monterrey, School of Engineering and Science, Puebla, Mexico

Virendra Kumar Vijay
Centre for Rural Development and Technology, Indian Institute of Technology, New Delhi, India

Garima Vishal
Department of Chemical Engineering, Indian Institute of Technology, New Delhi, India

Zularisam Abdul Wahid
Faculty of Civil Engineering Technology, Universiti Malaysia Pahang, Kuantan, Malaysia

Abu Yousuf
Department of Chemical Engineering and Polymer Science, Shahjalal University of Science and Technology, Sylhet, Bangladesh

Suzana Yusup
Chemical Engineering Department, Universiti Teknologi PETRONAS, Ipoh, Malaysia; Department of Chemical Engineering, Biomass Processing Cluster, HICOE-Centre for Biofuel and Biochemical Research, Institute of Self-Sustainable Building, Universiti Teknologi PETRONAS, Seri Iskandar, Perak, Malaysia

Nur Azmira Zainuddin
Centre for Environmental Sustainability and Water Security (IPASA), Research Institute of Sustainable Environment (RISE), School of Civil Engineering, Faculty of Engineering, Universiti Teknologi Malaysia (UTM), Skudai, Johor, Malaysia; School of Civil Engineering, Faculty of Engineering, Universiti Teknologi Malaysia, Skudai, Malaysia

CHAPTER 1

Microalgae biofuel bioreactors for mitigation of industrial CO_2 emissions

Corey A. Laamanen and John A. Scott
Bharti School of Engineering, Laurentian University, Sudbury, ON, Canada

1.1 Introduction

The reliance on fossil fuels continues to be an unsustainable option due to the depletion of reserves and the production of greenhouse gases, in particular carbon dioxide (CO_2), released by their combustion. This has led to a significant interest in renewable, sustainable energy sources, one of which is biofuels.

Biofuel sources have been extensively researched for the past several decades, and their production can be divided into three generations, namely food crops, energy crops, and microalgae [1]. The first generation is based on food crops such as corn being diverted into energy. In light of the food versus fuel debate, the second generation transitioned into the growth of dedicated energy crops such as *Jatropha*. While this marked a step change in areal energy production (from 172 L/ha for corn to 1892 L/ha) [2], arable land was still required to grow nonfood crops. To move away from arable land requirements, aquaculture represents the third generation of biofuels. This includes the cultivation of phototrophic microalgae, which use CO_2 as their carbon source. Using microalgae to produce biodiesel represents another order of magnitude increase in areal energy production (58,700 L/ha for microalgae at 30% oil by weight) [2].

For the production of biodiesel, microalgae cells are selected that accumulate between 15% and 85% lipid content. These lipids (triglycerides) can be directly converted into fatty acid methyl esters (biodiesel) through a transesterification reaction. The triglycerides are reacted with methanol in the presence of a catalyst (typically sulfuric acid) to produce glycerol and biodiesel. Transesterification can be shown as:

Bioreactors
DOI: https://doi.org/10.1016/B978-0-12-821264-6.00001-2

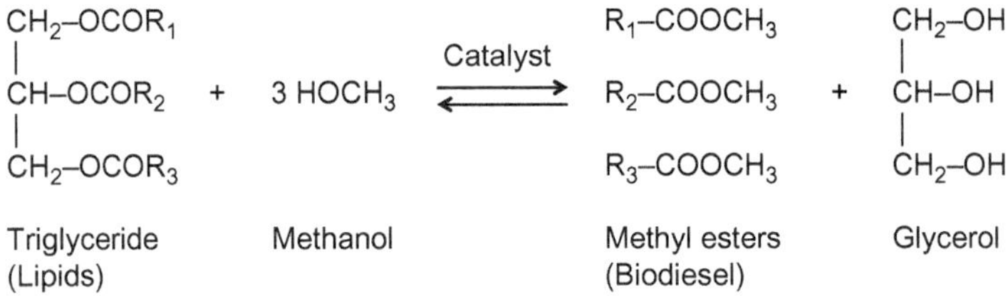

(1.1)

Phototrophic microalgae, in addition to not requiring arable land, grow faster and are able to capture more CO_2 per unit area than terrestrial plants [3]. The capture of anthropogenic CO_2 can be further enhanced by passing through aqueous medium industrial off-gases that have an elevated CO_2 content. That is, industrial CO_2 can be mitigated by microalgae using it as their carbon source [4]. As it has been shown, microalgal biomass is approximately 50% carbon [3], which translates to:

$$\left(0.5\frac{\text{g C}}{\text{g algae}}\right)\left(\frac{44\text{ g CO}_2/\text{mol}}{12\text{ g C/mol}}\right)=\frac{11\text{ g CO}_2}{6\text{ g algae}}=1.83\frac{\text{g CO}_2}{\text{g algae}} \tag{1.2}$$

That is, for every gram (dry weight) of microalgae, 1.83 g of CO_2 is taken up. The concept of carbon capture using microalgae and subsequent biofuel (biodiesel) production is shown in Fig. 1.1.

1.2 Microalgae

Microalgae can be biologically classified as single-celled plants that accumulate different products as a response to their environmental conditions, which is central to biodiesel production. Through the control of the process parameters and the selection of microalgae strains, the outcome can be aimed toward the production of lipids, which are cellular energy storage compounds [5]. Microalgal cells can accumulate lipids to up to 85% of total cell dry weight [6].

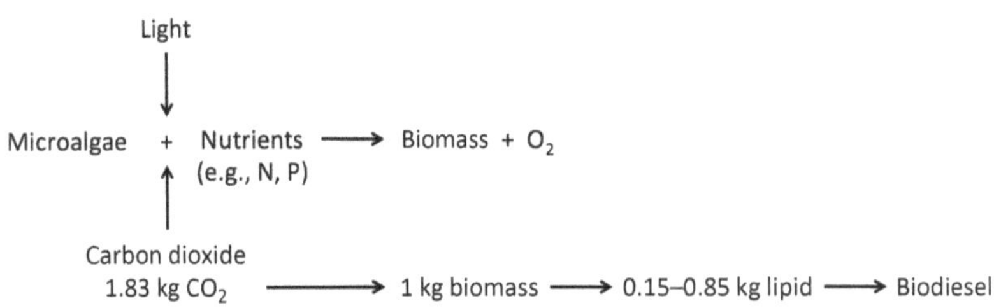

Figure 1.1 Carbon dioxide (CO_2) capture and biofuel production using microalgae.

To be able to best utilize different metabolic pathways a two-stage cultivation strategy is often utilized, the photosynthetic production of biomass and the accumulation of lipids. In the first stage, ideal growth conditions are provided to allow for the production of microalgal biomass. For any microalgal strain, these ideal conditions will be related to pH, temperature, and CO_2 and nutrient levels. A second stage will then be used to apply stress by changing one or more of these conditions, thereby triggering the accumulation of lipids [5].

Another benefit of microalgae cultivation for the production of biofuels is that the biomass remaining after lipid extraction can also be of significant value. This biomass has been examined for human dietary benefits, a protein source for fish farms or livestock, cosmetics, fertilizers, pharmaceuticals, and nutraceuticals [7–11]. While biofuel is a relatively low-value bulk product, additional high value coproducts can provide the economic benefit to make the entire operation profitable.

1.3 Microalgae growth parameters

Microalgae growth depends on a number of different parameters for the operation including the availability of nutrients (in particular the macronutrients nitrogen and phosphorus), the concentration of dissolved CO_2, pH, light intensity and photoperiod, and temperature. The typical ranges for operations are 15°C–30°C, a pH value between 4 and 11, and light intensities of 1000–10,000 lux [12]. These can be modified to promote biomass production, or to move away from ideal growth conditions in order to trigger other pathways that lead to the accumulation of lipids for biodiesel production [13].

1.3.1 Light

Microalgae use light as their source of energy, and both the intensity and duration (photoperiod) must be optimized. If the intensity is above the saturation limit (around 6500 lux) [14], then photoinhibition can occur, which reduces the efficiency of the system [15]. Wahidin et al. [14] found that for a *Nannochloropsis* sp., both the light intensity and photoperiod needed to be optimized, rather than maximized, to achieve the best growth rate.

Microalgae can only utilize wavelengths between 400 and 700 nm, which accounts for about 50% of sunlight. This limits their potential photosynthetic efficiency, which is further reduced through reflection and cellular respiration [15]. On a cellular basis, the maximum photoconversion efficiency that light is used for photosynthesis and biomass production is approximately 9% [16]. This represents, however, a significant increase compared to terrestrial plants, which have maximums in the 4.6%–6.0% range [16].

Light delivery, intensity, spectra, photoperiods, the frequency of light/dark cycles, and the amount of light exposed surface area have all been reported to have a significant impact on microalgal biomass formation. Light characteristics have also been shown to influence the lipid content of microalgal cells, as well as the production of carbohydrates, proteins, and other cellular components [17].

To provide light, it is ideal to utilize only sunlight, as this reduces the capital, maintenance, and energy costs associated with artificial lighting systems. This, however, limits the depth or length of the light path for systems, due to microalgal self-shading as biomass concentrations increase [18,19].

With respect to artificial lighting, assuming a culture is lit over a 3-week cycle for 18 h a day [14], then for every 100 W of light supplied and with a 100% efficient light bulb, the amount of energy used is:

$$\left(100\frac{\mathrm{J}}{\mathrm{s}}\right)\left(3\frac{\mathrm{week}}{\mathrm{cycle}}\right)\left(7\frac{\mathrm{day}}{\mathrm{week}}\right)\left(18\frac{\mathrm{h}}{\mathrm{day}}\right)\left(3600\frac{\mathrm{s}}{\mathrm{h}}\right)\left(10^{-6}\frac{\mathrm{MJ}}{\mathrm{J}}\right)=136\mathrm{MJ} \tag{1.3}$$

Based on the above energy requirement, with a biomass composed of 25% lipid, a biodiesel energy density of 37.8 MJ/kg, and a 90% conversion rate of lipids to biodiesel [20], the amount of biomass (dry weight) required to break even would be:

$$\frac{136\ \mathrm{MJ}}{(37.8\ \mathrm{MJ/kg})(0.25)(0.9)}=16\ \mathrm{kg} \tag{1.4}$$

That is, over a 3-week cycle, the addition of 100 W of lighting would require an additional 16 kg of microalgae to break even on an energy balance.

The above example highlights a major issue with traditional indoor and artificially lit bioreactors for microalgae. By moving a production system outdoors, whilst losing the ability to exactly control temperature, by utilizing sunlight and avoiding heating costs, overall electricity consumption can be reduced by 90% [20]. To overcome self-shading when utilizing only sunlight, microalgal cultures can be continually vertically circulated, resulting in individual microalgae cells cycling between light and dark areas [21].

In moves toward utilizing sunlight exclusively in bioreactors, work is being done on increasing the efficiency of their utilization. This largely involves different capture and distribution mechanisms such as fiber optics or light guides [17]. Sun et al. [22] used 3 cm long hollow horizontal tubes on a flat plate reactor, which led to a 23% increase in biomass productivity. More complex designs have included the use of lenses to focus sunlight into a distribution system [23]. However, as complexity increases so do capital and maintenance costs. Furthermore, any additional components added to a bioreactor, especially for the distribution of light, run the risk of accumulating surface biofilms that reduce efficiency. The design of light transmitting devices must, therefore, consider preventing biofouling to avoid frequent maintenance [9].

An approach for avoiding the accumulation of biofilms is the use of simple designs such as light tubes [17]. Light tubes are vertical tubes containing water that are inserted into the culture solution. As the interior of the tube does not contain microalgae, it provides a path for sunlight to pass into the culture medium. By providing light deeper into the system, the use of these tubes has resulted in a 33.6% increase in biomass production [17]. They are also ideally located above gas spargers that bubble in off-gas CO_2, with rising gas bubbles "scrubbing" the light tube surfaces.

To further increase their effectiveness, light tubes could also be combined with other sunlight focusing systems such as the lenses described in Ref. [23]. Low-cost solutions for improving the utilization of sunlight, and thereby increasing bioreactor performance, hold significant promise for more economic production of algal biofuel and other products.

1.3.2 Nutrients

A known limitation in the large-scale production of biofuels is a lack of sustainable sources of low-cost nutrients, but relatively little research has been done regarding this issue.

At the laboratory scale, most experiments give nutrients in excess to promote microalgae growth, reducing them only to induce stress in order to trigger lipid accumulation. Seyed et al. [13] did not, however, find a significant effect from varying media concentrations. This is in accordance with results from [24], which showed, with the exception of potassium phosphate, no significant changes when diluting 3N-Bold Basal medium sixfold. This suggests that projections for nutrient cost for large-scale production could be too high, and if too many nutrients are provided, the economics of the operation could be negatively impacted.

To economically provide microalgal macronutrients, namely nitrogen and phosphorus, a widely proposed solution is to adapt current microalgal systems used to treat municipal or industrial wastewaters to remove these nutrients. The combination of wastewater nutrient removal and CO_2 capture from industrial off-gases is a promising approach, therefore, for microalgal biofuel production. This route could potentially reduce microalgal production costs, as it has been estimated that nutrients, water, and CO_2 contribute to 10%–30% of the total production costs of commercial microalgae cultivation [25].

1.3.3 Carbon dioxide

Microalgae are photosynthetic microorganisms that utilize light energy to capture carbon from CO_2 and release oxygen as a by-product. This oxygen by-product is important in the design of bioreactors, as high concentrations of dissolved oxygen (DO) can trigger photorespiration (where oxygen is consumed and CO_2 is released), and high DO concentrations can be toxic to microalgae [15].

Carbon capture by microalgae relies on atmospheric CO_2 (approximately 380 ppmv) dissolving into water, but this can be enhanced through the use of industrial off-gas. The use of atmospheric CO_2 alone limits biomass productivity due to its low concentration and the high surface tension of water [26]. Therefore the direct application of industrial off-gas containing elevated levels of CO_2 is a promising strategy to promote greater gas mass transfer into the culture medium and improved biomass productivity. However, to achieve the necessary mass transfer into the liquid, an appropriate surface area and residence time must be provided.

To enhance the surface area for mass transfer, the gas is commonly sparged into the medium to produce fine bubbles. For a given gas flow rate, the smaller the bubbles the greater the surface area per unit gas bubble volume. To achieve this goal, work has been done on fluidic oscillation to decrease the bubble sizes out of a sparger, thereby increasing mass transfer rates [26].

The gas residence time is a function of the length of the bubble path through the culture medium. For existing commercial open pond systems, this relates to the depth, which is typically limited due to self-shading to

15–35 cm [27]. This means that the bubble residence will be short and the mass transfer will be low. This results in an 80%–90% loss of CO_2 to the atmosphere from open pond systems [28].

There are several industrial processes that emit significant amounts of CO_2 including natural gas combustion (9% CO_2 [29]), fossil fuel power plants (5%–15% [30,31]), steel and iron production (30% [29]), cement production (15%–25% [32,33]), smelter furnaces (6%–7% [4]), and waste incineration (11% [34]).

These off-gasses can be utilized as a carbon source for microalgae growth, but they can also contain components that are not necessarily beneficial to microalgae. These include acid gasses such as sulfur dioxide (SO_2) and nitrogen oxides (NO_x), which reduce the pH of culture media. Despite this, many reported experiments have been carried out using simulated off-gas, which is only air mixed with CO_2. Whilst low concentrations, in the range of 100 ppm NO_x and SO_2, may not affect microalgae growth [34,35], it is nevertheless important to consider them and how they may affect microalgal cultures.

1.3.4 Temperature

The temperature ranges for the majority of microalgae strains used for biodiesel production is 15°C–30°C [36], which has limited outdoor production in tropic and subtropic areas worldwide. Warm regions are often ones in which freshwater is limited, and as a consequence microalgae production may be restricted to salt (sea) water species. However, the majority of worldwide regions fall below 15°C for a portion of the year.

In hot climates, it has been shown that outdoor bioreactor temperatures can reach 10°C–30°C above the ambient temperature [15]. To mitigate any loss of productivity, there is need for cooling such as is achieved by submerging the entire bioreactor in a water pool, spraying the bioreactor surface with water, shading the surface, or using a heat exchanger. Using water to cool bioreactor systems adds significant additional pumping costs, whilst shading decreases the amount of solar energy available for biomass production [15]. In cold climates, the reverse issue is present as temperatures fall below the optimal range, thereby requiring heating. Temperature issues can be addressed by moving operations indoors, but building size can significantly reduce the production volume, and increased operational costs can arise from factors such as artificial lighting.

Many industries that produce off-gasses do, however, produce large amounts of waste heat. It has been proposed for a smelter, that an integrated system could use CO_2 in the furnace off-gas [37], and waste heat from the roaster off-gas [4,38]. The capture of waste heat provides a potential solution to expanding outdoor cultivation into cold climates.

1.4 Microalgae cultivation systems for CO_2 capture

Current large-scale operations mainly utilize shallow (30–50 cm) open pond systems due to their relatively low cost. Studies have shown that both temperature and contamination have a significant influence over the productivity of open pond systems [39]. These systems also provide limited control in the maintenance of microalgae species (e.g., avoidance of invasive species) and produce low final biomass concentrations [16], which makes downstream processing (the separation of biomass from the liquid medium) expensive. Furthermore, due to the short residence time of any sparged-in industrial off-gas, CO_2 adsorption will be low. Therefore other types of bioreactors have been examined for the capture of CO_2 emissions and the subsequent production of microalgal biofuel.

Singh and Sharma [9] listed several requirements for a microalgal bioreactor that improves on the performance of shallow open ponds:

- minimized contamination to allow for a single strain culture;
- increased control over process conditions (e.g., pH, light, temperature, and CO_2 concentration);
- reduced CO_2 losses;
- reduced water evaporative losses;
- increased microalgal cell concentrations; and
- allowance for the production of biopharmaceuticals and other specialty chemicals.

The major trade-off is between improved conditions and the resulting increase in operational costs due largely to the increased complexity leading to greater construction and maintenance costs, as well as auxiliary energy needs [16]. These costs can be reduced through the optimization of bioreactor geometry, orientation, and the utilization of available resources such as waste heat.

In consideration of bioreactor design, several areas should be optimized to achieve maximized biomass and biofuel production including [15]:

- harvesting as much sunlight as possible and efficiently distributing it to allow biomass formation;

- convenient and precise control of operational parameters;
- minimizing capital and operational costs; and
- minimizing energy consumption.

Numerous types of bioreactors have been developed including vertical column (airlift and bubble column), flat plate, and horizontal tubular bioreactors. However, no single type of bioreactor satisfies all of the necessary requirements of an ideal bioreactor [25].

One approach is to employ a hybrid system, that is, a system that has the characteristics of both an open pond and a bioreactor. The simplest of these is an open pond that is covered to provide a contained gaseous space above the culture surface. This allows for the increased capture (adsorption) of CO_2, the reduction of contamination, and an increased ability to control culture conditions [25]. Another example is the use of deeper ponds [21], thereby increasing the off-gas residence time, which can be achieved using a gas-lift system to promote culture circulation.

Bioreactors can be run in batch (discontinuous), fed-batch, semicontinuous, or continuous modes. Most common, due to the time requirements for growing microalgae to a desired biomass level, are batch or fed-batch systems. These allow for a simple operation, but are not necessarily fully efficient. Whereas a continuous system, in which some of the culture is continuously bled-off and processed, provides many benefits. These include a constant product stream, easy automation, and the ability to keep the system at its most productive. Continuous operation can allow for better volumetric productivities over time, but as most are the result of laboratory studies, significant work still needs to be done to translate research outcomes into full-scale operations [12].

For any CO_2 capture system, the configuration and operational mode of the bioreactor fundamentally determine the system's performance. These factors contribute to the operational costs of the reactor from energy consumption for any applied lighting to achieving mixing and gas sparging. This auxiliary energy must be considered in the design as it directly affects the net energy return of the operation. For bubble plates and airlift reactors, the energy inputs are in the range of 50–70 W/m^3, while for a horizontal tubular reactor it can be nearly 2000 W/m^3 [16].

The direct application of off-gas (bubbling in) to the culture medium in a vertical column is likely to provide the best capture rate of CO_2. However, the relatively small top surface-to-volume ratio reduces overall solar energy capture, thereby limiting maximum potential growth rates (albeit reducing the risk of photoinhibition). It is important, therefore, to obtain mixing from the bubbled in off-gas in order to provide an even distribution of microalgae and to promote their exposure to light through circulating them from the bottom to the top. Providing light–dark exposure cycles can improve photosynthetic efficiency [13,25]. Furthermore, off-gas bubbles can help remove oxygen from the medium produced during photosynthesis and to avoid accumulation issues [9].

Column bioreactors can be divided into two similar operations, namely bubble and gas-lift columns (Fig. 1.2).

In both operations, industrial off-gas could be injected into the bottom of the column and allowed to bubble up through the microalgal medium. This provides an optimum residence time for CO_2 mass transfer and varying degrees of mixing. The main difference between the two types is that with a gas-lift, the gas bubbles ascend through a riser (a vertical tube above the gas sparger), some of which can then be entrained in the flow back down the downcomer (Fig. 1.2). Gas-lift systems, therefore, provide more effective solution mixing and mass transfer [7,13]. Bubble rising and circulation in both types of column bioreactor need to be gentle enough that the shear created causes no cell damage [15]. A table of results from various column bioreactors is given in Table 1.1.

Zimmerman et al. [26] used fluidic oscillation to make smaller bubbles in an airlift column operation connected directly to steel plant exhaust gas (~20% CO_2 and unmeasured quantities of NO_x and SO_x). The results showed a carbon capture rate for Dunaliella *salina* of 0.1 g/L/day, with no

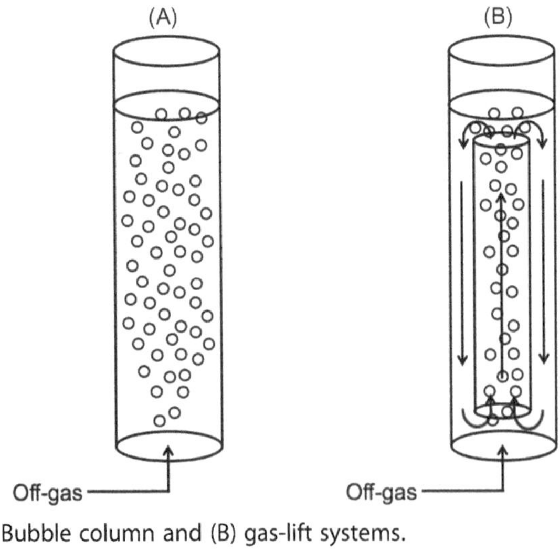

Figure 1.2 (A) Bubble column and (B) gas-lift systems.

Table 1.1 Column bioreactor results used for CO_2 capture.

Microalgae strain	Reactor	CO_2 source	CO_2 content (%)	Capture rate (g/L/day)	Reference
Dunaliella salina (culture collection)	Airlift	Steel plant off-gas	20	0.1	[26]
Spirulina sp. (culture collection)	Bubble column	Synthetic	12	0.27	[40]
Scenedesmus obliquus (bioprospected)	Bubble column	Synthetic	12	0.22	[40]
Chlorella sp. MTF-15 (bioprospected)	Bubble column	Steel plant off-gas	24–26	0.942 (calculated)	[41]
Chlorella sp. MTF-15 (bioprospected)	Bubble column	Steel plant off-gas	12–13 (50% diluted)	0.966 (calculated)	[41]
Scenedesmus sp. (culture collection)	Airlift	Synthetic	6	0.110 (calculated)	[21]
Chlorella sp. (culture collection)	Bubble column	Synthetic	10	0.249	[42]
Chlorella sp. (culture collection)	Bubble column	Coke oven	10–15	0.175	[42]

inhibition due to the presence of NO_x or SO_2. It should be noted, however, that the use of bubble columns, especially with small bubbles, whilst increasing surface area for carbon capture, can limit light capture effectiveness due to bubble reflection [43].

Different sources of off-gas from steel plants (24%–26%) were examined by Kao et al. [41]. Coke oven off-gas showed particularly promising results. Using *Chlorella* sp. MTF-15 in a bubble column, a capture rate of 0.942 g/L/day was calculated. The capture rate increased to 0.966 g/L/day when using half concentration coke oven flue gas, which was in line with the more commonly tested CO_2 percentage used for microalgae cultivation.

Morais et al. [40] utilized a three-stage sequential bubble column using synthetic coal combustion off-gas (12% CO_2) to increase carbon capture efficiency. The results showed a capture rate of 0.27 g/L/day for a species of *Spirulina* sp. and 0.22 g/L/day using a strain of *Spirulina obliquus* that was isolated from near a coal power station. Interestingly, the isolated *S. obliquus* strain, when exposed to NO and SO_2 (up to 100 ppm and 60 ppm, respectively), showed no significant effect regarding carbon capture.

Lee et al. [44], using 2%–4% CO_2 and four bubble columns in series with *Haematococcus pluvialis*, saw a 3.63-fold increase in capture efficiency (up to 49.37%) as compared to a single column. Cheng et al. [45] used a 14-stage sequential bubble column system, and with a synthetic off-gas (15% CO_2), obtained a capture efficiency of 85.6%. However, the experimental results may be of limited value in terms of use in scaling-up to a full-sized system, as they were obtained using columns with a working volume of only 300 mL.

Further experiments on the impact of potentially hindering components of industrial off-gas were carried out by Duarte et al. [46]. Using a bubble column fed with simulated coal-fired power station off-gas (10% CO_2, NO_x, SO_2, and ash), cultures of *Chlorella fusca* LEB 111 (isolated from a thermoelectric plant) showed significant resistance. Increasing the concentrations of both NO_x and SO_2 to 200 ppm and the concentration of ash to 40 ppm did not show any negative impacts on growth.

Yadav et al. [42] tested both a synthetic off-gas with 10% CO_2, and coal-fired flue gas with 10%–15% CO_2. The results showed a capture rate of 0.249 and 0.175 g/L/day for the synthetic and industrial gases, respectively. The strain of *Chlorella* sp. used was not, however, isolated from the region and had not, therefore, been acclimatized to the elevated CO_2 levels or acid gas components.

Most experimental systems are operated with continuous exposure to CO_2-enhanced gases, but some have used intermittent exposure to try and increase the capture percentage (i.e., the solution would not be continuous CO_2 saturation). Chiu et al. [47] tested a gas-lift column with a porous riser using a synthetic off-gas (10% CO_2). Semicontinuous harvesting was employed in which a portion of the culture was bled-off and its volume replenished. When 25% of the volume was bled-off every two days, the maximum carbon capture rate of 1.098 g/L/day was considerably higher than that achieved by Yadav et al. [42].

To increase the temperature resilience of column bioreactors for outdoor operation in cold climates, research has been carried out on the concept of burying them with only the top exposed [4]. Burying the columns provides a trade-off between the increase in thermal insulation and the decrease in illuminated area. However, deep top-lit columns with adequate vertical mixing can provide better areal productivities than traditional shallow open pond systems [13]. This is significant in terms of the required overall bioreactor footprint when located on an industrial site.

1.5 Conclusion

Currently the capture and mitigation of industrial CO_2 emissions through the production of microalgal biofuels is not considered economically viable as a competitive energy source [48–50]. However, with increasing fossil fuel costs, improvements in microalgal production technologies, and potential carbon credit benefits from CO_2 capture, the tipping point for economic viability is likely in the near future.

There are many promising results from column bioreactors, which can be adapted for specific scenarios. That said, further work needs to be done to test other designs for the direct capture of CO_2 from industrial off-gases, and toward taking designs from the laboratory to the industrial scale.

There are certain areas requiring additional research including the application of real off-gases rather than simulated gases that use air–CO_2 mixtures (i.e., the inclusion of acid gasses), the identification of more algal strains (i.e., acid gas–resistant) for different industrial operations, the scaling-up of laboratory technologies (e.g., CO_2 adsorption and light transfer), and the understanding of how process conditions, particularly for outdoor systems, affect operational performances (e.g., variable solar light and convective heat losses).

The carbon capture of industrial off-gas to produce microalgae biofuels has a bright future, and research should continue in order to help combat increasing anthropogenic CO_2 emissions into the atmosphere.

References

[1] L. Brennan, P. Owende, Biofuels from microalgae—a review of technologies for production, processing, and extractions of biofuels and co-products, Renew. Sustain. Energy Rev. 14 (2010) 557−577.

[2] Y. Chisti, Biodiesel from microalgae, Biotechnol. Adv. 25 (2007) 294−306.

[3] M. Daroch, S. Geng, G. Wang, Recent advances in liquid biofuel production from algal feedstocks, Appl. Energy 102 (2013) 1371−1381.

[4] C.A. Laamanen, H. Shang, G.M. Ross, J.A. Scott, A model for utilizing industrial off-gas to support microalgae cultivation for biodiesel in cold climates, Energy Convers. Manag. 88 (2014) 476−483.

[5] R. Münkel, U. Schmid-Staiger, A. Werner, T. Hirth, Optimization of outdoor cultivation in flat panel airlift reactors for lipid production by *Chlorella vulgaris*, Biotechnol. Bioeng. 110 (2013) 2882−2893.

[6] L. Rodolfi, G. Chini Zittelli, N. Bassi, G. Padovani, N. Biondi, G. Bonini, et al., Microalgae for oil: strain selection, induction of lipid synthesis and outdoor mass cultivation in a low-cost photobioreactor, Biotechnol. Bioeng. 102 (2009) 100−112.

[7] E. Jacob-Lopes, S. Revah, S. Hernández, K. Shirai, T.T. Franco, Development of operational strategies to remove carbon dioxide in photobioreactors, Chem. Eng. J. 153 (2009) 120−126.

[8] F. Bumbak, S. Cook, V. Zachleder, S. Hauser, K. Kovar, Best practices in heterotrophic high-cell-density microalgal processes: achievements, potential and possible limitations, Appl. Microbiol. Biotechnol. 91 (2011) 31.

[9] R.N. Singh, S. Sharma, Development of suitable photobioreactor for algae production − a review, Renew. Sustain. Energy Rev. 16 (2012) 2347−2353.

[10] G.N.A. Senhorinho, G.M. Ross, J.A. Scott, Cyanobacteria and eukaryotic microalgae as potential sources of antibiotics, Phycologia 54 (2015) 271−282.

[11] G.N.A. Senhorinho, C.A. Laamanen, J.A. Scott, Bioprospecting freshwater microalgae for antibacterial activity from water bodies associated with abandoned mine sites, Phycologia 57 (2018) 432−439.

[12] M. Koller, A. Salerno, P. Tuffner, M. Koinigg, H. Böchzelt, S. Schober, et al., Characteristics and potential of micro algal cultivation strategies: a review, J. Clean. Prod. 37 (2012) 377−388.

[13] N. Seyed Hosseini, H. Shang, J.A. Scott, Optimization of microalgae-sourced lipids production for biodiesel in a top-lit gas-lift bioreactor using response surface methodology, Energy 146 (2018) 47−56.

[14] S. Wahidin, A. Idris, S.R.M. Shaleh, The influence of light intensity and photoperiod on the growth and lipid content of microalgae *Nannochloropsis* sp., Bioresour. Technol. 129 (2013) 7−11.

[15] B. Wang, C.Q. Lan, M. Horsman, Closed photobioreactors for production of microalgal biomasses, Biotechnol. Adv. 30 (2012) 904−912.

[16] F. Lehr, C. Posten, Closed photo-bioreactors as tools for biofuel production, Curr. Opin. Biotechnol. 20 (2009) 280−285.

[17] N.S. Hosseini, H. Shang, J.A. Scott, Increasing microalgal lipid productivity for conversion into biodiesel by using a non-energy consuming light guide, Biochem. Eng. J. 134 (2018) 60−68.

[18] H.C. Greenwell, L.M.L. Laurens, R.J. Shields, R.W. Lovitt, K.J. Flynn, Placing microalgae on the biofuels priority list: a review of the technological challenges, J. R. Soc. Interface 7 (2010) 703–726.
[19] D. Vandamme, I. Foubert, K. Muylaert, Flocculation as a low-cost method for harvesting microalgae for bulk biomass production, Trends Biotechnol. 31 (2013) 233–239.
[20] E. Sevigné Itoiz, C. Fuentes-Grünewald, C.M. Gasol, E. Garcés, E. Alacid, S. Rossi, et al., Energy balance and environmental impact analysis of marine microalgal biomass production for biodiesel generation in a photobioreactor pilot plant, Biomass Bioenergy 39 (2012) 324–335.
[21] N. Seyed Hosseini, H. Shang, G.M. Ross, J.A. Scott, Microalgae cultivation in a novel top-lit gas-lift open bioreactor, Bioresour. Technol. 192 (2015) 432–440.
[22] Y. Sun, Y. Huang, Q. Liao, Q. Fu, X. Zhu, Enhancement of microalgae production by embedding hollow light guides to a flat-plate photobioreactor, Bioresour. Technol. 207 (2016) 31–38.
[23] J.-W.F. Zijffers, M. Janssen, J. Tramper, R.H. Wijffels, Design process of an area-efficient photobioreactor, Mar. Biotechnol. 10 (2008) 404–415.
[24] C. Welter, J. Schwenk, B. Kanani, J. Van Blargan, J.M. Belovich, Minimal medium for optimal growth and lipid production of the microalgae *Scenedesmus dimorphus*, Environ. Prog. Sustain. Energy 32 (2013) 937–945.
[25] R.P. Rastogi, A. Pandey, C. Larroche, D. Madamwar, Algal green energy – R&D and technological perspectives for biodiesel production, Renew. Sustain. Energy Rev. 82 (2018) 2946–2969.
[26] W.B. Zimmerman, M. Zandi, H.C. Hemaka Bandulasena, V. Tesař, D. James Gilmour, K. Ying, Design of an airlift loop bioreactor and pilot scales studies with fluidic oscillator induced microbubbles for growth of a microalgae *Dunaliella salina*, Appl. Energy 88 (2011) 3357–3369.
[27] B. Zhao, Y. Su, Process effect of microalgal-carbon dioxide fixation and biomass production: a review, Renew. Sustain. Energy Rev. 31 (2014) 121–132.
[28] J.C. Weissman, R.P. Goebel, J.R. Benemann, Photobioreactor design: mixing, carbon utilization, and oxygen accumulation, Biotechnol. Bioeng. 31 (1988) 336–344.
[29] R. Bounaceur, N. Lape, D. Roizard, C. Vallieres, E. Favre, Membrane processes for post-combustion carbon dioxide capture: a parametric study, Energy 31 (2006) 2556–2570.
[30] M.E. Huntley, D.G. Redalje, CO_2 mitigation and renewable oil from photosynthetic microbes: a new appraisal, Mitig. Adapt. Strateg. Glob. Chang. 12 (2007) 573–608.
[31] B. Wang, Y. Li, N. Wu, C.Q. Lan, CO_2 bio-mitigation using microalgae, Appl. Microbiol. Biotechnol. 79 (2008) 707–718.
[32] C.G. Borkenstein, J. Knoblechner, H. Frühwirth, M. Schagerl, Cultivation of *Chlorella emersonii* with flue gas derived from a cement plant, J. Appl. Phycol. 23 (2011) 131–135.
[33] J.A. Lara-Gil, M.M. Álvarez, A. Pacheco, Toxicity of flue gas components from cement plants in microalgae CO_2 mitigation systems, J. Appl. Phycol. 26 (2014) 357–368.
[34] L.M. Mortensen, H.R. Gislerød, The growth of *Chlorella sorokiniana* as influenced by CO_2, light, and flue gases, J. Appl. Phycol. 28 (2016) 813–820.
[35] E.M. Radmann, F.V. Camerini, T.D. Santos, J.A.V. Costa, Isolation and application of SO_x and NO_x resistant microalgae in biofixation of CO_2 from thermoelectricity plants, Energy Convers. Manag. 52 (2011) 3132–3136.
[36] M. Ras, J.-P. Steyer, O. Bernard, Temperature effect on microalgae: a crucial factor for outdoor production, Rev. Environ. Sci. Biotechnol. 12 (2013) 153–164.

[37] H. Shang, J.A. Scott, S.H. Shepherd, G.M. Ross, A dynamic thermal model for heating microalgae incubator ponds using off-gas, Chem. Eng. Sci. 65 (2010) 4591–4597.
[38] C.A. Laamanen, H. Shang, G.M. Ross, J.A. Scott, Smelter off-gas heat and carbon dioxide sequestration to promote production of biodiesel, CIM J. 8 (2017).
[39] C. Nurra, C. Torras, E. Clavero, S. Ríos, M. Rey, E. Lorente, et al., Biorefinery concept in a microalgae pilot plant. Culturing, dynamic filtration steam explosion fractionation, Bioresour. Technol. 163 (2014) 136–142.
[40] G. Morais Michele, M. Radmann Elisângela, A.V. Costa Jorge, Biofixation of CO_2 from synthetic combustion gas using cultivated microalgae in three-stage serial tubular photobioreactors, Z. Naturforsch. C: J. Biosci. 66 (2011) 313–318.
[41] C.-Y. Kao, T.-Y. Chen, Y.-B. Chang, T.-W. Chiu, H.-Y. Lin, C.-D. Chen, et al., Utilization of carbon dioxide in industrial flue gases for the cultivation of microalga *Chlorella* sp., Bioresour. Technol. 166 (2014) 485–493.
[42] G. Yadav, A. Karemore, S.K. Dash, R. Sen, Performance evaluation of a green process for microalgal CO_2 sequestration in closed photobioreactor using flue gas generated in-situ, Bioresour. Technol. 191 (2015) 399–406.
[43] S.-H. Ho, C.-Y. Chen, D.-J. Lee, J.-S. Chang, Perspectives on microalgal CO_2-emission mitigation systems – a review, Biotechnol. Adv. 29 (2011) 189–198.
[44] J.Y. Lee, M.-E. Hong, W.S. Chang, S.J. Sim, Enhanced carbon dioxide fixation of *Haematococcus pluvialis* using sequential operating system in tubular photobioreactors, Process. Biochem. 50 (2015) 1091–1096.
[45] J. Cheng, Y. Huang, J. Feng, J. Sun, J. Zhou, K. Cen, Improving CO_2 fixation efficiency by optimizing *Chlorella* PY-ZU1 culture conditions in sequential bioreactors, Bioresour. Technol. 144 (2013) 321–327.
[46] J.H. Duarte, L.S. Fanka, J.A.V. Costa, Utilization of simulated flue gas containing CO_2, SO_2, NO and ash for *Chlorella fusca* cultivation, Bioresour. Technol. 214 (2016) 159–165.
[47] S.-Y. Chiu, M.-T. Tsai, C.-Y. Kao, S.-C. Ong, C.-S. Lin, The air-lift photobioreactors with flow patterning for high-density cultures of microalgae and carbon dioxide removal, Eng. Life Sci. 9 (2009) 254–260.
[48] S. Dutta, F. Neto, M.C. Coelho, Microalgae biofuels: a comparative study on techno-economic analysis and life-cycle assessment, Algal Res. 20 (2016) 44–52.
[49] O.M. Adeniyi, U. Azimov, A. Burluka, Algae biofuel: current status and future applications, Renew. Sustain. Energy Rev. 90 (2018) 316–335.
[50] S. Rezania, M.F.M. Din, S.F. Kamaruddin, S.M. Taib, L. Singh, E.L. Yong, et al., Evaluation of water hyacinth (*Eichhornia crassipes*) as a potential raw material source for briquette production, Energy 111 (2016) 768–773.

CHAPTER 2

Microbiology and biochemistry of anaerobic digesters: an overview

Spyridon Achinas[1], Vasileios Achinas[2] and Gerrit Jan Willem Euverink[1]
[1]Faculty of Science and Engineering, University of Groningen, Groningen, The Netherlands
[2]Institute for Life Science and Technology, Hanze University of Applied Sciences, Groningen, The Netherlands

2.1 Introduction

This chapter provides an overview of the current understanding of the microbiological and biochemical environment in anaerobic digesters. This review is timely because of the rapid progress in "omics" technologies, which include transcriptomics, proteomics, metabolomics, and reverse genetic tools [1–4]. All these new technologies help researchers unravel complex regulatory and biochemical processes. Thus the field of the microbiology and biochemistry of digesters has advanced dramatically in recent times.

Nowadays, fossil fuels are being replaced by renewable sources and organic waste materials that are widely available. The production and processing of renewable sources and the additional reuse of organic waste make up a significant share of the bioeconomy [5,6]. Carbon is recovered from organic sources in the form of biogas in a cost-effective and environment-friendly process. Biogas can be used for heating, as a source of electricity, or as direct fuel, for example, in transportation vehicles. Residue from the anaerobic digestion (AD) process, so-called digestate, can be used as fertilizer in agricultural activities [7]. AD is a complex biological process that involves the reduction and oxidation of organic molecules through microbial activities. The AD of organic waste is a promising technology for the production of energy, the reduction of greenhouse gas emissions, and the application of sustainable waste management practices [8–10].

The biology and biochemistry of anaerobic digesters have been considerably investigated over the past 10–15 years [2–4,11].

Bioreactors
DOI: https://doi.org/10.1016/B978-0-12-821264-6.00002-4

Although much focus is still given to many classical and fundamental questions, microbial species can now be explored with a vast toolbox and with a more substantial comparative overview because of the advances made in genomics and related technologies such as bioinformatics. In this chapter, first the different stages of AD are described and then the microbial ecology and dynamics thereof are presented.

2.2 Anaerobic digestion steps

The AD process is conceptually divided into three stages, namely hydrolysis and fermentation, acidogenesis and acetogenesis, and methanogenesis; which illustrate the sequence of microbial activities that occur in the process biogas production (Fig. 2.1). During the first step, (in)soluble and complex organic molecules in the substrate are hydrolyzed and converted into simple end-products (soluble sugars, amino acids, glycerol, and long-chain carboxylic acids). The degradation reactions are catalyzed by exoenzymes excreted by hydrolytic and fermentative bacteria and mainly constitute a mixture of cellulases, amylases, proteases, and lipases. In the intermediate stage, the catabolism of long-chain carbon compounds to volatile fatty acids (VFA) by acidogens is characteristic. Acidogenic bacteria facilitate two biochemical substeps, namely fermentation and acetogenesis. Acidogenic bacteria convert soluble monomers into acetic and propionic, ethanol, carbon dioxide (CO_2), and hydrogen (H_2). In the first substep, soluble matter is converted into VFA, ketones, and alcohols. Acetogenesis is a substep of the acid-forming stage and is completed through carbohydrate fermentation, resulting in acetate, CO_2, and H_2 that can be utilized by methanogens to form methane. The final step in

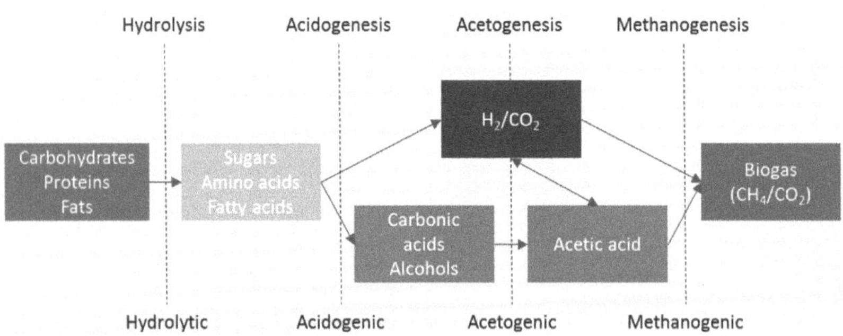

Figure 2.1 Anaerobic digestion networks to produce biogas from biomass.

the degradation of biomass is methanogenesis. Methanogenic species, which transform the end-products of acetogenesis into biogas, facilitate the final stage of AD. Acetoclastic and methylotrophic methanogens reduce acetate and methyl compounds to methane and CO_2. Hydrogen-oxidizing methanogens reduce CO_2 with H_2 into methane. Approximately 70% of the total methane produced is derived from acetate consumption, while the remaining 30% is the result of the reduction of CO_2 by H_2.

The AD process is widely used around the world at a commercial scale. AD is a green technology that aims to confront the challenges of sustainable energy production. Academic research, in collaboration with technology development industries, has contributed significantly to current innovations. The evolution of biofuels in the context of type and availability of feedstocks, codigestion, and conversion rates, has been widely discussed and published over the past few decades by researchers from the European Union, China, and the United States [12,13]. Considering that methane has a high climate change potential, the goal is to find an alternative to lower the environmental footprint of organic waste treatment.

Leading innovators include Germany and Italy, where the potential of biogas is demonstrated, not only using agricultural waste, but also other kinds of waste as well. The EU enlargement brought new members into the family of European biogas producers. The European Union will also benefit from implementing biogas technologies for renewable energy production while mitigating environmental problems and enhancing sustainable development. However, the number of innovations in the global landscape was limited in the period between 2004 and 2008, a fact that declares an uncertain future for advanced biofuels [14].

2.3 Challenges in anaerobic digesters operation

During the past two decades, researchers have focused on the optimization of AD technology through reactor redesign and the application of pretreatment technologies to enhance the biodegradation of substrates [15–18]. Many other factors including the type of feedstock or process conditions also affect the performance of AD [19–23].

The performance of the AD of lignocellulosic materials depends on the operational conditions used such as temperature, organic load, hydraulic retention time, and pH. Previous studies have shown that degradation and biogas production rates are higher at elevated temperatures. Cavinato et al. [24] reported the efficient AD of high lignocellulose-content materials at thermophilic conditions. However, this is not appropriate for energy conservation, and the process is relatively unstable. The pH is also a crucial parameter for the stable operation of a digester. Several studies conclude that a pH between 6.8 and 7.8 is optimal for biogas production [25]. Another factor that has been extensively examined and affects the start-up phase of the AD process is the inoculum-to-substrate ratio. The inoculum is crucial as it provides a consortium of microorganisms for the decomposition of the organic material [26].

The advances in reactor engineering are mainly focused on process integration and intensification with the objective of increasing the overall energy production and carbon decomposition, reducing the number of required process steps, and decreasing the required reactor volume. Up to now, biogas production through AD, upgrading to the quality of natural gas, and the necessary compression to inject green gas into the national grid are three separate procedures. Recently, an emerging technique based on high-pressure AD was developed and favors biogas production reaching 95% methane [27]. However, AD technology has suffered from the perception of it being periodically unstable. Mostly, this is due to a limited understanding of the relationship between microbial dynamics and process functions. Several studies report on the effect of environmental parameters on the microbial dynamics of digesters, primarily focusing on the versatility of methanogenic species and their resistance to toxic compounds since they play a crucial role in digester performance [3,28–32]. Currently, researchers are examining the influence of microbial community structures on digester function and stability. They use next generation sequencing (NGS) techniques and bioinformatic tools to produce a vast amount of data. They aim to understand or predict digester performance based on transcriptomic, proteomic, and metabolomic data [33–35]. Functional instability often occurs in engineered biological systems such as anaerobic digesters. Therefore studies to look for a correlation between microbial communities, process parameters, digester performance, and functional stability are frequently reported [1,36,37].

2.4 Microbial ecology

Microbial activities are responsible for the conversion of organic matter into methane and CO_2. Hydrolytic, acidogenic, and acetogenic bacteria and archaea (methanogens) form complex microbial communities in AD systems. Syntrophic relationships between microorganisms affect the metabolism in AD significantly [38]. The action of hydrolytic bacteria initiates the AD process. These organisms are facultative and catalyze the hydrolysis reaction with the help of extracellular enzymes. Proteins and polysaccharides are converted into monomeric sugars, fatty acids, amino acids, and alcohols [11,39]. Further fermentation of the simple and soluble products results in a wide variety of end-products including acetate, methanol, H_2, and CO_2 [2] (Tables 2.1 and 2.2).

Acetogenesis is the intermediate stage, occurring in well-established methanogenic systems when the products of hydrolysis are transformed into biogas (60% CH_4 and 40% CO_2) and acetate. Despite the significant importance of acetogens, which are obligatory H_2 producers, the knowledge of their taxonomic position, diversity, and physiology is insufficient. The microorganisms involved in the last phase of the AD process are known as methanogens. These microbes use specific coenzymes such as F420, F430, methanopterin, methanofuran, HS-HTP, and M, for the

Table 2.1 Groups of hydrolytic bacteria.

Substrate	Species	Products
Cellulose	*Bacteroides succinogenes, Clostridium lochhadii, Clostridium cellobioporus, Ruminococcus flavefaciens, Ruminococcus albus, Butyrivibrio fibrosolvens, Clostridium thermocellum,* and *Clostridium stercorarium*	Soluble oligosaccharides, cellobiose, which on further hydrolysis results in D-glucose
Hemicellulose	*Bacteroides ruminicola, B. fibrosolvens, R. flavefaciens,* and *R. albus*	Xylose, arabinose, mannose, galactose
Proteins	*Clostridium perfringens, Clostridium bifermentans, Clostridium histolyticum,* and *Clostridium sporogenes* *Bacteroides, Butyrivibrio, Fusobacterium, Slelnomonas, Peptococcus, Campylobacter,* and *Streptococcus*	Peptides, amino acids, ammonia and carbon dioxide, simple fatty acids such as acetic, propionic and butyric acid

Table 2.2 Groups of methane-forming microorganisms [39].

Order	Family
Methanobacteriales	Methanobacteriaceases
Methanococcales	Methanococcaceae
Methanomicrobials	Methanomicrobiaceas
	Methanosarcinaceae

synthesis of methane. These coenzymes are unique to these microorganisms [40].

Methanogens are strictly anaerobic *Archaea* that can be sub-divided into two groups, namely (1) hydrogenotrophic and (2) acetoclastic. Methanogens are characterized by their substrate particularity, slow growth rate, and sensitivity to environmental changes (pH, temperature, and ammonia) [41]. Acetoclastic and hydrogenotrophic methanogens contribute 70% and 30% respectively to the production of biogas and their growth and activity are crucial for the efficient functioning of the AD process [42].

Numerous studies report that the most common methanogenic species found in anaerobic digesters are *Methanobacterium*, *Methanothermobacter*, *Methanobrevibacter*, *Methanosarcina*, and *Methanosaeta* [4,42,43]. Methanogens are strict anaerobes, and they live in terrestrial and aquatic habitats such as swamps, volcanic vents, and deep sediments or the black mud of lakes, as well as in the digestive tract of humans and animals. Methane-forming bacteria are oxygen-sensitive, and they live in environments where other microbial activities rapidly remove oxygen. In nature, methanogens produce methane from the degradation of organic compounds [39].

2.5 Microbiological dynamics

AD requires an equal rate of degradation of all the intermediates due to the sensitivity of the process. However, the dynamics of the separate microbes are complex and interactive, therefore, equal degradation is hard to achieve. In particular, disproportionate amounts of microbial groups influence the degradation stability [44]. For example, complete degradation during hydrolysis is complex because organic compounds such as fats and proteins are depolymerized into monomers over several days, whereas carbohydrate polymers are hydrolyzed within a few hours. Additionally, if hydrolysis is too fast, acid accumulation will occur, which decreases pH and results in process failure. As mentioned before, the microorganisms involved in the three steps of AD are engaged in syntrophic interrelationships. To illustrate, if the growth rate of hydrolytic bacteria is low, then the rates of the other three steps will also decrease, resulting in a lesser biogas yield per unit time. Additionally, if the acetoclastic archaea do not metabolize acetate into methane and CO_2 sufficiently fast, the AD process will halt because of acidification due to acetate accumulation.

Thus the microbial population dynamics and metabolic activity influence the stability and rate of the degradation steps. In return, microbial population dynamics are affected by the presence and concentration of substrates, products, process conditions (e.g., alkalinity, VFA, total ammonia nitrogen, and total organic carbon), operating parameters (e.g., organic loading rate, pH, hydraulic retention time, and temperature), and substrate characteristics (type of lignocellulosic biomass). It has been reported that among the microbial groups involved in AD, methanogens are key microbes for biogas production, but they are also the most sensitive microorganisms to changes in process conditions and operating parameters. In general, methanogens become a rate-limiting step in the whole process when the right conditions are not precisely met [4].

AD also depends on metabolic interactions between microbial species. The classification of microbial species and comprehension of metabolic networks are critical in improving digestion efficacy [2]. Knowledge of the metabolomics of the two types of methanogens is currently characterized and engineered by genetic and bioinformatic tools, that is, quantitative polymerase chain reaction and its many variants. NGS and fluorescence in situ hybridization, among others, have been applied in the characterization of the microbial community structures and the expression of metabolic pathways in AD.

2.6 Conclusion

The study of microbial activities in anaerobic digesters has attracted the attention of researchers, and this chapter is merely a snapshot of the current state of affairs. As a derivative scientific discipline, researchers in microbiology and biochemistry embrace and apply new technologies taken from a wide array of disciplines including biological and genetic engineering and computer science to study AD in more detail than was possible before. A massive amount of data, obtained through advances in

DNA sequencing and bioinformatics has resulted in the accurate identification and classification of the microorganisms involved in AD. Recent advances in genomic techniques are now used to identify metabolic pathways in microbial communities (transcriptomics). Gene expression data provide insight into the metabolic versatility and the potential for the engineering of microorganisms to further improve the AD process. Future efforts aim to better understand the complexity of the ecosystems within biogas reactors by answering physiology and ecology related research questions by applying a combination of omics techniques.

References

[1] M. Loreau, S. Naeem, P. Inchausti, J. Bengtsson, J. Grime, A. Hector, et al., Biodiversity and ecosystem functioning: current knowledge and future challenges, Science 294 (2001) 804–808.

[2] V. O'Flaherty, G. Collins, T. Mahony, The microbiology and biochemistry of anaerobic bioreactors with relevance to domestic sewage treatment, Rev. Environ. Sci. Bio/Technol. 5 (2006) 39–55.

[3] J.J. Werner, D. Knights, M.L. Garcia, N.B. Scalfone, S. Smith, K. Yarasheski, et al., Bacterial community structures are unique and resilient in full-scale bioenergy systems, Proc. Natl. Acad. Sci. U.S.A. 108 (2011) 4158–4163.

[4] A.M. Ziganshin, E.E. Ziganshina, S. Kleinsteuber, M. Nikolausz, Comparative analysis of methanogenic communities in different laboratory-scale anaerobic digesters, Archaea 2016 (2016).

[5] F. Valenti, S. Porto, G. Chinnici, G. Cascone, C. Arcidiacono, Assessment of citrus pulp availability for biogas production by using a GIS-based model: the case study of an area in southern Italy, Chem. Eng. Trans. 58 (2017) 529–534.

[6] D. dell'Antonia, S.R. Cividino, A. Carlino, R. Gubiani, G. Pergher, Development perspectives for biogas production from agricultural waste in Friuli Venezia Giulia (Nord-East of Italy), J. Agric. Eng. (2013).

[7] S. Menardo, P. Balsari, An analysis of the energy potential of anaerobic digestion of agricultural by-products and organic waste, BioEnergy Res. 5 (2012) 759–767.

[8] S. Achinas, V. Achinas, G.J.W. Euverink, A technological overview of biogas production from biowaste, Engineering 3 (2017) 299–307.

[9] S. Achinas, A. Vasileios, Biogas: production, applications and global developments, in: N.A. Agostino Vico (Ed.), Biogas Combustion: An Introductory Briefing, Nova Science Publishers, USA, 2017, pp. 179–193.

[10] R. Kumar, L. Singh, Z.A. Wahid, M.F.M. Din, Exoelectrogens in microbial fuel cells toward bioelectricity generation: a review, Int. J. Energy Res. 39 (8) (2015) 1048–1067.63.

[11] J.D. Murphy, T. Thamsiriroj, Fundamental science and engineering of the anaerobic digestion process for biogas production, The Biogas Handbook, Elsevier, 2013, pp. 104–130.

[12] L. Singh, Z.A. Wahid, Methods for enhancing bio-hydrogen production from biological process: a review, J. Ind. Eng. Chem. 21 (2015) 70–80.

[13] P. Azadi, R. Malina, S.R. Barrett, M. Kraft, The evolution of the biofuel science, Renew. Sustain. Energy Rev. 76 (2017) 1479–1484.

[14] S.C. Albers, A.M. Berklund, G.D. Graff, The rise and fall of innovation in biofuels, Nat. Biotechnol. 34 (2016) 814.
[15] D. Bolzonella, L. Innocenti, F. Cecchi, Biological nutrient removal wastewater treatments and sewage sludge anaerobic mesophilic digestion performances, Water Sci. Technol. 46 (2002) 199−208.
[16] C. Sambusiti, E. Ficara, M. Rollini, M. Manzoni, F. Malpei, Sodium hydroxide pretreatment of ensiled sorghum forage and wheat straw to increase methane production, Water Sci. Technol. 66 (2012) 2447−2452.
[17] D. Boscaro, A. Pezzuolo, S. Grigolato, R. Cavalli, F. Marinello, L. Sartori, Preliminary analysis on mowing and harvesting grass along riverbanks for the supply of anaerobic digestion plants in north-eastern Italy, J. Agric. Eng. 46 (2015) 100−104.
[18] G. Mancini, S. Papirio, P.N. Lens, G. Esposito, Increased biogas production from wheat straw by chemical pretreatments, Renew. Energy 119 (2018) 608−614.
[19] D. Scaglione, S. Caffaz, E. Ficara, F. Malpei, C. Lubello, A simple method to evaluate the short-term biogas yield in anaerobic codigestion of WAS and organic wastes, Water Sci. Technol. 58 (2008) 1615−1622.
[20] E. Dinuccio, F. Gioelli, D. Cuk, L. Rollè, P. Balsari, The use of co-digested solid fraction as feedstock for biogas plants, J. Agric. Eng. 44 (2013) 153−159.
[21] F. Perazzolo, G. Mattachini, F. Tambone, A. Calcante, G. Provolo, Nutrient losses from cattle co-digestate slurry during storage, J. Agric. Eng. 47 (2016) 94−99.
[22] D. Bolzonella, P. Pavan, P. Battistoni, F. Cecchi, Mesophilic anaerobic digestion of waste activated sludge: influence of the solid retention time in the wastewater treatment process, Process. Biochem. 40 (2005) 1453−1460.
[23] D. Coppolecchia, D. Gardoni, C. Baldini, F. Borgonovo, M. Guarino, The influence on biogas production of three slurry-handling systems in dairy farms, J. Agric. Eng. 46 (2015) 30−35.
[24] C. Cavinato, D. Bolzonella, P. Pavan, F. Fatone, F. Cecchi, Mesophilic and thermophilic anaerobic co-digestion of waste activated sludge and source sorted biowaste in pilot-and full-scale reactors, Renew. Energy 55 (2013) 260−265.
[25] L.C. Grady, G. Daigger, H. Lim, Biological Wastewater Treatment, CRC Press, 1992.
[26] A. Fabbri, S. Serranti, G.J.W.M. Bonifazi, Biochemical methane potential (BMP) artichoke waste: inoculum effect, Waste Manag. Res. 32 (2014) 207−214.
[27] W. Merkle, K. Baer, N.L. Haag, S. Zielonka, F. Ortloff, F. Graf, et al., High-pressure anaerobic digestion up to 100 bar: influence of initial pressure on production kinetics and specific methane yields, Environ. Technol. 38 (2017) 337−344.
[28] A.S. Fernandez, S.A. Hashsham, S.L. Dollhopf, L. Raskin, O. Glagoleva, F.B. Dazzo, et al., Flexible community structure correlates with stable community function in methanogenic bioreactor communities perturbed by glucose, Appl. Environ. Microbiol. 66 (2000) 4058−4067.
[29] A. Briones, L. Raskin, Diversity and dynamics of microbial communities in engineered environments and their implications for process stability, Curr. Opin. Biotechnol. 14 (2003) 270−276.
[30] S.D. Allison, J.B. Martiny, Resistance, resilience, and redundancy in microbial communities, Proc. Natl. Acad. Sci. USA 105 (2008) 11512−11519.
[31] A. Schievano, G. D'Imporzano, F. Adani, Substituting energy crops with organic wastes and agro-industrial residues for biogas production, J. Environ. Manag. 90 (2009) 2537−2541.
[32] L. Wittebolle, M. Marzorati, L. Clement, A. Balloi, D. Daffonchio, K. Heylen, et al., Initial community evenness favours functionality under selective stress, Nature 458 (2009) 623.
[33] S. Naeem, S. Li, Biodiversity enhances ecosystem reliability, J. Nat. 390 (1997) 507.

[34] D. Traversi, V. Romanazzi, R. Degan, E. Lorenzi, E. Carraro, G.J.B.T. Gilli, Microbial-chemical indicator for anaerobic digester performance assessment in full-scale wastewater treatment plants for biogas production, Bioresour. Technol. 186 (2015) 179–191.

[35] V. Ventorino, I. Romano, G. Pagliano, A. Robertiello, O. Pepe, Pre-treatment and inoculum affect the microbial community structure and enhance the biogas reactor performance in a pilot-scale biodigestion of municipal solid waste, Waste Manag. 73 (2018) 69–77.

[36] G. Merlino, A. Rizzi, F. Villa, C. Sorlini, M. Brambilla, P. Navarotto, et al., Shifts of microbial community structure during anaerobic digestion of agro-industrial energetic crops and food industry byproducts, J. Chem. Technol. Biotechnol. 87 (2012) 1302–1311.

[37] K. Venkiteshwaran, B. Bocher, J. Maki, D. Zitomer, Relating anaerobic digestion microbial community and process function: supplementary issue: water microbiology, Microbiol. Insights (8)(2015). MBI S33593.

[38] J.-L. Garcia, B.K. Patel, B. Ollivier, Taxonomic, phylogenetic, and ecological diversity of methanogenic Archaea, Anaerobe 6 (2000) 205–226.

[39] D. Deublein, A. Steinhauser, Biogas From Waste and Renewable Resources: An Introduction, John Wiley & Sons, 2011.

[40] K.C. Costa, J.A. Leigh, Metabolic versatility in methanogens, Curr. Opin. Biotechnol. 29 (2014) 70–75.

[41] N. Serrano-Silva, Y. Sarria-Guzmán, L. Dendooven, M. Luna-Guido, Methanogenesis and methanotrophy in soil: a review, Pedosphere 24 (2014) 291–307.

[42] B. Schink, Syntrophic associations in methanogenic degradation, Molecular Basis of Symbiosis, Springer, 2005, pp. 1–19.

[43] L.M. Steinberg, J.M.J.A.E.M. Regan, Phylogenetic comparison methanogenic communities an acidic, oligotrophic fen an anaerobic digester treating municipal wastewater sludge, Appl. Environ. Microbiol. 74 (2008) 6663–6671.

[44] M.E. Griffin, K.D. McMahon, R.I. Mackie, L. Raskin, Methanogenic population dynamics during start-up of anaerobic digesters treating municipal solid waste and biosolids, Biotechnol. Bioeng. 57 (1998) 342–355.

CHAPTER 3

Process intensification for the production of canola-based methyl ester via ultrasonic batch reactor: optimization and kinetic study

Awais Bokhari[1,2], Suzana Yusup[1], Saira Asif[3], Lai Fatt Chuah[4] and Leow Zi Yan Michelle[1]

[1]Department of Chemical Engineering, Biomass Processing Cluster, HICOE-Centre for Biofuel and Biochemical Research, Institute of Self-Sustainable Building, Universiti Teknologi PETRONAS, Seri Iskandar, Perak, Malaysia
[2]Chemical Engineering Department, Biomass Conversion Research Center (BCRC), COMSATS University Islamabad (CUI), Lahore, Pakistan
[3]Department of Botany, PMAS Arid Agriculture University, Rawalpindi, Pakistan
[4]Malaysia Marine Department Northern Region, Gelugor, Penang, Malaysia

3.1 Introduction

Due to fossil fuel depletion and environmental concerns, there should foster an intense awareness in a diesel fuel substitution presently all over the world with a clean, viable, and renewable energy, accelerating international efforts to achieve both economic development and ecological protection [1]. Hence, energy sources such as solar, wind, geothermal, and biofuels are considered as renewable fuel alternatives. They produce lower or negligible levels of greenhouse gases and other pollutants when compared to fossil energy sources. Globally, biofuels production has received much attention in the past two decades [2]. The industry is projected to produce more than 130 Mt of oil equivalent by 2035, which represents an 18-fold increase in biofuels production from 1990 [3].

Renewable liquid biofuels such as bioethanol, biohydrogen, biodiesel and bio-oil have been reported as a highly modern and technological area of study for researchers; in particular, their application for transportation has attracted worldwide attention due to their renewability, sustainability, widespread availability, and biodegradability [2]. Among these, biodiesel shows great potential as a sustainable energy resource for the mitigation of

Bioreactors
DOI: https://doi.org/10.1016/B978-0-12-821264-6.00003-6

global warming due to it being derived from renewable and domestic vegetable oils, thereby relieving reliance on petroleum fuel and reducing concentrations of carbon dioxide (CO_2) in the atmosphere during the growth of these plants [4]. Located in South-East Asia, Malaysia has tropical weather with uniform temperatures throughout the year and an average rainfall of 200–250 cm per year, which are excellent conditions for the growth of palm trees and other plants. On one hand, it has been reported that Malaysia relies heavily on oil and natural gas to sustain its economic development and is expected to get into trouble due to fossil fuel shortages in future [4]. Palm trees are grown easily in Malaysia and have helped Malaysia become the second largest producer of palm oil in the world, contributing 40% of the total worldwide demand for crude palm oil. This proves that palm oil has the potential capability to be utilized as a source of renewable feedstock and Malaysia has a high biodiesel production capacity, which would reduce overdependence on fossil fuels and their harmful effects on the environment [5].

Furthermore, as mentioned above, the majority of energy consumption in Malaysia is from the transportation sector. Together with the widespread use of energy in the industrial sector, these contribute to global climate change, an inevitable consequence of emission production from fossil fuel combustion [6]. However, through the application of biodiesel, greenhouse gas emissions can be alleviated as expected [4]. From an energy demand point of view, it has been described that Malaysians produced 8.196 Mt of solid waste in 2010, from which the potential energy extractable is projected to reach 360 MW by 2022 [7]. Besides palm oil, municipal solid waste and other nonedible oils can be considered as alternative renewable energy sources for biodiesel production and Malaysia has the potential to become one of the major contributors to the biodiesel industry [8]. All the positive steps taken toward alternative and renewable energies were based on the awareness created by environmental aspects, affordability, and limited fossil fuel resources. To overcome the negative effects of fossil fuels on the environment, Brazil, United States, and the European Union, as well as other countries took the initiative to create policies and regulations toward alternative energy utilization and technologies. The major producers and users of biodiesel have been the United States, European Union, and Brazil. Since the 1970s, South Asia and South East Asia have also developed biodiesel production technologies to reduce agricultural and environmental issues. China, India, Japan, South Korea, Taiwan, Malaysia, Singapore, Thailand, and Indonesia are the main countries in Asia that have shown great interest in biodiesel technologies to make positive contributions to their environmental and economic futures [9–11].

Biodiesel is the most important and essential technology for the government of China. China has carried out strategic policies for biodiesel production technologies in terms of renewable legislation, tax support, and excise exemption. In this way, China's biodiesel production has increased from 12,000 t/a to 200,000 t/a between 2002 and 2006. Currently China has five large plants with 30,000 t/a capacities and some smaller plants with 10,000 t/a capacities. India promotes the utilization and production of biodiesel to reduce its fossil fuel imports as +70% of the energy demand of India is fulfilled by imported fossil fuels, and this affects their economy negatively. India uses around 55.3 million hectares for jatropha cultivation for biodiesel production and this is targeted to increase by 11.2 million hectares in the coming years [12]. Japan developed biodiesel production technology in 1997 to comply with the Kyoto Protocol. The main aim of the agreement was to reduce the environmental issues associated with the utilization of fossil fuels. Japan achieved its goal of reducing 6% of its CO_2 emissions in 2010 as compared to those of 1990, with almost 30 small plants producing 3000–4000 t/a of biodiesel using recycled cooking oil [12,13].

The main focus for Taiwan was to produce biodiesel to increase energy sources for the betterment of farmers' incomes. In Taiwan, five biodiesel plants operated using waste cooking oil with a capacity of 5900 t/a. Thailand's biodiesel production is around 18,000 t/a and B5 is currently implemented in Bangkok. Thailand's main focus is to reduce its fossil fuel imports. Philippines, being a major exporter of coconut products, produces biodiesel from coconut oil and currently produces 144000 t/a of biodiesel. Malaysia and Indonesia are major global exporters of palm oil. Indonesia has many biodiesel plants with a capacity of 820,000 t/a. Singapore established its biodiesel industry in 2008, with a capacity of 250,000 t/a, and a targeted capacity of 3 million t/a by 2015 [12].

Malaysia started research and development in the biodiesel sector at the beginning of the 1980s, utilizing their huge palm oil plantations for the production of biodiesel. In collaboration with the Malaysian Palm Oil Board (MPOB), Malaysia took the first step to set up a small scale production unit with a capacity of 3000 t/a in 1985. The small-scale plant produced biodiesel from palm oil [14,15]. Between 1986 and 1994 many tests were conducted on different diesel vehicles and engines operated using palm oil biodiesel [16]. The study areas of the tests included engine performance, environmental benefits, fuel quality, and wear and tear. One test was conducted on 30 buses without any engine modifications, using a Mercedes Benz engine, model OM352, and covering a distance of 300,000 km. Malaysia produced its biodiesel commercially by setting up three biodiesel plants in 2006 and two more plants by 2008. This commercial biodiesel successfully met the criteria of European Union and United States standards [12,17,18].

The Malaysian government announced the National Biofuel Policy (NBP) in August, 2005, in collaboration with all stakeholders' policy makers by keeping the facts and figures provided by research conducted by MPOB. The NBP was implemented in 2006 to promote research and development in the palm oil-based biodiesel sector. The main objective of the policy is to create awareness in society about the benefits and positive effects of biofuels. The main vision of this policy is to utilize environment-friendly, renewable, alternative energy sources to control dependency on fossil fuels. In addition, it is also focused on controlling the fluctuating prices of palm oil and fossil fuels. The government announced the short-term policy, which illustrated B5 biodiesel pumps being launched in 2007 at selected sites of Malaysia. The extensive demand for palm oil for both food and fuel resulted in dramatically increasing palm oil prices. However, in February, 2010, the government took action and decided to postpone the B5 biodiesel implementation until 2011 [14,15].

Currently B5 biodiesel blends are implemented in the Central Peninsula Region of Malaysia. Current policy makers are trying to minimize the effect of the fluctuating palm oil and petroleum diesel prices by keeping in line with the views of the consumer market. Government plans to give subsidies to the biodiesel industry and make new policies that will provide positive change in Malaysia's economic and energy sectors. From the early history of biodiesel, this sector is aimed at obtaining advantages in terms of environmental, social, political, economic, and energy security benefits [9,12]. In summary, global energy consumption trends report renewables as an undeniable fact. An efficiently developed biodiesel industry with technological innovations would safeguard Malaysia's energy supply, warrantee socioeconomic development, improve the standard of living for every citizen, and foster a cleaner environment. There are several technologies available for the transformation of oils into fuels having properties comparable to diesel. These technologies include pyrolysis, microemulsion, dilution, and transesterification [19]. It is of vital importance to understand that viscosity is a crucial factor in evaluating the possibility of fuel compositions for compression–ignition engines. Due to the high viscosity and low volatility of oils, it is unfeasible to directly use them in an engine. The process of transesterification (alcoholysis) replaces the glycerol moiety of the triglyceride molecule with an alkali radical of the alcohol used, leading to a 10- to 15-fold reduction of viscosity [20]. Being a simple reaction, the lowering of viscosity and enhancing of physicochemical properties to gain a good yield of better-quality biodiesel are apparent benefits of transesterification over other processes. This has resulted in the widespread use of alcoholysis in the development of biodiesel.

3.2 Feedstocks for biodiesel production

Being nutritional, vegetable oils have a great impact on almost every human being. Essential vitamins and fats for healthy living are provided by vegetable oils. Historically, these oils were not only famous for nutritional purposes, but people utilized them to meet their energy requirements by using them in lamps, lanterns, and for fire. In ancient civilizations, vegetable oils were the only available resource used to produce heat and light. Some of the most common oils used include fish oil, corn oil, and nut oil, but their utilization was and is dependent on the availability of the feedstock.

The initiation of the use of vegetable oils in an engine dates to the beginning of the 1900s, when a French company ran an engine on peanut oil. During World War I and II, vegetable oils were extensively used to run army trucks and cars. The oil produced from energy crops is known as biodiesel [21,22]. More than approximately 4000 vegetable species were found on Earth that contain these oils [23]. Vegetable oils can be obtained by either chemical or mechanical means. The most commonly used traditional and alternative oil crops are shown in Table 3.1.

Vegetable oils and fats are most commonly known as triglycerides or triacylglycerols. The chemical structure of triglycerides consists of a three-carbon backbone with a long hydrocarbon chain attached to each of the carbons. Oxygen atoms and carbonyl carbon are attached to these chains. The chemical and physical properties vary with origin. The properties of these oils differ from one another by length of the fatty acid chains attached to the backbone and the influence of the carbon–carbon double

Table 3.1 Properties of canola oil feedstock.

Parameter	This work	[24]
Acid value (mg KOH/g oil)	0.076	0.64
FFA value (%)	0.038	0.32
Iodine value (g I_2/g oil)	112	–
Water content (%)	0.22	0.19
Kinetic viscosity at 40°C	37.75	

FFA, Free fatty acid.

bond on the chain. Oils derived from vegetables or fats mostly contain 18 carbon-long chains with 0–3 double bonds on each chain. Those chains of fatty acid that contain double bonds are known to be unsaturated and those remaining without double bonds are known as saturated. Biodiesel methyl esters are dramatically influenced by chain length and the presence of a double bond. Unsaturated fatty acid is liquid at ambient temperatures because of the deformation of the molecule due to the presence of a double bond. Saturated fatty acid becomes solid at room temperature due to similar reasons [22,25].

The use of vegetable oils and fats as an alternative resource for energy production exhibit the following useful and beneficial advantages to humans along with the environment.

- Being domestically produced, vegetable oils give economic benefits by reducing fossil fuel-based diesel imports [26].
- They are biodegradable in nature, nontoxic, and are less hazardous to human health than fossil fuel-based diesel [9].
- They are renewable in nature, and can be produced from oil-bearing crops or other feedstocks that would be considered as waste [26,27].
- They exhibit higher heating values, that is, almost 80% of that of fossil fuel-based diesel [26,28].
- They contain lesser aromatic compounds.
- Their reasonable cetane number results in a lesser knocking tendency in engines [29].
- They are environment-friendly as they contain low sulfur contents [26,29].
- They show enhanced lubricity, so no engine modifications are required [26].

Many types of vegetable crops, oils, and fats are widely available throughout the world. To meet the demand and supply for vegetable oils for nutritional and energy reasons the number of such plantations is growing worldwide. People are interested in investing in this sector to harvest more oil-bearing crops. The Food and Agricultural Organization (FAO) in the United States reported that the production of oil-bearing crops increased million tons in the year 2000 to around 127 million by year 2006. According to FAO statically database, almost 633 crops and livestocks, harvested and processed, give quite excellent yield of oil [30].

Feedstock availability, harvesting, and plantations are dependent on growing conditions and climate. For example, for palm oil plantations, the most suitable climate for its cultivation is in tropical regions. Currently Malaysia is the second largest producer and exporter of palm oil after Indonesia. Similarly, Brazil and United States have a large number of soybean oil plantations. Furthermore, rapeseed is widely available in the European Union.

Besides edible oils, nonedible oils are also found to be a promising alternative energy production source, and these show many advantages over edible oils. Developing countries focus more on nonedible oils due to the tremendous demand for edible oils for nutritional purposes, and edible oils are too expensive to be used as an alternative fuel source. Nonedible oils, as a renewable energy source, have many benefits including economic, environmental, and social benefits, as well as their contribution to reducing expenditure over imports, the fuel-versus-food controversy, reducing the rate of deforestation, and contributing toward developing countries' gross domestic product.

Biodiesel produced from nonedible crops has been extensively investigated by researchers over the past few years. Nonedible oils that have been investigated by researchers include jatropha, karanja, tobacco seed, rice bran, mahua, neem, castor, linseed, microalgae, and rubber seed. The results reported by researchers successfully met the international standards proscribed for biodiesel methyl esters. Some of the methyl esters are suitable for use as fuel in cold climates [28,31].

3.3 Current research

Ultrasonic irradiation is a new cavitating technique that has been widely exploited for its high methyl ester yields in a relatively short reaction time. It is capable of overcoming mass transfer between the immiscible reactant by generating the cavitational phenomena (process intensification) [32].

Canola seed oil is used in the production of biodiesel. There have been no reports of ultrasonic-assisted transesterification using canola-based oil. The aim of this research is to explore the feasibility of canola in the production of biodiesel through an alcoholysis reaction in an ultrasonic, irradiated environment. The aim of the current research is to study ultrasonic-assisted action and transesterification processes of canola seed oil by investigating the effect of four input parameters including methanol-to-oil molar ratio, catalyst amount, reaction time, and ultrasonic frequency on the conversion of methyl ester.

3.4 Materials and methodology

3.4.1 Materials

In the present work, canola oil was procured from Felcura oil Sdn. Bhd. Malaysia. The characterization of the canola oil feedstock is given Table 3.1. All the chemicals and reagents such as potassium hydroxide (99%), potassium carbonate (99%), methanol (99%), and propanol (98%) were attained through Merck (Malaysia). Methyl ester standards for gas chromatography—flame ionization detector was purchased from Sigma Scientific Chemicals.

3.4.2 Experimental

Fig. 3.1 depicts the batch ultrasonic reactor setup. The details of the experimental procedure followed for the batch ultrasonic reactor have been described in previous research work.

3.5 Results and discussion

3.5.1 Effect of methanol-to-oil ratio on methyl ester content

Fig. 3.2 depicts the effect of methanol-to-oil molar ratio on methyl ester content. Transesterification is known as a reversible reaction. The alcohol-to-oil molar ratio of stoichiometry for the transesterification reaction is 3:1. Nonetheless, Encinar, González, Martínez, Sánchez, and Pardal [33] revealed that using a 3:1 methanol-to-oil molar ratio could only achieve a less than 75% conversion. From the theory of Le Chatelier's principle, an increase in alcohol may shift the position of the equilibrium in the opposite direction, offsetting a change; in this case, to the right (forward reaction), which produces more biodiesel. The work of

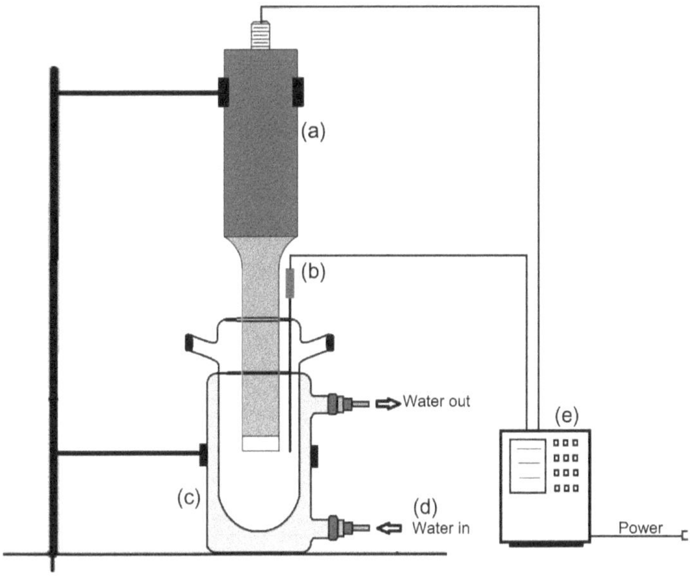

Figure 3.1 Batch ultrasonic reactor setup.

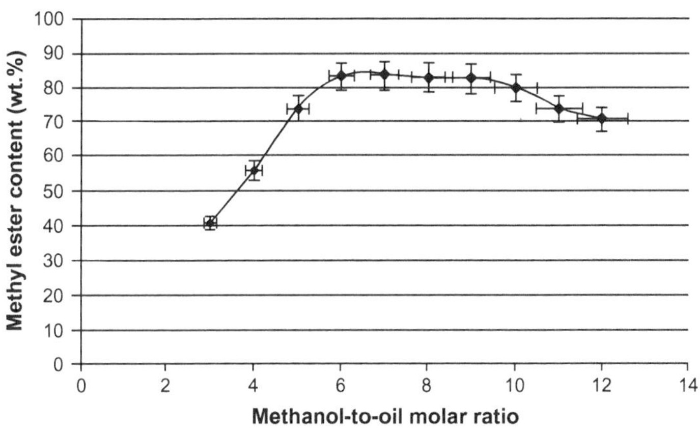

Figure 3.2 Effect of methanol-to-oil ratio on methyl ester content (3.0 wt.% of KOH, 50% of amplitude, and 40 min of reaction).

previous researchers revealed that an optimum molar ratio of alcohol to oil of between 6:1 and 9:1 has a positive effect on converting triglyceride to methyl esters [34]. However, greater amounts of alcohol may cause some negative effects in an ultrasonic-assisted reactor. The researches of

Chen et al. [35] and Priambodo et al. [36] indicated that an increase of the methanol-to-oil molar ratio from 6:1 to 12:1 caused a decrease of conversion from 97.9% to 91.6%, and from 93% to less than 90%, respectively. It is important to note that alcohols (especially methanol) possess dipole moments, thus, the use of alcohols in excess can absorb ultrasonic energy resulting in the loss of the solvent, and also may reduce the intensification field, as well as increase the miscibility of biodiesel and glycerol. In addition, a high volume of alcohol also reduces the catalyst concentration in a given medium. One should bear in mind that the amount of alcohol used for biodiesel production in an ultrasonic system is much lower than that of conventional heating (where the methanol-to-oil molar ratio can range from 9:1 to 30:1) and the role of alcohol is not dependent on temperature [34].

3.5.2 Effect of catalyst amount on methyl ester content

Fig. 3.3 demonstrates the effect of the amount catalyst on the methyl ester content under ultrasonic cavitation. With regards to the catalyst concentration, it is evident that insufficient amounts of catalyst result in an incomplete conversion of triglycerides into alkyl esters, but excess catalyst may rise the bulk viscosity of the reactants and lead to the formation of gels, an obstacle in glycerin separation, hence, diminishing the reaction yields [37]. As reported in the work of Tan et al. [34] an increased concentration of CH_3ONa from 1 wt.% to 1.5 wt.% in the microwave-assisted continuous-flow transesterification of crude jatropha oil resulted in

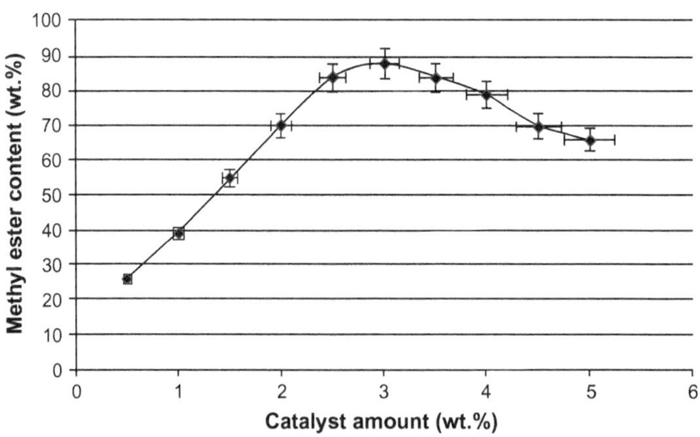

Figure 3.3 Effect of catalyst amount on methyl ester content (50% amplitude, 6:1 methanol-to-oil ratio, and reaction time of 40 min).

a slight decrease in the biodiesel yield from 96.5% to about 92%. Another publication also indicated that there was no significant increase in biodiesel yield when raising the NaOH catalyst concentration from 1 wt.% to 1.5 wt.% in the transesterification of palm oil in an ultrasonic reactor.

3.5.3 Effect of reaction time on methyl ester content

Fig. 3.4 depicts the effect of reaction time on the ultrasonic-based synthesis of canola methyl ester. One of the overwhelming advantages of using an ultrasonic batch reactor is the dramatic reduction in reaction time. Conventional or laboratory-scale processes usually take 90–180 min for the transesterification process to occur. Asif et al. [37] reported for their findings that ultrasonic transesterification of nonedible oil could achieve the conversion of 91.3% after 30 min of reaction time at 60°C in the presence of 1.5 wt.% KOH and 6:1 methanol-to-oil molar ratio, as compared to the conventional heating 91.4% of methyl ester was produced after 90 min of reaction time with the same conditions. It could be concluded that by using ultrasonic technology, the reaction time may be reduced five times compared with conventional heating.

On the other hand, increasing the reaction time may cause negative effects on the reaction yield, causing a reduction in the transesterification reaction. This may be attributed to the formation of soap during overheating, which can destroy some organic molecules, and the cracking followed by oxidization of the biodiesel to aldehydes, ketones, and lower chained organic fractions [38].

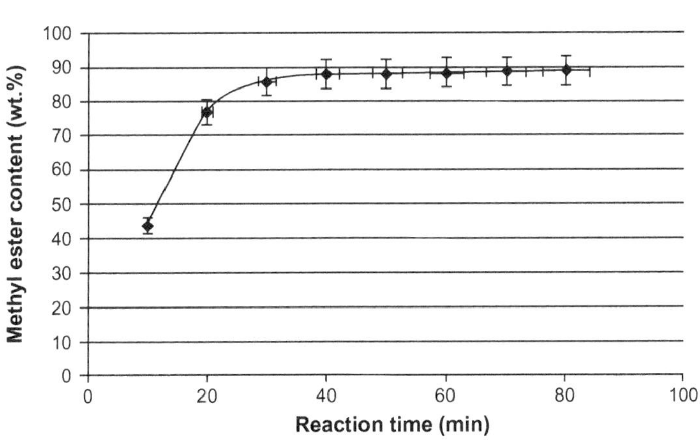

Figure 3.4 Effect of reaction time on methyl ester content (6:1 methanol-to-oil ratio, 3.0 wt.% of KOH, and 50% amplitude).

3.5.4 Effect of amplitude on methyl ester content

Fig. 3.5 depicts the effect of amplitude on methyl ester content. A 50% ultrasonic amplitude successfully increased the methyl ester content. However, a high ultrasound amplitude may bring a nonaffirmative effect on the reaction rate of transesterification [39]. Salamatinia et al. [40] studied the ultrasonic (having an amplitude and power of 20 kHz and 200 W, respectively) technique with heterogeneous catalyst using numerous sources of feedstock oils. Different ultrasonic pulses and power levels were studied and optimization was achieved by response surface methodology. The work reported a similar effect to the current study.

3.5.5 Reaction kinetics of canola oil methyl ester via ultrasonic-assisted technique

Chemical kinetics is the study of the rates of chemical reactions, factors which are influential in the rates and the explanation of the rates with respect to the reaction mechanisms of chemical processes [34]. Chemical kinetics and reactor design are of primary concern in the exploitation of chemical reactions in industrial production [38]. From an economic point of view, it is also a crucial factor in the success or failure of a chemical plant. Unlike in thermodynamics, where the concern is expressed in the change of reactions without considering the intermediate states or time, in chemical kinetics, the rate of change of the concentration of reactants or products is followed by time. Therefore the rate of chemical reactions, as

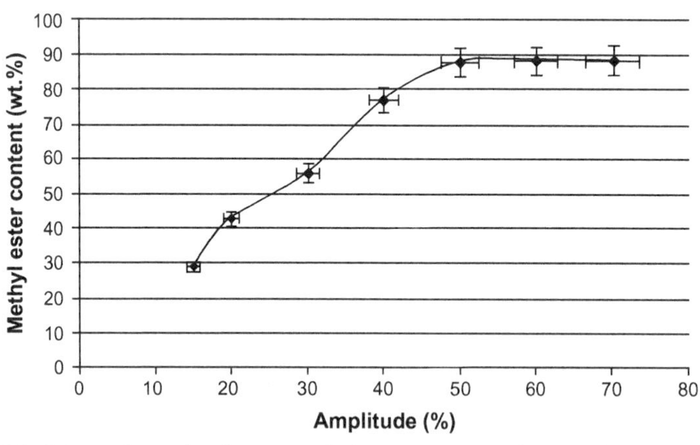

Figure 3.5 Effect of amplitude on methyl ester content (6:1 methanol-to-oil ratio, reaction time of 40 min, and 3.0 wt.% of KOH).

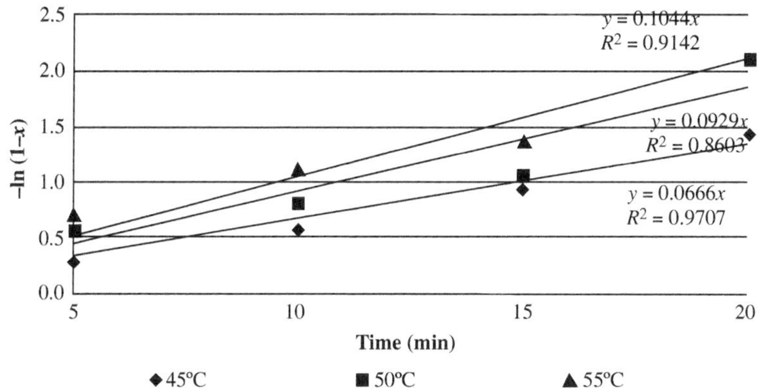

Figure 3.6 Pseudo first-order reaction kinetics fitting for canola-based methyl ester.

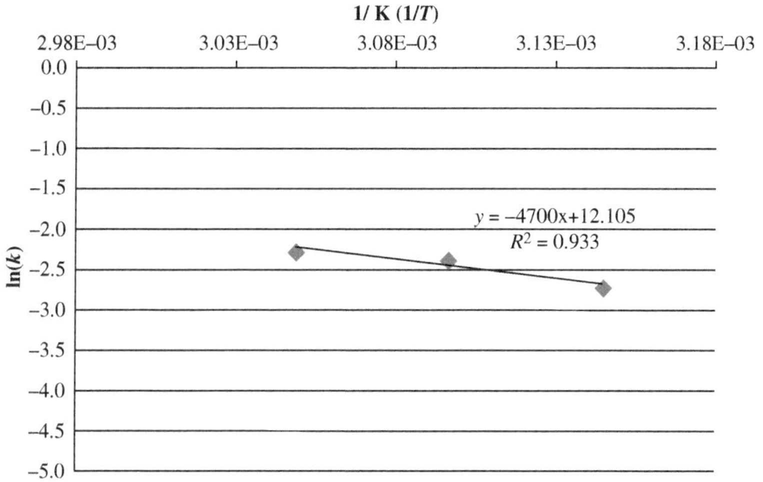

Figure 3.7 Activation energy and frequency identification plot.

well as the effect of various factors such as concentration, temperature, catalyst, and so forth, are determined quantitatively [39]. Fig. 3.6 shows the reaction kinetics of the pseudo first order for canola methyl ester synthesis under ultrasonic-assisted conditions. The kinetics has been performed at three different temperatures. The reaction rate constants were found to be 0.0666 min^{-1} at 45°C, 0.0929 min^{-1} at 50°C, and 0.1044 min^{-1} at 55°C. Fig. 3.7 reveals the activation energy of 39.05 kJ/mol and frequency factor of 12×10^7 min^{-1}.

3.6 Conclusion

For the transesterification process, a 90% methyl ester content was achieved using a 3 wt.% potassium hydroxide catalyst concentration and a 6:1 methanol-to-oil molar ratio in 40 min at 50% of amplitude. From the viewpoint of energy efficiency, ultrasonic-assisted process intensification proved its innovation in the significant reduction of energy consumption as well as CO_2 emission. In terms of chemical kinetics, the obtained results showed that the ultrasonic-accelerated transesterification reactions with catalyst were irreversible pseudo-homogeneous first-order reactions. The activation energy of the transesterification reaction was found to be 39.05 kJ/mol. The high pre-exponential factor A with the value of 12×10^7/min for the transesterification reaction has proven the energy efficiency of ultrasonic for one-step biodiesel production.

Acknowledgments

The authors would like to express their gratitude to University Teknologi PETRONAS, MyRA research grant, COMSATS University Islamabad, Lahore Campus, PMAS Arid Agriculture University, and Marine Department Malaysia for the support given while conducting this research work.

References

[1] J. Ahmad, S. Yusup, A. Bokhari, R.N.M. Kamil, Study of fuel properties of rubber seed oil based biodiesel, Energy Convers. Manag. 78 (2014) 266–275.
[2] R.P.B. Costa-Felix, M.K.K. Figueiredo, A.V. Alvarenga, An ultrasonic method to appraise diesel and biodiesel blends, Fuel 227 (2018) 150–153.
[3] T. Lieu, S. Yusup, M. Moniruzzaman, Kinetic study on microwave-assisted esterification of free fatty acids derived from Ceiba pentandra Seed Oil, Bioresour. Technol. 211 (2016) 248–256.
[4] S.E. Hosseini, M.A. Wahid, Necessity of biodiesel utilization as a source of renewable energy in Malaysia, Renew. Sustain. Energy Rev. 16 (2012) 5732–5740.
[5] A. Johari, B.B. Nyakuma, S.H. Mohd Nor, R. Mat, H. Hashim, A. Ahmad, et al., The challenges and prospects of palm oil based biodiesel in Malaysia, Energy 81 (2015) 255–261.
[6] K.W. Cheah, S. Yusup, L.F. Chuah, A. Bokhari, Physio-chemical studies of locally sourced non-edible oil: prospective feedstock for renewable diesel production in Malaysia, Procedia Eng. 148 (2016) 451–458.
[7] M.H.M. Ashnani, A. Johari, H. Hashim, E. Hasani, A source of renewable energy in Malaysia, why biodiesel? Renew. Sustain. Energy Rev. 35 (2014) 244–257.
[8] A. Bokhari, S. Yusup, L.F. Chuah, R.N.M. Kamil, Vacuum distilled post blend crude palm and rubber seed oil methyl esters: optimisation and parametric studies, Procedia Eng. 148 (2016) 479–486.
[9] S. Lim, L.K. Teong, Recent trends, opportunities and challenges of biodiesel in Malaysia: an overview, Renew. Sustain. Energy Rev. 14 (2010) 938–954.

[10] M.H. Jayed, H.H. Masjuki, M.A. Kalam, T.M.I. Mahlia, M. Husnawan, A.M. Liaquat, Prospects of dedicated biodiesel engine vehicles in Malaysia and Indonesia, Renew. Sustain. Energy Rev. 15 (2011) 220–235.
[11] N.N.A.N. Yusuf, S.K. Kamarudin, Z. Yaakub, Overview on the current trends in biodiesel production, Energy Convers. Manag. 52 (2011) 2741–2751.
[12] Y.M.C. Mohd, B. Wahid, C.W. Puah, Biodiesel in Asia, in: J.K. Gerhard Knothe, J.V. Gerpen (Eds.), The Biodiesel Handbook, AOCS Press, Urbana, IL, 2010, pp. 355–363.
[13] S. Lim, K.T. Lee, Implementation of biofuels in Malaysian transportation sector towards sustainable development: a case study of international cooperation between Malaysia and Japan, Renew. Sustain. Energy Rev. 16 (2012) 1790–1800.
[14] M.P.O.B. (MPOB) (Ed.), Malayisian Oil Palm Statistics, Malaysian Palm Oil Board, Bandar Baru Bangi, Selangor, Malaysia, 2006.
[15] A.Z. Abdullah, B. Salamatinia, H. Mootabadi, S. Bhatia, Current status and policies on biodiesel industry in Malaysia as the world's leading producer of palm oil, Energy Policy 37 (2009) 5440–5448.
[16] Y.M. Choo, A.N. Ma, Y. Basiron, Production and evaluation of palm oil methyl esters as diesel substitute, Elaeis (1995) 5–25 (special issue).
[17] Y.M. Choo, W. Mohd Basri, Latest development in palm biodiesel in Malaysia, in: PIPOC "Palm Oil: Empowering Change", Kuala Lumpur Convention Centre, Malaysia, 2007.
[18] Y.M. Choo, C.W. Puah, W. Mohd Basri, Biofuel: greening the future, in: BIMP-EAGA Investment Conference Davao City, Philippines, 2007.
[19] E.M. Shahid, Y. Jamal, Production of biodiesel: a technical review, Renew. Sustain. Energy Rev. 15 (2011) 4732–4745.
[20] B.L. Salvi, N.L. Panwar, Biodiesel resources and production technologies – a review, Renew. Sustain. Energy Rev. 16 (2012) 3680–3689.
[21] G. Knothe, Historical perspectives on vegetable oil-based diesel fuels, Ind. Oils 12 (2001) 1103–1107.
[22] R.D. Misra, M.S. Murthy, Straight vegetable oils usage in a compression ignition engine—a review, Renew. Sustain. Energy Rev. 14 (2010) 3005–3013.
[23] G. Santori, G. Di Nicola, M. Moglie, F. Polonara, A review analyzing the industrial biodiesel production practice starting from vegetable oil refining, Appl. Energy 92 (2012) 109–132.
[24] M.G. Jang, D.K. Kim, S.C. Park, J.S. Lee, S.W. Kim, Biodiesel production from crude canola oil by two-step enzymatic processes, Renew. Energy 42 (2012) 99–104.
[25] S.P. Singh, D. Singh, Biodiesel production through the use of different sources and characterization of oils and their esters as the substitute of diesel: a review, Renew. Sustain. Energy Rev. 14 (2010) 200–216.
[26] A.S. Ramadhas, S. Jayaraj, C. Muraleedharan, Use of vegetable oils as I.C. engine fuels—a review, Renew. Energy 29 (2004) 727–742.
[27] J. Tickell, From the Fryer to the Fuel Tank: The Complete Guide to Using Vegetable Oil as an Alternative Fuel, third ed., Joshua Tickell Publications, 2000.
[28] A. Demirbas, Progress and recent trends in biodiesel fuels, Energy Convers. Manag. 50 (2009) 14–34.
[29] S.S. L. Ranganathan, A review on biodiesel production, combustion, emissions and performance, Int. J. Adv. Sci. Tech. Res. 1 (2011) 86–106.
[30] Food and agricultural organization of United Nations, Statistics Division, 2012 ISSN 0081-4539, URL : http://www.fao.org/3/a-i3028e.pdf.
[31] M. Balat, Potential alternatives to edible oils for biodiesel production – a review of current work, Energy Convers. Manag. 52 (2011) 1479–1492.

[32] R.M. Baesso, P.A. Oliveira, G.C. Morais, A.V. Alvarenga, R.P.B. Costa-Felix, Using ultrasonic velocity for monitoring and analysing biodiesel production, Fuel 226 (2018) 389–399.
[33] B.T. Mohammad, M. Al-Shannag, M. Alnaief, L. Singh, E. Singsaas, M. Alkasrawi, Production of multiple biofuels from whole camelina material: a renewable energy crop, BioResources 13 (3) (2018) 4870–4883.
[34] S.X. Tan, S. Lim, H.C. Ong, Y.L. Pang, State of the art review on development of ultrasound-assisted catalytic transesterification process for biodiesel production, Fuel 235 (2019) 886–907.
[35] K.-S. Chen, Y.-C. Lin, K.-H. Hsu, H.-K. Wang, Improving biodiesel yields from waste cooking oil by using sodium methoxide and a microwave heating system, Energy 38 (2012) 151–156.
[36] R. Priambodo, T.-C. Chen, M.-C. Lu, A. Gedanken, J.-D. Liao, Y.-H. Huang, Novel technology for bio-diesel production from cooking and waste cooking oil by microwave irradiation, Energy Procedia 75 (2015) 84–91.
[37] S. Asif, L.F. Chuah, J.J. Klemeš, M. Ahmad, M.M. Akbar, K.T. Lee, et al., Cleaner production of methyl ester from non-edible feedstock by ultrasonic-assisted cavitation system, J. Clean. Prod. 161 (2017) 1360–1373.
[38] A. Bokhari, S. Yusup, L.F. Chuah, J.J. Klemeš, S. Asif, B. Ali, et al., Pilot scale intensification of rubber seed (*Hevea brasiliensis*) oil via chemical interesterification using hydrodynamic cavitation technology, Bioresour. Technol. 242 (2017) 272–282.
[39] S. Asif, M. Ahmad, A. Bokhari, L.F. Chuah, J.J. Klemeš, M.M. Akbar, et al., Methyl ester synthesis of *Pistacia khinjuk* seed oil by ultrasonic-assisted cavitation system, Ind. Crop. Products 108 (2017) 336–347.
[40] B. Salamatinia, H. Mootabadi, I. Hashemizadeh, A.Z. Abdullah, Intensification of biodiesel production from vegetable oils using ultrasonic-assisted process: optimization and kinetic, Chem. Eng. Process.: Process. Intensif. 73 (2013) 135–143.

CHAPTER 4

Conversion of rubber seed oil to biodiesel using continuous ultrasonic reactor

Huong Trinh and Suzana Yusup
Chemical Engineering Department, Universiti Teknologi PETRONAS, Ipoh, Malaysia

4.1 Introduction

Because fossil fuel is being depleted and causing serious harm to the environment, biodiesel production through a transesterification process has attracted attention from governments around the world. Overall global biodiesel production was increased by 5 billion liters from 2001 to 2006 [1]. Biodiesel production is currently underway at an industrial scale all over the world. The rapid growth of the biodiesel industry has caused the "food versus fuel" argument. Moreover, feedstock sources, especially edible oils, are limited and more than 70% of the total production cost are accounted by feedstock cost [2]. Thus finding a sustainable nonedible feedstock and developing an innovative, intensification process is necessary.

4.2 Rubber seed oil as feedstock

Rubber tree (*Hevea brasiliensis*) has its natural home in the Amazon. Thus the hot, wet climate of Malaysia is ideal for its growth. Malaysia is one of the top producers of natural rubber in the world [3]. In Malaysia, rubber plantations account for the second-largest portion of land after palm tree plantations. Normally, the lifetime of rubber is about one hundred years or more [4]. It is estimated that 800−1200 kg seed/ha can be harvested annually. Nevertheless, these are normally regarded as waste. According to the Department of Statistics Malaysia, Malaysia had an estimated acreage of 1,078,630 ha of rubber plantation in 2015 [5]. Therefore it is estimated that there are one million metric tons of rubber seed produced annually. Eka et al. [4] estimated that Malaysia can produce

Bioreactors
DOI: https://doi.org/10.1016/B978-0-12-821264-6.00004-8

355,200,000 kg of oil per year from rubber seed. A rubber seed contains 40%–50% of oil [6]. Compared to palm seed oil content, which is about 45%–50%, rubber seed oil (RSO) has the potential to be a feedstock source in biodiesel production in Malaysia.

In general, transesterification is used for vegetable oils with low free fatty acid (FFA) values because alkali catalysts can react with FFA to form soap [7]. An extra step involving an acid catalyzed reaction is required to convert the high amount of FFA into fatty acid methyl ester (FAME) before transesterification. Typically, the high FFA content in RSO has to be pretreated to reduce it to less than 1%. The pretreatment of FFA in RSO can be done by several methods such as distillation, acid esterification with methanol and sulfuric acid, esterification with ion exchange resins, or neutralization with alkali followed by soap separation in a decanter and extraction along with acid esterification [8]. Acid esterification is reported to be the most efficient method for FFA pretreatment. Basically, esterification is a reaction between FFA and alcohol in the presence of an acid catalyst [9]. In the esterification reaction, excess alcohol is also required because esterification is reversible.

4.3 Ultrasonic method in biodiesel production

Conventionally, the yield of biodiesel produced through transesterification is mainly affected by alcohol-to-oil molar ratio, catalyst concentration, reaction temperature, and stirrer speed [10]. In a chemical reaction, the diffusion of reactants plays an important role in the reaction rate. In the transesterification reaction, mass transfer of reactants is limited because oils and methanol are immiscible. In the case of a heterogeneous catalyzed transesterification, there are three different phases (oil, alcohol, and solid catalyst). The reaction usually occurs on the catalyst surface and inside the catalyst pores where reactant molecules make contact with the active sites of the catalyst [11]. It has been known that ultrasonic irradiation can dramatically increase the interface mixing and, thus, enhance the overall reaction rate. In addition, the ultrasonic technique has been mentioned to be commonly applied for a mass transfer intensification process between a solid and a liquid medium [12].

Ultrasonic or ultrasound is sound waves with frequencies greater than human audibility limits. It provides mechanical energy for vigorous mixing as well as the initial activation energy for transesterification [13]. Therefore it results in a faster reaction and higher biodiesel yield. The phenomenon called cavitation has the ability to create physical and chemical effects on the reaction [14]. A huge amount of cavitation bubbles is generated and these increase in size over many ultrasonic cycles. At the moment the collapse occurs, each bubble acts as a hotspot. Consequently, high temperature and pressure (up to 5000K and 1000 atm, respectively) occur throughout the system followed by cooling at a fast rate (up to 10^9 K/s) [15]. This results in vigorous mixing between the immiscible reactants and rapid heating throughout the system [15,16]. Several studies have shown benefits of using the ultrasonic method such as increase in yield, reaction time reduction, and saving energy [17–20]. Two common ultrasonic methods used in biodiesel production are indirect (bath-type) and direct (probe-type) methods. In an ultrasonic bath, ultrasonic waves are indirectly transmitted to the sample in a container through the liquid (normally water) in the bath [21]. In a probe-type ultrasonic reactor, ultrasonic energy is directly supplied into the sample through a probe that is submerged in the sample [21]. The ultrasonic bath method was used in early lab-scale synthesis, but was reported to be less efficient compared to the probe-type method [22]. Deng et al. [23] conducted a two-step biodiesel production process from *Jatropha curcas* L. oil (10.47 mg KOH/g acid value) using an ultrasonic bath, and the acid value was reduced to 1.2 mg KOH/g in 1 h with a 96.4% yield achieved in 30 min with a catalyst loading of 0.32 mg KOH/g. Worapun et al. [18] used a probe-type ultrasonic reactor to produce biodiesel from *J. curcas* L. oil (12.5% FFA). The FFA content was reduced to 3% in 20 min of acid esterification and a 98% biodiesel conversion was obtained in 30 min for base transesterification. The transesterification process in both these studies was twice as fast as the conventional process conducted by Berchmans and Hirata [24]. Reshad et al. [19] used a probe-type reactor to produce biodiesel from RSO. In the transesterification step using $Ba(OH)_2 \cdot 8H_2O$ as a catalyst, a 96% FAME conversion was achieved in 15 min at 30°C. On the other hand, the mechanical stirring method only obtained a 44.44% conversion in 1 h. In a previous study, a probe-type ultrasonic reactor (frequency of 20 kHz) was used to convert RSO to biodiesel. In the first step, an FFA conversion of 98% was found at an amplitude of 50% with 7.50 wt.% of H_2SO_4, a methanol-to-oil molar ratio of 23:1, a temperature of 50°C, and a reaction time of 30 min [20]. In the second step, a FAME content of 89.03% was found at a 33% amplitude, 9.8:1 molar ratio of methanol to oil, 3.16 wt.% of K_2CO_3/nano-MgO, a temperature of 60°C, and a 11.50 min reaction time [20]. Both processes showed positive influences on reaction time compared to mechanical stirring. When mechanical stirring was used, only 75% FFA and 79.83% FAME were obtained in esterification and transesterification, respectively. Batch processing has been commonly studied for biodiesel production at both laboratory and industrial scales. However, batch processing has several drawbacks compared to continuous processing. It requires larger reactor volumes, longer reaction times, and higher investment costs for the same product capacity [25]. Studies [26,27] have shown that the continuous flow reactor has the potential to be up-scaled for biodiesel production. Cintas et al. [27] reported that a transesterification of 1.0 L of soybean oil was completed in 1 h with a flow rate of 55 mL/min. Energy consumption was estimated around 0.28 kWh/L. Boffito et al. [26] reported that a 90% biodiesel yield was obtained within 18 s with 3.6 kJ of energy input. The aim of the study described in Sections 4.4 and 4.5 was to test the scale-up potential of a continuous ultrasonic reactor using previous optimum conditions in a batch process of biodiesel production from RSO.

4.4 Material and methodology

4.4.1 Materials

In the present work, RSO was imported from Vietnam by Kinetics Chemicals (M) Sdn. Bhd. Malaysia. The characterization of RSO as a feedstock is given in Table 4.1. All the chemicals and reagents such as sulfuric acid (98%), potassium carbonate (99%), methanol (99%), and magnesium oxide were attained through Merck (Malaysia). The base solid catalyst, K_2CO_3/nano-MgO, for the transesterification step has a BET surface area of 17.10 m^2/g and an average crystallite size ranging between 15.5 and 41.2 nm. Methyl ester standards for gas chromatography were purchased from Sigma Scientific Chemicals.

Table 4.1 Properties of RSO feedstock.

Parameter	This work	Chuah et al. [28]
Acid value (mg KOH/g oil)	80.28	85
FFA value (%)	40.14	42.5
Iodine value (g I_2/g oil)	144	131.8
Peroxide value (mg KOH/g oil)	1.9	–
Saponification value (mg KOH/g oil)	205.5	201.5

FFA, Free fatty acid; *RSO*, rubber seed oil.

4.4.2 Experiments

Ultrasonic-assisted esterification and transesterification reactions were carried out in a Sonitube (Bench top 35 kHz SONITUBE). It outputs up to 800 W of acoustic power and operates at a frequency of approximately 35 kHz. Though the operation amplitude of the reactor ranges from 50% to 100%, low amplitudes are preferable as the operation must be strictly maintained under 60°C for safety reasons.

Fig. 4.1 describes the continuous ultrasonic reactor setup. A mixture of catalyst and methanol was mixed in a 2000 mL beaker, and then the required amount of oil was added. The sample was stirred by a magnetic stirrer for 10 min before being pumped through the ultrasonic reaction tube (70 mL capacity). After the reaction was carried out under irradiation, the flow of the sample was recirculated back to the beaker. The flow rate was set at 1000 mL/min. A 30 mL sample was taken out for analysis after an expected time. In order to evaluate the performance of the continuous ultrasonic reactor, the experiments were carried at the optimum conditions, which were obtained from the batch ultrasonic process. During the reaction, the reaction temperature was in uncontrolled operation and increased under the ultrasonic effect. A temperature probe was equipped to measure the temperature of the medium. Since it is recommended that the operating temperature should not exceed 60°C, the experimental amplitude was selected in the range of 50%–80% for safety reasons. As observed, the higher the amplitude, the faster the rise of reaction temperature. At an amplitude of 80%, the temperature reached 50°C within 4 min of reaction.

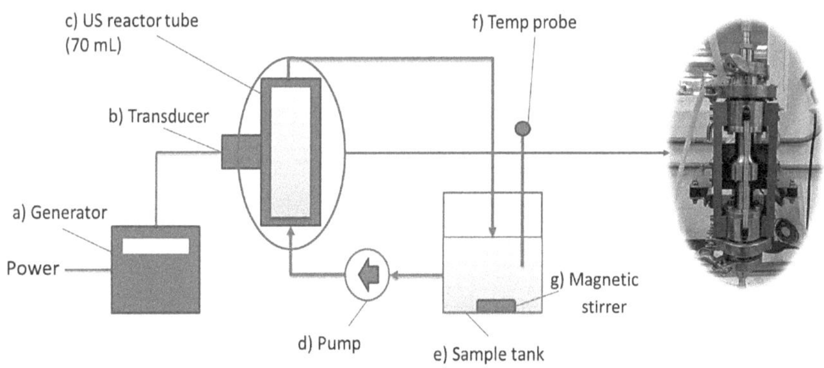

Figure 4.1 The continuous ultrasonic reactor setup.

In the acid esterification step using H_2SO_4 as the catalyst, the collected sample was put into a separating funnel and left for 6 h to separate the esterified oil and the excess of methanol and acid catalyst. The upper layer contained the methanol, while the acid catalyst was discarded after separating. The treated oil was washed a few times with warm water to remove impurities. After that, the remaining methanol and water were evaporated using a rotary evaporator for 1 h at 60°C. Anhydrous sodium sulfate was added to the product to remove the water and this was filtered using filter paper. Ultimately, the acid value of the product was determined based on AOCS-Cd 3d-63 [29] and calculated as:

$$\text{Acid value} = \frac{(A - B) \times M \times 56.1}{m}, \text{mg KOH/g oil} \qquad (4.1)$$

where A is the volume of titrant solution used in the titration of the sample (mL), B is the volume of titrant solution used in the titration of the blank (mL), M is the molarity of the titrant solution (mol/L), and m is the mass of the sample (g).

The FFA content was calculated using Eq. (4.2):

$$\text{FFA}(\%) = \frac{\text{Acid value}}{1.99} \qquad (4.2)$$

In base transesterification, the experimental setup was similar to the acid esterification, but the separation and purification procedure was different because at this stage a solid catalyst (K_2CO_3/nano-MgO) was used. The solid catalyst was mixed with methanol by agitation and pretreated oil was added to this mixture. The collected sample was centrifuged, and it formed three phases. The liquid was transferred into a separation funnel to separate the glycerol and methyl esters. After 2 h of separating, the upper layer (methyl esters) was distilled under vacuum (60°C, 1 h) to remove excess methanol. To remove traces of water, the oil was dried over anhydrous sodium sulfate and then filtered. The methyl esters were collected for chromatographic analysis to measure FAME content, Eq. (4.3).

$$\text{FAME}(\%) = \frac{\sum A - A_{\text{IS}}}{A_{\text{IS}}} \times \frac{W_{\text{IS}}}{W} \times 100 \qquad (4.3)$$

where $\sum A$ is the total peak area from the methyl ester including nonadecanoic acid methyl ester C19, A_{IS} is the peak area corresponding to nonadecanoic acid methyl ester C19, W is the weight of the sample (g), and W_{IS} is the weight of nonadecanoic acid methyl ester C19 (g).

4.5 Results and discussion

During the reaction, the reaction temperature was in uncontrolled operation and increased under the ultrasonic effect. A temperature probe was equipped to measure the temperature of the medium. Since it is recommended that the operating temperature should not exceed 60°C, the experimental amplitude was selected in the range of 50%–80% for safety reasons. As observed during the reactions, the higher the amplitude, the faster the rise of reaction temperature. At an amplitude of 80%, the temperature reached 50°C within 4 min of reaction.

4.5.1 Esterification process

Fig. 4.2 depicts the increase in FFA conversion when the amplitude was increased from 50% to 80% at a reaction time of 4 min. Maximum conversion was found at 80% due to the better mixing effect along with better temperature condition. Thus an amplitude of 80% was used for esterification process. Fig. 4.3 describes the reaction time from 20 to 240 s (time interval of 20 s). The FFA conversion sharply increased after the first 20 s and gradually reached the equilibrium stage at 95.95% within 240 s.

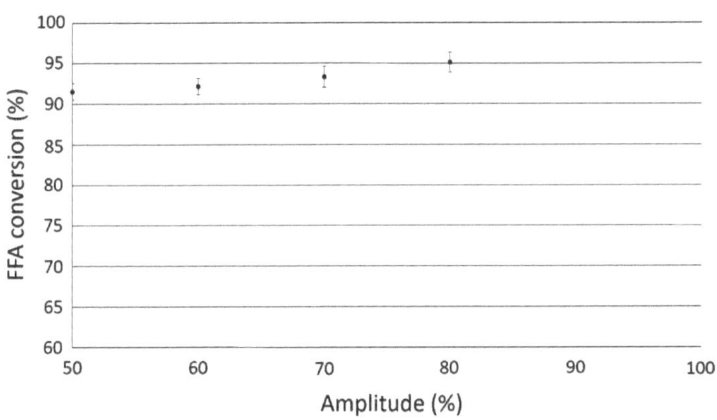

Figure 4.2 FFA conversion at different amplitudes (50%–80%) (23:1 methanol:oil, 7.5 wt.% H_2SO_4 and 4 min reaction). *FFA*, Free fatty acid.

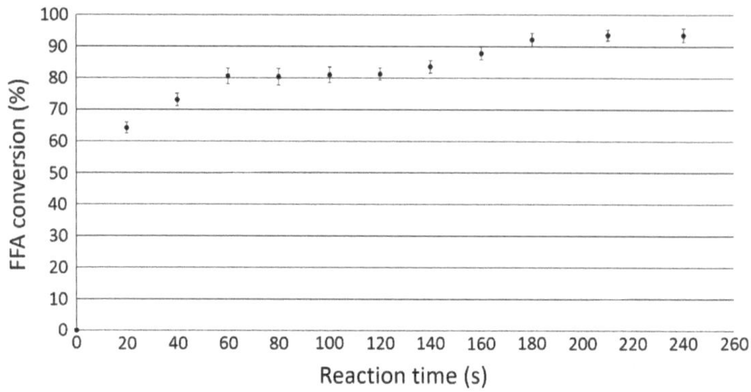

Figure 4.3 Increase in FFA conversion from 0 to 240 s reaction time (80% amplitude, 23:1 methanol:oil, and 7.5 wt.% H_2SO_4). *FFA*, Free fatty acid.

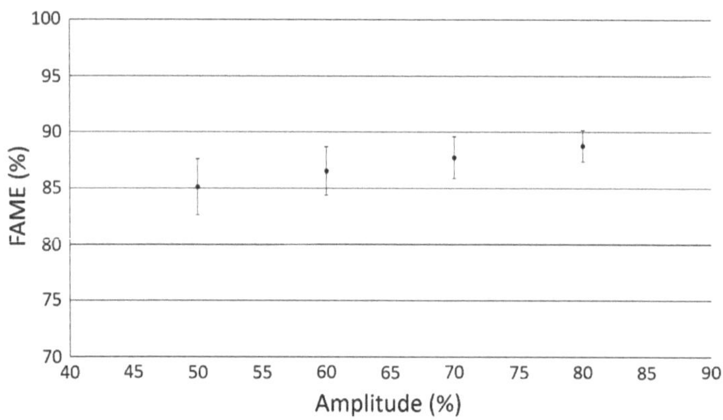

Figure 4.4 FAME content at different amplitudes (50%–80%) (9.8:1 methanol:oil, 3.16 wt.% 60%K_2CO_3/nano-MgO, and 4 min reaction). *FAME*, Fatty acid methyl ester.

4.5.2 Transesterification process

The FAME content at different amplitudes is described in Fig. 4.4. Clearly, the amplitude of 80% yielded the highest FAME percentage. The trend of FAME content over time at 80% amplitude is given in Fig. 4.5. The finding was found to be similar to the results of the esterification process. The FAME content quickly reached the highest value (89.23%) at 120 s and remained almost the same afterward. A more rapid reaction via a continuous ultrasonic system was also reported by Thanh et al. [30] and Mostafaei et al. [31] in the transesterification of waste oil. Mostafaei et al. [31] reported that a maximum reaction yield of 91.12% could be obtained within 1 min with a continuous reactor, while it took about 6 min with a

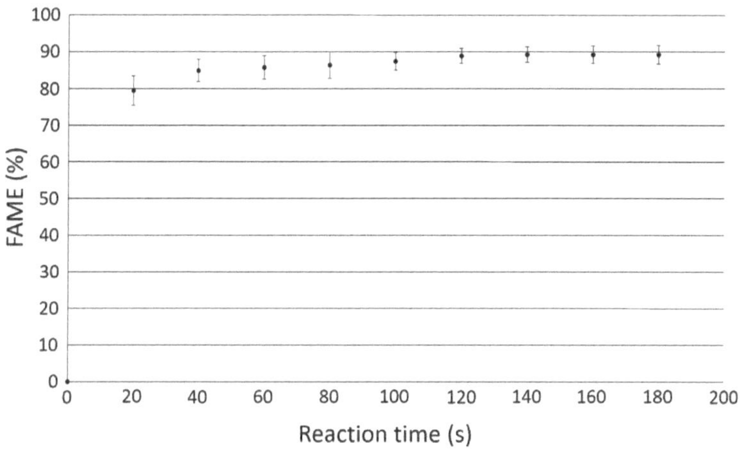

Figure 4.5 Increase in FAME content from 0 to 180 s reaction time (80% amplitude, 9.8:1 methanol:oil, and 3.16 wt.% 60%K_2CO_3/nano-MgO). *FAME*, Fatty acid methyl ester.

batch reactor [31]. Two-step transesterification conducted by Thanh et al. [30] took only 0.93 min for the entire process to obtain a 99% FAME yield.

4.5.3 Upscale potential of continuous sonitube reactor

As shown in Table 4.2, the reactions carried out in the continuous system could occur in a mild condition (45°C–50°C) at a very fast rate. Thus the continuous system was found to be more efficient in terms of reaction time and energy usage. Moreover, it can produce 20–40 times the capacity than the batch system. Although the optimum FFA content in the continuous process was slightly higher than that in the batch process, the maximum FAME content of the continuous process was found to be similar to that of the batch one. The potential cause of the lower conversion in the continuous process is the poor magnetic mixing between the initial feedstock and the output product in the mixing tank of the system. It can be concluded that the continuous Sonitube reactor can be used for upscaling the biodiesel production process at low temperatures.

4.6 Conclusion

Both the batch ultrasonic reactor and the continuous reactor showed an increased reaction rate compared to conventional methods. Experiments

Table 4.2 Results of the batch system and the continuous system.

Parameters	Esterification process		Transesterification process	
	Batch reactor (20 kHz) [20]	Continuous reactor (35 kHz)	Batch reactor (20 kHz) [20]	Continuous reactor (35 kHz)
Capacity (mL)	50	1000	25	1000
Amplitude (%)	50	80	33	80
Methanol:oil (mol:mol)	23:1	23:1	9.8:1	9.8:1
Catalyst concentration (wt.%)	7.5	7.5	3.16	3.16
Reaction time (min)	30	3.5	11.5	2.3
Temperature (°C)	50	50	60	45
Optimum FFA/FAME (%)	FFA = 0.75	FFA = 1.63	FAME = 89.03	FAME = 89.23
Power (W)	45.14	310	28.04	310
Energy consumption (kJ)	62.29	65.1	19.35	42.78
Energy consumption (kJ/mL)	1.26	0.07	0.77	0.04

FAME, Fatty acid methyl ester; *FFA*, free fatty acid.

in the continuous system demonstrated improvements with respect to reaction time and energy saving. In the esterification process, the continuous Sonitube reactor achieved a 95.95% FFA conversion in 3.5 min, while the batch reactor required 30 min to achieve a 98% conversion. Energy consumption was only 0.07 kJ/mL, 18 times less than when using the batch reactor. In the transesterification process, 89.23% FAME was obtained in 2.3 min using the continuous reactor, whereas the batch reactor took 11.5 min for 89.03% FAME. Similarly, energy consumption was 19 times less than in the batch reactor. In addition, the capacity of the Sonitube reactor is 20–40 times larger than in the batch reactor. The challenge with the Sonitube reactor was that it required to be used with care to ensure operation at a low temperature. Operating at low temperature is also safer for the operator since it can minimize the formation of methanol vapor, which is flammable and toxic.

Acknowledgment

The authors would like to express gratitude to University Teknologi PETRONAS and Exploring Research Grant Scheme (ERGS) for the support given to conduct the research work.

References

[1] W. Institute, Biofuels for Transport Global Potential and Implications for Sustainable Energy and Agriculture, Earthscan, London, UK, 2007.

[2] M. Tabatabaei, K. Karimi, R. Kumar, I.S.L. Horváth, Renewable energy and alternative fuel technologies, BioMed. Res. Int. 2015 (2015).

[3] C.R. Soccol, S.K. Brar, C. Faulds, L.P. Ramos, Recent developments and innovations in solid state fermentation, Green Fuels Technology: Biofuels, Springer, 2016.

[4] H. Eka, Y. Tajul Aris, W. Wan Nadiah, Potential use of Malaysian rubber (*Hevea brasiliensis*) seed as food, feed and biofuel, Int. Food Res. J. 17 (2010) 527–534.

[5] M.R. Board, Natural Rubber Statistics 2016, Malaysian Rubber Board, Malaysia, 2016.

[6] Y.B.A. Ghazali, Biobased Products From Rubber, Jatropha and Sunflower Oil, University of Groningen, 2015.

[7] M. Chai, Q. Tu, M. Lu, Y.J. Yang, Esterification pretreatment of free fatty acid in biodiesel production, from laboratory to industry, Fuel Process. Technol. 125 (2014) 106–113.

[8] N.S. El-Gendy, S.F. Deriase, Application of statistical approaches to optimize the productivity of biodiesel and investigate the physicochemical properties of the bio/petro-diesel blends, High-Performance Materials and Engineered Chemistry, Apple Academic Press, 2018, pp. 161–239.

[9] R.E. Babcock, E.C. Clausen, M. Popp and W.B. Schulte, Yield characteristics of biodiesel produced from chicken fat-tall oil blended feedstocks; Completion report project number MBTC-2092, 2007, Mack-Blackwell Rural Transportation Center, Fayetteville, AR, USA.

[10] F. Al Basir, P.K. Roy, Effects of temperature and stirring on mass transfer to maximize biodiesel production from jatropha curcas oil: a mathematical study, Int. J. Eng. Math. 2015 (2015).

[11] R. Klaewkla, M. Arend, W.F. Hoelderich, A review of mass transfer controlling the reaction rate in heterogeneous catalytic systems, Mass Transfer-Advanced Aspects, IntechOpen, 2011.

[12] A. Mulet, J. Carcel, C. Benedito, C. Rosselló, S. Simal, Ultrasonic mass transfer enhancement in food processing, Transp. Phenom. Food Process. 18 (2003) 265–278.

[13] J.C. Serrano-Ruiz, Advanced Biofuels: Using Catalytic Routes for the Conversion of Biomass Platform Molecules, CRC Press, 2015.

[14] C. Stavarache, M. Vinatoru, R. Nishimura, Y. Maeda, Fatty acids methyl esters from vegetable oil by means of ultrasonic energy, Ultrason. Sonochem. 12 (2005) 367–372.

[15] K.S. Suslick, Sonochemistry, Science 247 (1990) 1439–1445.

[16] D.R. Fernandez, S. Kuhn, Synergy of microfluidics and ultrasound: process intensification challenges and opportunities, Top. Curr. Chem. 374 (2016). 70-70.

[17] X. Deng, Z. Fang, Y.-H. Liu, Ultrasonic transesterification of *Jatrophacurcas L.* oil to biodiesel by a two-step process, Energy Convers. Manag. 51 (2010) 2802–2807.

[18] I. Worapun, K. Pianthong, P. Thaiyasuit, Two-step biodiesel production from crude *Jatropha curcas L.* oil using ultrasonic irradiation assisted, J. Oleo Sci. 61 (2012) 165–172.

[19] A.S. Reshad, D. Panjiara, P. Tiwari, V.V. Goud, Two-step process for production of methyl ester from rubber seed oil using barium hydroxide octahydrate catalyst: process optimization, J. Clean. Prod. 142 (2017) 3490–3499.

[20] H. Trinh, S. Yusup, Y. Uemura, Optimization and kinetic study of ultrasonic assisted esterification process from rubber seed oil, Bioresour. Technol. 247 (2018) 51–57.

[21] I. Mahbubul, Preparation, Characterization, Properties, and Application of Nanofluid, William Andrew, 2018.
[22] N. Dhanalakshmi, R. Nagarajan, Ultrasonic intensification of the chemical degradation of methyl violet: an experimental study, Int. J. Chem. Mol. Nucl. Mater. Metall. Eng. 59 (2011) 1019–1024.
[23] X. Deng, Z. Fang, Y.-H. Liu, Management, Ultrasonic transesterification of *Jatrophacurcas L.* oil to biodiesel by a two-step process, Energy Convers. 51 (2010) 2802–2807.
[24] H.J. Berchmans, S.J.B.T. Hirata, Biodiesel production from crude *Jatropha curcas L.* seed oil with a high content of free fatty acids, Bioresour. Technol. 99 (2008) 1716–1721.
[25] H.-S. Zou, J. Chai, A novel ultrasonic reactor for continuous production of biodiesel from waste acid oil, Korean J. Chem. Eng. 34 (2017) 353–359.
[26] D.C. Boffito, J.-M. Leveque, C. Pirola, C. Bianchi, R. Vibert, A. Perrier, et al., Batch and continuous ultrasonic reactors for the production of methyl esters from vegetable oils, Production of Biofuels and Chemicals withUltrasound, Springer, 2015, pp. 87–114.
[27] P. Cintas, S. Mantegna, E.C. Gaudino, G. Cravotto, A new pilot flow reactor for high-intensity ultrasound irradiation. Application to the synthesis of biodiesel, Ultrason. Sonochem. 17 (2010) 985–989.
[28] L.F. Chuah, S. Yusup, A.R. Abd Aziz, J.J. Klemeš, A. Bokhari, M.Z. Abdullah, Influence of fatty acids content in non-edible oil for biodiesel properties. J. Clean Technol. Environ. Pol. 18 (2016) 473–482
[29] A.O.C. Society, AOCS Official Method Cd 3d-63: Acid Value. In Official Methods and Recommended Practices of the AOCS, fifth ed., American Oil Chemists' Society, IL, 1999.
[30] L.T. Thanh, K. Okitsu, Y. Sadanaga, N. Takenaka, Y. Maeda, H. Bandow, A two-step continuous ultrasound assisted production of biodiesel fuel from waste cooking oils: A practical and economical approach to produce high quality biodiesel fuel, Bioresour. Technol. 101 (2010) 5394–5401.
[31] M. Mostafaei, B. Ghobadian, M. Barzegar, A. Banakar, Optimization of ultrasonic assisted continuous production of biodiesel using response surface methodology, Ultrason. Sonochem. 27 (2015) 54–61.

CHAPTER 5

Conversion of biomass into biofuel: a cutting-edge technology

Md. Saiful Alam and Md. Sifat Tanveer
Department of Petroleum & Mining Engineering, Shahjalal University of Science and Technology, Sylhet, Bangladesh

5.1 Introduction

Power plays an important role in today's modern society and the demand for power in our daily lives is increasing day by day. Currently about 70% of the energy demand is met by burning fossil fuels such as natural gas, oil, and coal. After being utilized, these fossil fuels cannot be reproduced in a relatively short time and are, therefore, known as nonrenewable resources. Due to the limited availability of these nonrenewable resources, the world is gradually moving toward severe energy crises [1]. Increasing environmental pollution, in addition to global warming caused by the use of fossil fuels and the depletion of these limited fuels, are leading researchers to search for alternative energy sources to meet future energy demands. Therefore the utilization of renewable energy sources is expected to play a key role in meeting future energy demands and creating a sustainable world for the next generation.

5.1.1 Biomass potential

From the beginning of civilization, biomass has been considered as the major sources of energy throughout the world. According to Svetlana et al., nearly 50% of the world's population meets their demand of primary energy sources from biomass [2]. The global demand for biomass-based energy is continuously increasing and has doubled in the past four decades. Biomasses available for energy production include agricultural and forest crops, agricultural and forest wastes and residues, and waste materials from food processing, fisheries, and other industries as well as households. Energy produced from biomass can easily be converted into

Bioreactors
DOI: https://doi.org/10.1016/B978-0-12-821264-6.00005-X

all the other forms of energy currently required in modern society. Besides the availability of the sources, compared to renewable energy, biomass is the most common and widespread of all other global renewable resources. Biomass has emerged as the fourth largest energy resource after fossil fuels such as natural gas, oil, and coal, and is estimated at about 10% of global primary energy resources [3].

Thus biomass is expected to be a potential source of renewable energy, for both small- and large-scale production around the world. But the future possibility of biomass-based energy depends to a great extent on the availability of cultivable land. Currently, the amount of land utilized to grow energy crops and raw materials for biofuels is only 0.19% of the world's total land area of 13.2 billion hectares. On the other hand, only 0.5%–1.7% of global agricultural land is used for the cultivation of raw materials for biofuels production [4]. To meet future demands for the production of biofuels, much larger areas must be considered for the cultivation of more energy crops. The World Biomass Association estimates that by 2035, about 5% of global agricultural land (240 Mha) can be used for growing dedicated energy crops for biofuels [5]. Moreover, agricultural productivity needs to be increased to cut down the use of additional forest land by employing new and advanced technologies. In addition, new energy crops that are not yet cultivated will need to be planted. Special attention needs to be given to ensure the sustainable production of new energy crops since the uncontrolled production of biomass may create detrimental effects on biodiversity, soil, and water, etc.

5.1.2 Carbon dioxide emissions and carbon cycle

Carbon dioxide (CO_2) is considered to be one of the major greenhouse gases (GHGs). It has now been proven that global warming is accelerating due to increasing GHG emissions. Emitted CO_2 produced from the combustion of fossil fuels remains trapped in the Earth's atmosphere for thousands of years, and is considered as one of the major causes of global warming. Although the combustion of biofuels produces CO_2 similar to that from fossil fuels, biofuels are considered to be CO_2 neutral because the plant feedstock used during the production of biofuels absorbs CO_2 from the atmosphere during photosynthesis. After the biomass is converted into biofuel and burned in a combustion process CO_2 is released back into the atmosphere. The carbon cycle therefore is closed.

5.2 Classification of biofuels

Biomass can be converted into biofuel through three main process technologies, namely biochemical, thermochemical, and physiochemical. Anaerobic digestion (to produce biogas and biomethane) and fermentation (to produce ethanol) are considered as biochemical conversion processes. Pyrolysis, gasification, combustion, and hydrothermal processing are categorized as thermochemical conversion processes. Extraction and esterification are treated as physiochemical conversion [6]. Unlike fossil fuels, biofuels can also be served in a number of different forms and meet all sorts of consumer requirements.

5.2.1 Classification based on feedstock

Depending on the type of feedstock used in biofuel production, biofuels can be subdivided into three categories. (http://biofuel.org.uk).

5.2.1.1 First generation biofuels

Biofuels said to be first generation are directly related to biomass produced from edible oils such as rapeseed, soybeans, sunflower, safflower, palm oil, coconut, and peanut [7,8]. If these feedstocks are utilized on a large scale, it would have a large impact on global food supply and, therefore, they are not considered to be sustainable or green. First-generation biofuels represent most of the biofuels currently in use.

5.2.1.2 Second generation biofuels

Biofuels that are produced from a variety of different feedstocks, ranging from lignocellulosic feedstocks to municipal solid wastes are known as second-generation biofuels. In this sense, these biofuels are "greener" as they are made from sustainable feedstocks and can be reproduced. The term sustainable is defined mainly by the availability of the source of the feedstock, the impact of its use on GHG emissions, and its ultimate effect on biodiversity. At this point, most second-generation biofuels are under various stages of experimentation and development, and are not widely available to consumers.

5.2.1.3 Third generation biofuels

The term third generation biofuel refers to biofuel mainly made from algae. Previously algae were lumped in with second-generation biofuels.

Now they have been moved into their own category due to their relatively high yields and low resource inputs compared to other feedstocks.

5.2.2 Classification based on biofuel nature

According to the chemical nature of biofuels, they can be classified into various categories including biodiesel, bioethanol, biomethane, biohydrogen, and so forth. Sections 5.2.2.1–5.2.2.5 describe how various biofuels are produced from biomass feedstock.

5.2.2.1 Biodiesel

5.2.2.1.1 Introduction

Biodiesel refers to any diesel fuel substitute that is derived from any renewable resources. More specifically, biodiesel refers to a family of products, called alkyl esters of fatty acids, made from vegetable oils or animal fats in combination with alcohol such as methanol or ethanol. One popular process for producing biodiesel is transesterification. Biodiesel is made from a variety of natural oils, especially rapeseed oil (a close cousin of canola oil) and soybean oil. There are many other feedstock candidates including recycled cooking oils, microalgae, animal fats, and other oilseed crops. The potentiality of fuel production from microalgae, animal fats, and waste oils is expected to be very high [9].

5.2.2.1.2 Biodiesel production

A simple process flow diagram of biodiesel production from biomass (rapeseed) feedstock is shown in Fig. 5.1. The whole process mainly consists of four important steps, namely feedstock preparation, oil extraction, oil refining, and transesterification, which are described here.

Feedstocks preparation. The preparation of seeds involves the removal of the outer layers of the fruit and the separation of kernels or seeds from the fruits. The kernels or seeds then have to be dried to make the moisture content optimum for high oil extraction [10]. For example, a 15% moisture content in the seeds or kernels of beauty leaf (Calophyllum *inophyllum*) provides the maximum oil yields in both mechanical and solvent extraction methods [11].

Oil extraction. The most important step in the production of biodiesel is oil extraction. Currently, different methods and techniques are available for oil extraction such as (1) mechanical extraction, (2) chemical or solvent extraction; (3) enzymatic or biological extraction, (4) accelerated solvent extraction, (5) supercritical fluid extraction, and (6)

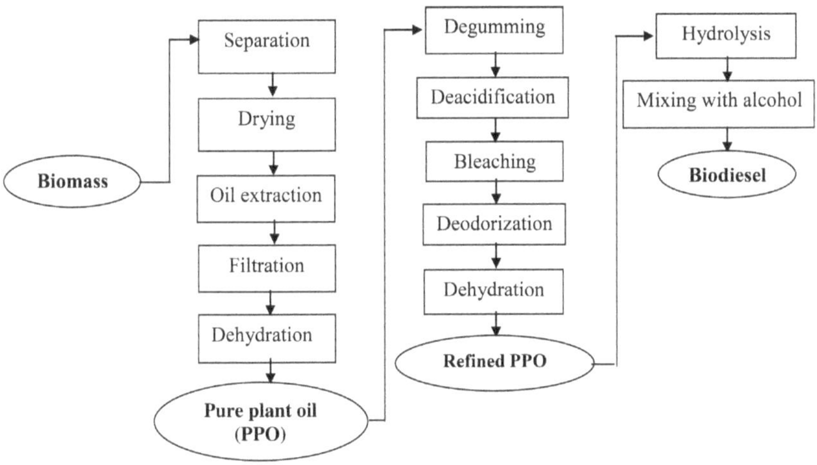

Figure 5.1 A simple process flow diagram of biodiesel production from biomass (rapeseed) feedstock.

microwave-assisted extraction [7,10]. Among these, mechanical extraction and solvent extraction are the most commonly used commercialized methods of oil extraction [12]. In mechanical extraction, the oil is extracted from the oleaginous material by applying high pressure. Various types of mechanical devices such as the Ram Press, Bridge Press, Hydraulic Press, Ghani, and Screw Presses or expellers, etc., are currently in use to extract oil [13]. In chemical extraction, also known as solvent extraction, a specific or desired component is isolated from solids or liquids using a solvent. Sometimes, the solvent extraction method is used to extract residual oil present in the seed cake due to an incomplete oil extraction by mechanical methods. Among different types of solvent extraction, liquid–liquid and solid–liquid extractions are found to be the most common [13]. The pressed cake left over after the extraction of oil can be used as fertilizer for soil enrichment and protein-rich feed for animals [6,12]. The extracted oil is then filtered and dehydrated to produce pure plant oil (PPO).

Oil refining. The produced PPO contains many undesired components such as phosphatides, free fatty acids (FFAs), and colorants, etc. These need to be removed through several refining steps. First, phosphatides are removed by a process called degumming. Degumming can be accomplished either by water degumming or by acid degumming. In the water degumming process, water is mixed with the oil at 60°C–90°C where soluble phosphatides are hydrated and separated by centrifugal separation. Acid degumming is done through the addition of acid substances such as citric or phosphoric acid to remove phosphatides that cannot be hydrated in water. The second refining step is deacidification where FFAs are removed. Several methods for deacidification such as neutralization with alkali, distillation, and esterification, etc., are applied today. After deacidification, colorants are removed through a bleaching step to enhance the storage life of the biofuel. In this step, bleaching earth, silica gel, or activated carbon are generally used as adsorbing substances. Oxygen, ozone, and hydrogen peroxide can also be used for bleaching. Colorants are also removed by simply heating the biofuel to 200°C. After bleaching, a deodorization step (steam distillation) is conducted to remove odorous substances such as ketone and aldehyde. Finally a dehydration step is conducted to remove water through a distillation process under reduced pressure [14].

Transesterification. According to Oh et al., oil produced from energy crops can be used directly for combustion in diesel engines [15]. However, a viscous plant oil always generates poor atomization in the combustion chamber creating other operational problems in the process. To solve this problem, transesterification has gained attention among all other methods due to the simultaneous dilution of the plant oil during the process [16]. In the transesterification process, the molecular structure of lipid molecules of refined oil is converted into methyl or ethyl esters (biodiesel) and glycerin as a byproduct. The lipid molecules are hydrolyzed to produce FFAs. Then the produced fatty acids are converted into fatty acid esters by mixing with methanol or ethanol. The final mixture is placed in a separatory funnel to settle down the glycerin and the biodiesel (methyl- or ethyl-ester) into two layers. After settling, the bottom layer of glycerin is removed and the remaining biodiesel is sent to the next purification process. The transesterification reaction is usually conducted using several catalysts such as alkaline material, acidic material, silicates, and lipases, etc. In most cases, transesterification is carried out using methanol for the production of biodiesel. Due to its relatively low cost and high reactivity, methanol is of a higher priority than other alcohols. However, the use of bioethanol in transesterification processes is considered to be an alternative to methanol due to its renewability and less-toxic nature. It also improves the quality of the resulting fuel with regards to heating value and cetane numbers [17]. A general chemical transesterification reaction is shown in Fig. 5.2.

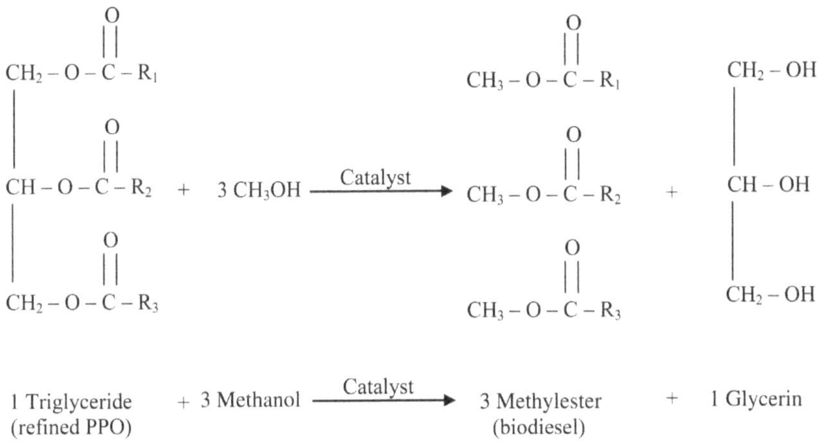

Figure 5.2 General chemical transesterification reaction using methanol.

5.2.2.1.3 Properties of pure plant oil and biodiesel

PPO exhibits a wide range of differences in primary fuel properties when compared with fossil diesel. A comparison of common fuel properties of diesel, PPO (rapeseed oil), and biodiesel is given in Table 5.1. The table indicates that the viscosity of PPO is 10 times higher than that of fossil diesel. This makes PPO difficult to blend with conventional diesel fuel. But for the higher flashpoint of PPO over diesel, it is safe for storage and transportation, and is easy to handle in gaseous environments. As shown in Table 5.1, the viscosity and other ignition properties of biodiesel are quite similar to that of the fossil diesel. Although the energy content of biodiesel is lower than that of diesel fuel, the complete combustion of biodiesel in combustion engines makes them advantageous over diesel fuel. This is due to the presence of an oxygen content in biodiesel. Since biodiesel contains practically no sulfur, emissions of sulfur oxides from biodiesel combustion is zero. However, some problems may occur during any long-term storage of biodiesel as it readily oxidizes, but these problems can be minimized using additives.

5.2.2.2 Bioethanol

5.2.2.2.1 Introduction

Bioethanol has evolved as a potential source for biofuel production at the current state of energy insecurity and in environmental safety challenges over fossil. Broad classes of biomass resources have been investigated and experimented on for bioethanol production, which can be categorized

Table 5.1 Some properties of biofuels in comparison with fossil fuels [6,18].

Fuels	Density (kg/L)	Viscosity (mm^2/s)	Flash point (°C)	Calorific value (at 20°C MJ/kg)	Octane number (–)	Cetane number (–)
Diesel	0.84	5	80	42.7	–	50
PPO (rapeseed oil)	0.92	74	240	37.6	–	40
Biodiesel	0.88	7.5	120	37.1	–	56
Petrol	0.76	0.6	<21	42.7	92	–
Bioethanol	0.79	1.5	<21	26.8	>100	–
Ethanol	0.785	–	13	27.0	108	2–12
Gasoline	0.791	–	–43	43.9	>90	–
Biobutanol	0.81	3.7	36	33.2	96	17

into sugar product, starch, and lignocellulosic biomass [19]. Different agricultural wastages, leaves, waste wood, and residues from forest, municipal solid wastages, and wastages from pulp/paper processing and energy crops are common cellulosic biomasses. Crop residues such as wheat and rice straw, corn leaves, stalks and cobs, sugarcane bagasse, and residues from sugar production are the main cellulosic agricultural wastes. Forestry waste includes the logs and roots of trees left in forests after wood production. Municipal solid wastes includes different paper packages, paper and cardboard are in higher percentages of cellulosic materials. On the other hand, dedicated energy crops that are planted specially for ethanol production include warm season grasses (switchgrass) and fast growing trees such as poplars and shrubs (willows). The cellulosic components of these energy crops range between 30% and 70%. The concept of utilizing cellulosic feedstock for bioethanol production has not yet started on a commercial scale, and intensive research of this subject is required. Although, the large-scale production of agricultural ethanol requires substantial amounts of cultivable land with fertile soils and a supply of fresh water for irrigation. These limitations of ethanol production could not attract entrepreneurs from densely populous regions and industrially occupied regions like Western Europe where desire for new cultivable land exploiting rainforests arose environmental concerns.

5.2.2.2.2 Production of bioethanol

A simple process flow diagram of bioethanol production from sugar, starch, and cellulose is shown in Fig. 5.3. Although the conversion routes

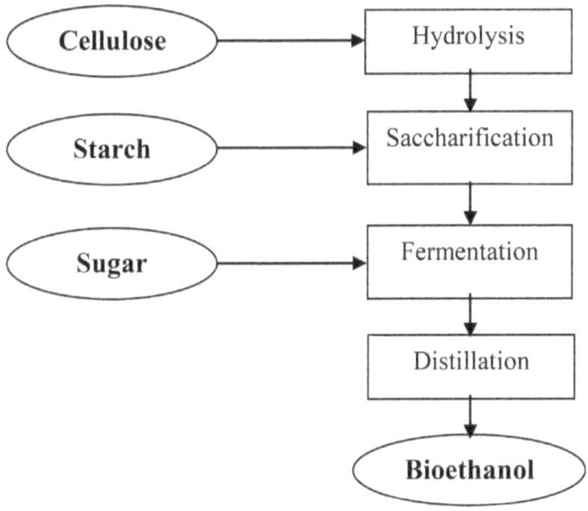

Figure 5.3 A simple process flow diagram of bioethanol production from sugar, starch, and cellulose.

of bioethanol production vary depending on the source of feedstock, the steps listed are considered to be the main steps.

- The reduction of particle size by milling/grinding of feedstock.
- Conversion of cellulose and starch into fermentable sugar (glucose) in the presence of an enzyme by a saccharification process.
- Dilution of the fermentable sugars with water.
- Addition of bacteria, yeasts, or other microorganisms.
- Conversion of sugar into bioethanol and CO_2 through a fermentation process.

Sugar-to-ethanol process. In this simple process, biomass containing six-carbon sugars can be fermented directly to ethanol. The extraction of sugar is achieved by grinding up the feedstock. Yeast is added to the extracted sugar for the fermentation process. Fungi, bacteria, or yeast microorganisms can be added to the extracted sugar for fermentation. In a closed anaerobic chamber, the yeast secretes enzymes that digest the sugar and produce mainly ethanol and CO_2 [20]. Sugarcane, sugar beet, sweet sorghum, and other plants containing a large proportion of simple sugars are considered to be the most common feedstocks for ethanol production.

Starch-to-ethanol process. The production of bioethanol from starch mainly consists of saccharification and fermentation processes. During the saccharification process long chains of glucose molecules in starchy material are broken into simple glucose molecules using enzymatic hydrolysis. Then the conversion of simple glucose molecules is accomplished by adding it to yeast in the fermentation process. The yeast secretes enzymes that digest the simple glucose, yielding mainly ethanol and CO_2 [20].

Cellulose-to-ethanol process. The process of ethanol production from cellulosic biomass is quite complicated compared to the process of ethanol production from sugar or starch. A large portion of cellulosic materials contain lignin, hemicellulose, and cellulose. Bioethanol production from these cellulosic materials requires an additional pretreatment process before the saccharification and fermentation processes. In the pretreatment process, cellulosic material is converted into glucose molecules either by enzymatic hydrolysis or acid hydrolysis. This treatment breaks down the lignin structure and disrupts the crystalline structure of cellulose [21]. The saccharification process aims to convert cellulose into glucose using enzymatic hydrolysis and then glucose is converted into ethanol using microorganisms through a fermentation process [22].

5.2.2.2.3 Purification of bioethanol

Bioethanol produced in a fermentation process is usually a dilute solution of ethanol in water. In order to be used as biofuel, the concentration of ethanol must be increased. The ethanol concentration can be increased using various separation routes such as the use of a distillation column to reach the azeotropic composition followed by vapor permeation membranes, a distillation column followed by an extractive distillation column, and a liquid–liquid extraction column followed by extractive distillation columns [23].

5.2.2.2.4 Properties of bioethanol

Ethanol, also known as "ethyl alcohol" or "grade alcohol," is a flammable, colorless, chemical compound, representing one of the most commonly found alcohols in alcoholic beverages. It is often referred to simply as alcohol. Its molecular formula is C_2H_6O, variously represented as EtOH or C_2H_5OH. Bioethanol and ethanol are chemically identical and the properties of both bioethanol and ethanol do not differ much in parameters. The properties of bioethanol are shown in Table 5.1 and compared to those of petrol. It is found that the octane number of ethanol is higher than that of conventional petrol. Ethanol is also increasingly used as an oxygenate additive for standard petrol as a replacement for methyl tertiary butyl ether (MTBE) to improve its octane number. Because MTBE has toxic properties and is responsible for considerable groundwater and soil contamination, MTBE is more and more frequently replaced by ethyl tertiary butyl ether, which is produced from bioethanol. Since the energy yield of ethanol is lower than that of petrol, it is recommended to utilize bioethanol in producing biodiesel rather than replacing conventional petrol.

5.2.2.3 Biobutanol

5.2.2.3.1 Introduction

Recently, the biofuel market has been primarily dominated by bioethanol, biobutanol, and biogas, relying on substrates such as sugars, starch, oil crops, agricultural and animal residue, and lignocellulosic biomass. Fuel properties like energy density and hygroscopicity are higher in butanol (C_4H_9OH) than in traditional fuels. However, due to its low production cost and availability over the seasons, lignocellulosic biomass is the most suitable raw material for butanol production [24].

5.2.2.3.2 Biobutanol production

The conversion route for biobutanol is similar to the that used in the production of bioethanol, but requires the use of different enzymes. Traditionally, biobutanol is produced through an acetone–butanol–ethanol (ABE) fermentation process in which *Clostridia* species, namely *Clostridium acetobutylicum*, *Clostridium beijerinckii*, *Clostridium saccharobutylicum*, and *Clostridium saccharoper butylacetonicum* are used [25,26]. ABE fermentation is a process that uses bacterial fermentation to produce three solvents, namely acetone, *n*-butanol, and ethanol, from carbohydrates such as starch and glucose. ABE fermentation produces solvents in a ratio of three parts acetone ($CH_3{-}CO{-}CH_3$), six parts butanol ($CH_3{-}CH_2{-}CH_2{-}CH_2OH$), and one part ethanol ($CH_3{-}CH_2{-}OH$). However, traditional butanol fermentation processes have some major challenges, and these are listed here [24].

- Increased operating costs due to high raw material cost.
- High cost of recovery due to low production yield.
- Increased capital and operating costs due to low volumetric productivity.
- High cost for product recovery with conventional distillation processes.

- Toxicity of the produced biobutanol to microorganisms at high concentration.
- Some by-products such as acetone and ethanol hinder the purification processes.
- Culture degeneration.
- Phage contamination especially in large-scaled units.

To produce sustainable biobutanol and to overcome the challenges mentioned above, some possible recommended ways are:

- Finding cheaper and sustainable lignocellulosic raw materials.
- Development of butanol selective microorganisms with improved yield.
- Development of novel reactive separation processes.
- Development of continuous fermentation processes toward reducing reaction times.
- Development of low-cost energy efficient methods for solvent purification and recovery.
- Avoiding the excessive acidification of the culture.
- Good culture practices, proper sterilization, and the use of phage resistant strains.

5.2.2.3.3 Properties of biobutanol

The chemical composition of biobutanol is identical to that of *n*-butanol and, therefore, it does not differ in other parameters. Having some intrinsic properties, biobutanol becomes an attractive source of biofuel. Comparisons of biobutanol to other fuels are shown in Table 5.1. The energy content of biobutanol is quite close to that of gasoline, and the energy density per liter of biobutanol is 25% greater than that of ethanol. Although the production costs of butanol are more than that of ethanol, it gives a better performance in engines. A further energy advantage is that about 18% more energy as hydrogen is produced in the production of biobutanol by fermentation from the same amount of fermentable substrate as ethanol [18]. In addition, the high flash point of biobutanol (36°C) over ethanol and gasoline makes it safer to use and handle.

5.2.2.4 Biomethane

5.2.2.4.1 Introduction

Present infrastructure for transport around the globe is still based on liquid and gaseous fossil fuels like compressed natural gas, gasoline, and diesel, etc. Biomethane is quite famous due to its high efficiency and clean burning properties compared to all other biofuels available today. Nearly all types of biomass can be used to produce biomethane. The raw materials for biomethane generation include dairy and swine farm manures, food processing residues and organic household wastes, vegetable oil residues, and energy crops, etc.

5.2.2.4.2 Biomethane production

The production of biomethane mainly consists of two steps. In the first step, biogas is produced from the raw materials previously mentioned. Then, the produced biogas is further processed and cleaned in order to get biomethane suitable for transport applications. A simple process flow diagram of biomethane production from biomass feedstock is shown in Fig. 5.4.

Biogas production. Biogas is produced by means of an anaerobic digestion process. In this digestion process, complex organic materials are broken down by symbiotic groups of bacteria at different stages of the process. There are four basic types of microorganisms involved in the four main steps, namely hydrolysis, acidogenesis, acetogenesis/dehydrogenation, and methanation [27,28]. First, hydrolyzing and fermenting microorganisms attack the complex biomass materials. Microorganisms taking part in this step are anaerobes such as *Bacteriocides*, *Clostridia*, and *Bifidobacteria*. During this step, acetate, hydrogen, and a mixture of various volatile fatty acids are produced. In the second step, other anaerobic bacteria change the resultant compounds into short-chain organic acids (e.g., butyric, propionic, and acetic acids), alcohols, hydrogen, and CO_2. At the third stage, acidogenic microorganisms convert the products into hydrogen, CO_2, and acetate. The high concentrations of hydrogen produced in this step can hinder the performance of the remaining bacteria. Thus proper control of the hydrogen concentration is important to achieve a maximum production rate. Finally, methanogenic bacteria such as *Methanosarcina barkeri*, *Metanon coccus mazzei*, and *Methanosaeta concilii* convert either one or all of the produced materials into biogas [27]. Since these bacteria are sensitive to temperature, this has to be considered in the digestion process. In order to promote bacterial activity, temperatures of at least 20°C are required. Generally, higher temperatures shorten the processing time and reduce the required volume of the digester tank by 25%–40%. Regarding the temperature, bacteria of anaerobic digestion can be divided into psychrophile (25°C), mesophile (32°C–38°C), and thermophile (42°C–55°C)

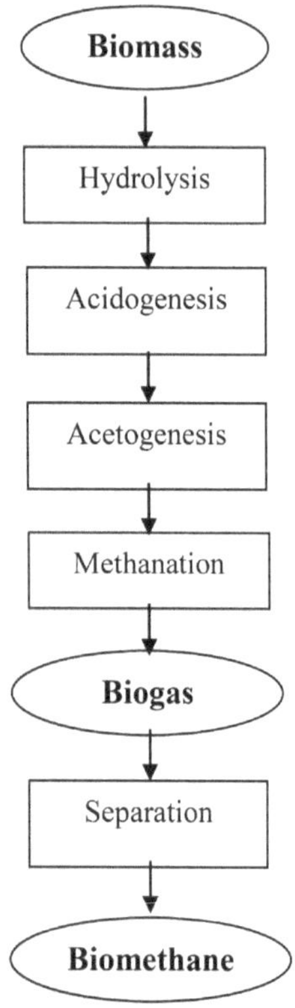

Figure 5.4 A simple process flow diagram of biomethane production from biomass feedstock.

have to be heated in colder climates in order to encourage the bacteria to carry out their function. Digestion times range from a couple of weeks to a couple of months depending on the feedstock used and the digester type as well as on the digestion temperature. The kinetics of the whole process is proportional to many parameters such as substrate, thermal and pH conditions of the fermenter, and the enzymes produced during the steps. A more detailed explanation of the metabolisms and kinetics of biomethane production can be found elsewhere [29,30].

Biogas purification. The produced biogas usually comprises of methane and CO_2, generally in a ratio of 6:4 (55%–80% methane). In addition, some associated gases such as hydrogen sulfide in small quantities and other trace gases are produced. To produce biomethane, methane has to be separated from CO_2 and the remaining components of the biogas. Biogas purification is normally performed in two steps where the produced CO_2 is removed from the gas in the main step. Produced biogas is made free of minor contaminants (e.g., sulfur compounds) before the CO_2-removal step, and the water dew point can be adjusted before or after the CO_2 removal process.

5.2.2.4.3 Types of biomethane

Based on the raw material used and transformation process involved in the production of biomethane, it can be categorized as:

First-generation biomethane. Anaerobic decomposition of organic wastes is involved in the production of biomethane, where the natural breakdown of organic matters occurs under bacterial action. Slightly preheated biogas produced this way can be used to mitigate household demand for cooking, with limited supply for the generation of electricity, heat, and combined cycle.

Second-generation biomethane. The production of second-generation biomethane comprises "thermochemical conversion" to gasify the lignocellulosic biomass (wood and straw). The thermochemical conversion process is carried out in two stages. In the first step, synthetic gas is produced from lignocellulosic biomass, then the produced synthetic gas is transformed into biomethane by introducing a catalyst into the process.

Third-generation biomethane. This is produced from the direct transformation of microalgae cultured in high-yield photosynthetic reactors where natural light, water, and minerals work as reagents and produced CO_2 is recycled in each step. This is an emerging technology to be developed on an industrial scale by 2020–30 (www.engie.com).

5.2.2.4.4 Properties of biomethane

Methane is the simplest of all hydrocarbons that can exists as gas at standard temperature and pressure. Its chemical formula is CH_4. Further, methane is a highly combustible and odorless gas. It is also a GHG with a global warming potential of 23 in 100 years (IPCC 2001). The methane content of biomethane is about 95%–100%. Thus biomethane produced from biogas is chemically similar to natural gas, and it also does not differ in other energy components and characteristic parameters from natural gas. Therefore biomethane is suitable for all the applications of natural gas.

5.2.2.5 Biohydrogen

5.2.2.5.1 Introduction

Hydrogen (H_2) is the most abundant and common element in the universe. H_2 barely exists naturally as free substance, but can easily be found chemically combined with other elements. It mainly exists as molecules chemically bound with water, biomass, or fossil fuels. It can be produced from all of these substances. However, the conversion of biomass to H_2 has gained special interest as it is a method to exploit renewable energy toward the reduction of GHG emissions.

5.2.2.5.2 Biohydrogen production

Currently, biohydrogen from biomass can be produced using two different methods, namely biomass gasification and biomass digestion. The gasification process is an important component of many current and up-coming low environmental impact energy systems [31]. The process is done by heating biomass material in the absence of air or oxygen. It produces synthesis gas primarily consisting of carbon monoxide (CO), CO_2, and H_2. The production of H_2 and CO from biomass gasification is reportedly higher than simple biomass combustion [32]. H_2 production can also be increased with the rise of the operation temperature in the gasification process [33]. On the other hand, in the digestion method, wet feedstock such as manure is converted to produce primarily CH_4 and CO_2 (as discussed in Section 5.2.2.4.2). CH_4 is then converted into H_2 through a steam reforming process. The endothermic steam reforming process converts CH_4 and water vapor into H_2 and CO. The product gas containing CO can be further separated into H_2 and CO_2. The gas produced from both methods is then sent to for a purification process to increase the H_2 concentration and is then used in various applications.

5.2.2.5.3 Properties of biohydrogen

Biohydrogen is chemically identical to conventional hydrogen gas and does not differ in characteristic parameters. It is just produced from the gasification or digestion of biomass. It is a colorless, highly flammable, gaseous element, and the lightest of all gases. The transportation and storage of hydrogen gas becomes difficult, since it occupies a large volume compared to other gaseous fuels. Safe handling during transportation and storage is another concern that requires many precautions. H_2 is mainly stored in a liquid form after cooling the gas to liquid phase using nearly cryogenic temperatures (−253°C). Another common storing option includes storing it as a constituent in other liquid such as $NaBH_4$ solutions, rechargeable organic liquids, or anhydrous NH_3 through the presence of a few catalysts [34].

5.3 Barriers of biofuels

Biofuels are regarded as one of the most viable options for reducing CO_2 emissions in the transport sector. However, conventional plant-based biofuels share only 4% of the total transportation fuel consumption due to several major limitations [35]. These barriers limit the development of the biofuel market. The main market constraints that restrain the commercial production and use of biofuels have been summarized here [17,36].

- Biomass sources may compete with food supply (feedstock barrier).
- The production of biofuels is still expensive (financial barrier).
- New technology or improvements in technologies for certain biofuels are still needed (technology barrier).
- The production of a specific biofuel requires new or modified infrastructures (infrastructure barrier).
- Existing laws and regulations are immature for biofuels (law and regulation barrier).
- Existing storage and transportation systems are inappropriate for biofuels (storage and transportation barrier).
- There is a lack of political will to promote biofuel market development (political barrier).
- No quality standards exist for some biofuels (trade barrier).
- There is a lack of knowledge on biofuels (knowledge barrier).

5.4 Conclusion

Biomass cannot realistically fulfill the world's future energy demands. On the other hand, the versatility of biomass with the diverse portfolio of conversion options makes it possible to meet the demand for secondary energy carriers. For avoiding CO_2 emissions, replacing natural gas is, at present, a highly selective way of using biomass. However, replacing natural gas with biomass for power generation, results in levels of CO_2 mitigation similar to second generation biofuels. In the future, using biomass for transport fuels will gradually become more attractive from a CO_2 mitigation perspective because of their lower GHG emissions.

Especially promising are the production via advanced conversion concepts biomass-derived fuels such as biodiesel, biomethanol, biohydrogen, and bioethanol. Ethanol produced from sugarcane is already a competitive biofuel in tropical regions and further improvements are on-going. The use of waste materials in biogas production contributes to the reduction animal wastes and odors. In addition, it destroys disease-causing pathogens existing in waste materials. The actual role of biofuel will depend on its competitiveness with fossil fuels and on agricultural policies worldwide. Biofuels are expected to be a significant and necessary component of the movement toward a low-carbon economy including for use in vehicles and particularly for aviation and heavy-duty trucks [37,38].

References

[1] A. Pandey, Biofuels: Alternative Feedstocks and Conversion Processes, Academic Press, 2011.
[2] S. Ladanai, J. Vinterbäck, Global potential of sustainable biomass for energy, Report 013, Swedish University of Agricultural Sciences, 2009.
[3] L. Singh, Z.A. Wahid, Methods for enhancing bio-hydrogen production from biological process: a review, J. Ind. Eng. Chem. 21 (2015) 70–80.
[4] A. Faaij, Developments in international bio-energy markets and trade, Biomass Bioenergy 32 (2008) 657–659.
[5] World Bioenergy Association, Report, Global Biomass Potential Towards 2035, 2016.
[6] N. Paul, D. Kemnitz, Biofuels–Plants, Raw Materials, Products, Fachagentur, Nachwachsende Rohstoffe eV (FNR), WPR Communications, Berlin, 2006, pp. 43.
[7] M. Bhuiya, M. Rasul, M. Khan, N. Ashwath, A. Azad, Prospects of 2nd generation biodiesel as a sustainable fuel—part: 1 selection of feedstocks, oil extraction techniques and conversion technologies, Renew. Sustain. Energy Rev. 55 (2016) 1109–1128.
[8] R.A. Lee, J.-M. Lavoie, From first-to third-generation biofuels: challenges of producing a commodity from a biomass of increasing complexity, Anim. Front. 3 (2013) 6–11.
[9] J. Sheehan, V. Camobreco, J. Duffield, H. Shapouri, M. Graboski, K. Tyson, An Overview of Biodiesel and Petroleum Diesel Life Cycles, National Renewable Energy Lab, Golden, CO (US), 2000.
[10] E. Yadessa, D. Tulu, A. Bogale, G. Mengistu, M. Aleme, S. Shiferawu, et al., Characterization of smallholder poultry production systems in Mezhenger, Sheka and Benchi-Maji zones of south western Ethiopia, Acad. Res. J. Agric. Sci. Res. 5 (2017) 10–19.
[11] M. Jahirul, J. Brown, W. Senadeera, N. Ashwath, C. Laing, J. Leski-Taylor, et al., Optimisation of bio-oil extraction process from beauty leaf (*Calophyllum inophyllum*) oil seed as a second generation biodiesel source, Procedia Eng. 56 (2013) 619–624.
[12] A. Atabani, A. Silitonga, H. Ong, T. Mahlia, H. Masjuki, I.A. Badruddin, et al., Non-edible vegetable oils: a critical evaluation of oil extraction, fatty acid compositions, biodiesel production, characteristics, engine performance and emissions production, Renew. Sustain. Energy Rev. 18 (2013) 211–245.
[13] G. Bhargavi, P.N. Rao, S. Renganathan, Review on the extraction methods of crude oil from all generation biofuels in last few decades, IOP Conference Series: Materials Science and Engineering, IOP Publishing, 2018, p. 012024.
[14] M. Mittelbach, C. Remschmidt, Biodiesel thé Comprehensive Handbook, M. Mittelbach, Gratz, Austria, 2004.

[15] P.P. Oh, H.L.N. Lau, J. Chen, M.F. Chong, Y.M. Choo, A review on conventional technologies and emerging process intensification (PI) methods for biodiesel production, Renew. Sustain. Energy Rev. 16 (2012) 5131–5145.
[16] F. Ferella, G.M. Di Celso, I. De Michelis, V. Stanisci, F. Vegliò, Optimization of the transesterification reaction in biodiesel production, Fuel 89 (2010) 36–42.
[17] D. Rutz, R. Janssen, Biofuel Technology Handbook, WIP Renewable Energies, ISBN Contract No. EIE/05/022/SI2. München, Germany, 2008.
[18] H.W. Doelle, J.S. Rokem, M. Berovic, Biotechnology-Volume VIII: Fundamentals in Biotechnology, EOLSS Publications, 2009.
[19] H. Zabed, J. Sahu, A. Suely, A. Boyce, G. Faruq, Bioethanol production from renewable sources: current perspectives and technological progress, Renew. Sustain. Energy Rev. 71 (2017) 475–501.
[20] Biofuels for Transportation, Global Potential and Implications for Sustainable Agriculture and Energy in the 21st Century, Submitted Report Prepared for BMELV in Cooperation With GTZ and FNR, 2006.
[21] J.D. McMillan, Pretreatment of Lignocellulosic Biomass, ACS Publications, 1994.
[22] E. Triwahyuni, S. Hariyanti, D. Dahnum, M. Nurdin, H. Abimanyu, Optimization of saccharification and fermentation process in bioethanol production from oil palm fronds, Procedia Chemistry 16 (2015) 141–148.
[23] C. Conde-Mejía, A. Jiménez-Gutiérrez, F.I. Gómez-Castro, Purification of bioethanol from a fermentation process: alternatives for dehydration, Computer Aided Chemical Engineering, Elsevier, 2016, pp. 373–378.
[24] A. Prakash, R. Dhabhai, V. Sharma, A review on fermentative production of biobutanol from biomass, Curr. Biochem. Eng. 3 (2016) 37–46.
[25] P. Dürre, Fermentative production of butanol—the academic perspective, Curr. Opin. Biotechnol. 22 (2011) 331–336.
[26] B. Ndaba, I. Chiyanzu, S. Marx, *n*-Butanol derived from biochemical and chemical routes: a review, Biotechnol. Rep. 8 (2015) 1–9.
[27] P. Weiland, Biogas production: current state and perspectives, Appl. Microbiol. Biotechnol. 85 (2010) 849–860.
[28] M. Tabatabaei, A. Sulaiman, A.M. Nikbakht, N. Yusof, G. Najafpour, Influential Parameters on Biomethane Generation in Anaerobic Wastewater Treatment Plants, AlternativeFuel, IntechOpen, 2011.
[29] S. Rana, L. Singh, Z. Wahid, H. Liu, A recent overview of palm oil mill effluent management via bioreactor configurations, Curr. Pollut. Rep. 3 (2017) 254–267.
[30] C. Grieder, G. Mittweg, B.S. Dhillon, J.M. Montes, E. Orsini, A.E. Melchinger, Kinetics of methane fermentation yield in biogas reactors: genetic variation and association with chemical composition in maize, Biomass Bioenergy 37 (2012) 132–141.
[31] N. Gnanapragasam, B. Reddy, M. Rosen, Reducing CO_2 emissions for an IGCC power generation system: effect of variations in gasifier and system operating conditions, Energy Convers. Manag. 50 (2009) 1915–1923.
[32] P. Gilbert, C. Ryu, V. Sharifi, J. Swithenbank, Tar reduction in pyrolysis vapours from biomass over a hot char bed, Bioresour. Technol. 100 (2009) 6045–6051.
[33] A.T. Wijayanta, M.S. Alam, K. Nakaso, J. Fukai, M. Shimizu, Optimized combustion of biomass volatiles by varying O_2 and CO_2 levels: a numerical simulation using a highly detailed soot formation reaction mechanism, Bioresour. Technol. 110 (2012) 645–651.
[34] International Energy Agency, Hydrogen Production and Storage, OECD/IEA, 2006.
[35] Y.-K. Oh, K.-R. Hwang, C. Kim, J.R. Kim, J.-S. Lee, Recent developments and key barriers to advanced biofuels: a short review, Bioresour. Technol. 257 (2018) 320–333.

[36] S. Krishnan, M.F.M. Din, S.M. Taib, Y.E. Ling, H. Puteh, P. Mishra, et al., Process constraints in sustainable bio-hythane production from wastewater, Bioresour. Technol. Rep. 5 (2019) 359–363.
[37] International Energy Agency, Technology Roadmap: Delivering Sustainable Bioenergy, OECD/IEA, 2017.
[38] J. Littlejohns, L. Rehmann, R. Murdy, A. Oo, S. Neill, Current state and future prospects for liquid biofuels in Canada, Biofuel Res. J. 5 (2018) 759–779.

CHAPTER 6

Dry fermenters for biogas production

Abu Yousuf[1], Ahasanul Karim[2], M. Amirul Islam[3], Shefa Ul Karim[4], Md. Maksudur Rahman Khan[3] and Che Ku Mohammad Faizal[2]
[1]Department of Chemical Engineering and Polymer Science, Shahjalal University of Science and Technology, Sylhet, Bangladesh
[2]Faculty of Engineering Technology, Universiti Malaysia Pahang, Pahang, Malaysia
[3]Faculty of Chemical and Natural Resource Engineering, Universiti Malaysia Pahang, Pahang, Malaysia
[4]Department of Geological Sciences, Chiang Mai University, Chiang Mai, Thailand

6.1 Introduction

Nowadays, organic waste is of great concern to municipal as well as local authorities as it can become a threat, causing water pollution, land occupation, soil contamination, and public health deterioration. In addition, the exhaustion of fossil fuel resources and global warming have been driving governments and researchers to inspect alternative sources of fuels. For securing the clean energy demand along with the safety of the environment, a feasible renewable energy source could be biogas, which is generally produced by the anaerobic digestion (AD) of organic wastes [1]. AD is considered to be a convincing method for treating organic wastes and producing methane-rich biogas simultaneously [2]. The anaerobic fermentation (AF) process can be operated at relatively low temperature ranges, and also requires less energy input compared to other biomass processing technologies [3]. It is significant to mention that the term AF is often used interchangeably with AD when describing the physical decomposition of organics. Therefore fermentation is a biological process in which organic material is degraded and decomposed by microorganisms resulting in the release of biogas. The biogas fermentation processes are classified into three major categories, namely submerged or wet fermentations [with total solids (TS) contents considerably lower than 10%] [4], semidry fermentations (with TS ranges of 10%–20%) [5,6], and dry or solid-state fermentations (with TS contents higher than 20%) [7,8]. However, solid-state anaerobic fermentation (SSAF) is generally performed at solid concentrations of more than 15% [9,10].

Bioreactors
DOI: https://doi.org/10.1016/B978-0-12-821264-6.00006-1

SSAF or dry fermentation (DF) is defined as a fermentation process in which digestion occurs in the absence of water while the substrate possesses enough moisture content to support the growth as well as the metabolism of microbes [11]. The usage of SSAF has been increasing since 2005 as the technique gives both economic and environmental benefits compared to other solid waste treatment techniques such as incineration, composting, and landfilling. The SSAF technique is considered as superior to typical liquid-state anaerobic fermentation (LSAF) because it requires a comparatively smaller volume of digester in the absence of water, does not require mechanical stirring, dewatering, or drying of the effluent, shows a higher solid loading capacity with TS contents of between 20% and 50%, high yields, better energy recovery, less wastewater generation with lesser risk of bacterial contamination, less material management, and less loss of total parasitic energy [12–18]. However, SSAF systems also have challenges, for example, large inoculum amounts are required along with long retention times. Therefore different kinds of bioreactors (Fig. 6.1) have been developed to mechanize the process to lower retention time and to make the process more cost-effective and environment-healthy [1].

In the past decade, several fermentation processes such as batch fermentation, continuous fermentation, and two-phase fermentation have been studied to produce biogas through a DF process [7,8]. However, the batch fermentation process is not attractive since it cannot be applied to large- or industrial-scale production [19,20]. The two-phase fermentation process is able to generate several organic acids during the methane producing phase, thus increasing the rate of methane production by removing significant amounts of volatile fatty acids (VFAs). However, it is difficult to separate acid production and methane generation phases in a two-phase fermentation process. Moreover, the process is high cost due to it containing dual phases, hence narrowing its practical applications. Generally, plug-flow

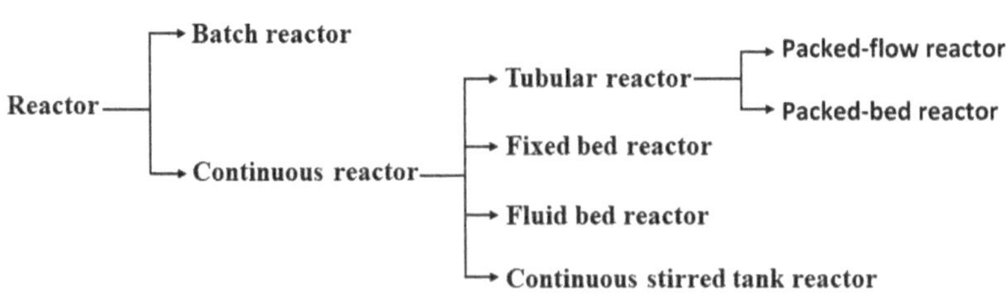

Figure 6.1 Different kinds of reactors that are explored for fermentation processes.

digesters are effective in treating ruminant substrate that contains high percentages of TS [21]. It is worthy to note that continuous fermentation processes are more economical compared with other fermentation systems [22,23] since the process allows high volumes of substrate to be treated per unit time. However, it gives very low concentrations of final products because of the complicated recovery process [22,24].

Several studies have been performed to find some elucidations to the technical difficulties of DF processes, to lower retention time, and to make them more cost effective and environment-friendly considering the architecture of digesters, the degradation efficiency of organic waste, and heterogeneous microbial growth within substrates. This chapter summarizes several types of dry fermenters along with influential factors, major findings, and the benefits and limitations of different fermenters.

6.2 Different kinds of dry fermenters for biogas production

6.2.1 Continuous plug flow reactor

The plug flow reactor (PFR) is used to describe the continuous chemical reactions, its shape is cylindrical or tubular and also known as continuous tubular reactor or piston flow reactor. Nutrients (and sometimes microorganisms) are introduced into the reactor continuously and move through the reactor as a "plug" in a PFR. The reactor system can either be open or confined. Although there is no mixing of the medium along with the long axis, that is, the *X*-axis but there may be lateral mixing in the medium at any point along with the long axis, that is, the *Y*-axis in an ideal PFR. PFRs are commonly used in large-scale and continuous production, and for rapid, homogeneous or heterogeneous, and high-temperature reactions [25].

Patinvoh et al. [26] designed and developed a horizontal PFR (Fig. 6.2) for continuous DF in Sweden, and studied the efficiency of this fermenter using untreated cattle manure as a substrate. The cattle manure was bedded with straw (TS concentration of 22%), which was obtained from a local cattle farm and used in a continuous digestion process. The anaerobic sludge (used as inoculum) was collected from a digester that was operating to treat wastewater sludge under mesophilic conditions. The major findings of the study are listed here.

- The newly developed reactor is a PFR designed for continuous DF.
- This PFR has operated smoothly for 230 days using raw cattle manure bedded with straw (22% TS).

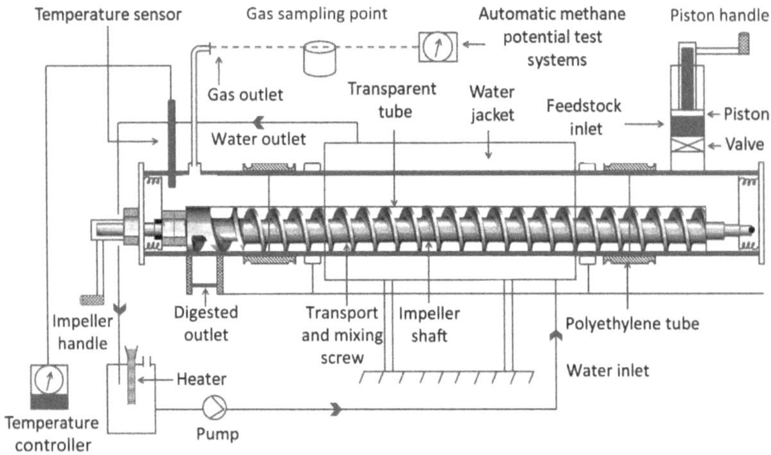

Figure 6.2 A schematic diagram of the newly developed plug flow reactor and the experimental setup with other fixtures such as the water bath, heater, and temperature sensor for controlling the temperature, the composition analysis of biogas through a sampling point, and the automated methane potential test system for measuring the rate of biogas production. *R.J. Patinvoh, A.K. Mehrjerdi, I.S. Horváth, M. J. Taherzadeh, Dry fermentation of manure with straw in continuous plug flow reactor: reactor development and process stability at different loading rates, Bioresour. Technol. 224 (2017) 197–205.*

- Theoretically, a 56% methane (CH_4) yield was obtained with a 57% removal efficiency of volatile solids (VS) at a 4.2 gVS/L/day organic loading rate (OLR).
- The process was unstable at 6 gVS/L/day with a VS removal efficiency of 41%.

PFRs are considered as efficient for DF processes since this type of reactor is easy to build and is cost effective [27]. PFRs have been reported to have the highest success rate in the United States, where 42% out of the 242 anaerobic fermenters were operating at local livestock farms in 2015 with plug flow designs [26]. However, different drawbacks have been listed by researchers including thermal stratification, solid sedimentation problems, and low mass transfer rates as a consequence of insufficient mixing [27]. Researchers have found that the implementation of impellers in PFRs can lessen these kinds of limitations since reactor performance is highly sensitive to the amount of mixing. Less mixing is related to better performance. It was also observed that continuous mixing causes an unstable performance at high OLRs, but continuously mixed unstable reactors could be made stable by reducing the mixing level [28].

The effectiveness of PFRs using substrates with 11%–14% TS and manure have also been investigated by other researchers [29].

6.2.2 Garage-type dry fermenter

A study was conducted by Qian et al. [30] on the first industrial-scale garage-type dry fermenter (GTDF) plant for treating municipal solid waste (MSW) located in the city of Harbin in Heilongjiang province, China. The first GTDF was developed to treat MSW for producing biogas and monitored for 180 days in their study. The percolate was obtained from the codigestion of the highly biodegradable organic fraction of MSW and cattle manure, which was used as liquid inoculum. The plant has been operating to treat MSW of 100 Mt as per designed capacity and producing 8000 m^3 of biogas. Evaluating the AD efficiency and investigating the process stability were the main objectives of this study. The major findings of GTDF study are listed here.

- This reactor was the first industrial-scale GTDF designed and operated for treating MSW in China.
- After six months of observation, it was found that specific biogas production was about 270 m^3 CH_4 t/VS.
- It was reported that the biogas production for the digester and percolate were 0.72 and 2.22 m^3/day, respectively.

GTDF is widely used for the treatment of stackable non-free flowing materials like energy crop or MSW in Europe because of its several special features, such as greater acceptance of input materials with higher dry matter content, and lower energy consumption [31]. A number of GTDF plants have been designed and developed for operation in many European countries, especially in Germany, during the past decade [30]. The process perception for a batch system that was implemented by BEKON Energy Technologies GmbH & Co. KG is shown in Fig. 6.3. It was operated through the simultaneous inoculation of the feedstock with the digestate before each feeding cycle and the percolate was run in the circuit. Inoculum usually contains methanogenic bacteria that can produce CH_4 from the organic acids in a digester. Another process that was carried out by GICON Holding GmbH is presented in Fig. 6.4, which was operated without any inoculation with the digestate, but by the recirculation of percolates. It can be seen that the percolate is stored in a percolate take that is connected to an external digester and most of the acids are degraded into biogas there.

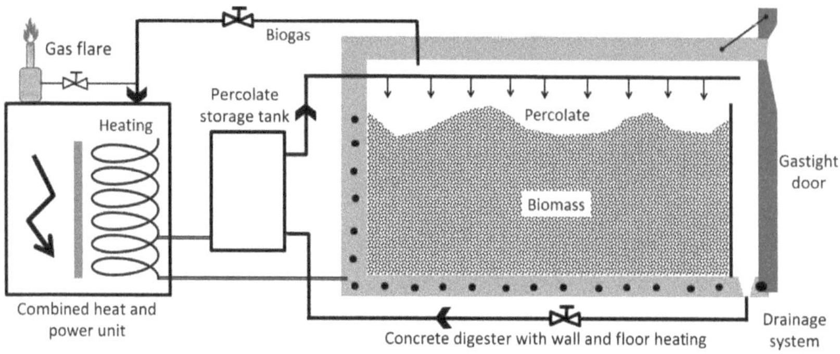

Figure 6.3 A schematic diagram of the process concept for the batch mode with percolate tank that was developed by BEKON. *M. Qian, R. Li, J. Li, H. Wedwitschka, M. Nelles, W. Stinner, et al., Industrial scale garage-type dry fermentation of municipal solid waste to biogas, Bioresour. Technol. 217 (2016) 82–89.*

Good contact between feedstock and microorganism is essential for the proper degradation of a substrate. For this purpose, the movement of the substrate in the liquid digester, is mainly performed by agitating/mixing the whole substrate (especially, in continuous stirred-tank reactor digesters) or by flowing/passing liquid substrate to microorganism surface (such as in upflow anaerobic sludge blanket reactor and fixed-bed reactors). On the contrary, this connection between feedstock and microorganism is mainly achieved by the percolate in GTDF systems. However, biogas production can be stopped soon after stopping the percolate. The solid–liquid mass transfer rate can be accelerated by means of a high flow rate percolate recirculation and, hence, the production of biogas can be increased dramatically [32]. On the other hand, it was observed that the yield of biogas can be lessened significantly due to a low content of moisture and low flow rate of feedstock. The results of the study by Qian et al. [30] indicated that the two-phase method, where hydrolysis and acetogenesis occur in the digester and methanogenesis takes place in the percolate tank, has its advantages in the digestion of MSW. To separate the AD process into a two-phase system, the recirculation of the percolate is an attractive and feasible inoculation approach. This two-phase system of separating the acetogenesis from the methanogenesis offers an improved process stability since the risk of process acidification is reduced [33]. Therefore it provides an enhanced process control since the methanogenesis process can be detached if problems occur [34].

It was observed that the digestion efficiency of the fermenter was much lower than that of the percolate tank. The efficiency can be

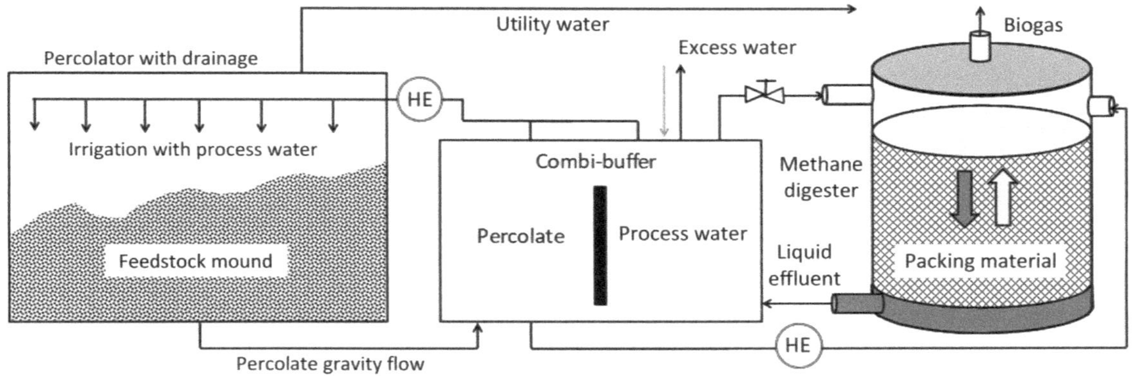

Figure 6.4 A schematic diagram of the process concept of the batch system with external methanogenesis of GICON. *M. Qian, R. Li, J. Li, H. Wedwitschka, M. Nelles, W. Stinner, et al., Industrial scale garage-type dry fermentation of municipal solid waste to biogas, Bioresour. Technol. 217 (2016) 82–89.*

enhanced by shortening the lag phase through increased inoculation, and the process stability can be increased by operating the digestion process in accordance with a preset schedule. Increasing the percolate recirculation or mixing the digestate at the start-up phase can enhance the inoculation to shorten the lag phase, and this process is considered as a potential approach to boost the AD efficiency of a digester [30].

6.2.3 Upflow anaerobic solid-state reactor

A laboratory-scale upflow anaerobic solid-state (UASS) reactor test system was designed and constructed by Mumme et al. [35]. The system was equipped with a UASS reactor and an anaerobic filter consisting of two reactors with working volumes of 26.5 L and 79.2 L, respectively (Fig. 6.5). The substrates used for the experiment were maize and barley straw, which were collected from local farms. The target of the study was to evaluate the practicability of a novel UASS reactor consisting of liquor recirculation. A performance analysis was done based on the operation output over a time span of 68 days of the UASS reactor operating in thermophilic conditions with a total 26.5 L working volume. The substrate

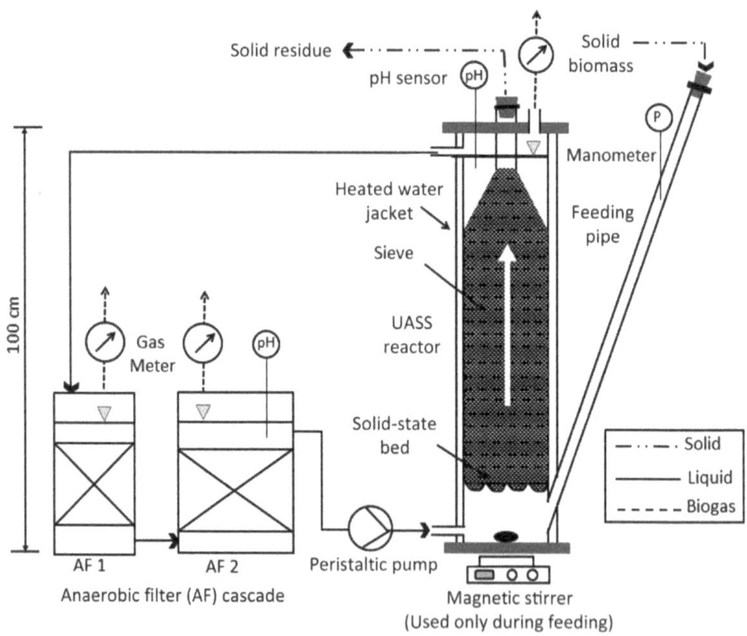

Figure 6.5 A schematic diagram of a laboratory-scale upflow anaerobic solid-state reactor and its experimental setup. *Mumme et al. [35].*

used in this UASS reactor was the mixture of straw and maize silage. The UASS reactor was divided into three parts, namely a lower liquid zone, an upper liquid zone, and the solid-state bed (SSB) in between. The major findings of the study on UASS reactors are listed here.

- A novel UASS reactor was designed for a continuous operation and solid-state process, which is considered as an optimistic solution for digesting different sorts of organic materials.
- The process liquor was recirculated continuously to avoid the accumulation of VFAs in this system.
- The overall methane yield was decreased from 384 to 312 per kg_{vs} by increasing the VS loading rate from 7.1 to 17 g_{vs}/L/day, while the yield contribution of the anaerobic filters was boosted from 12% to 70%.
- The chemical oxygen demand (COD) removal efficiency of the organic matter obtained was 86-93% with the highest hydrolysis rate of 16.4 g_{COD}/L/day.

The experimental results showed that the hydrolytic and methanogenic performance were the highest for solid biomass assimilation reported to date, which proves the feasibility of the UASS concept. The CH_4 yield was stabilized at OLRs as high as 17 g_{vs}/L/day while an anaerobic filter was connected to the UASS reactor. A notable disadvantage of the UASS process is that it is not suitable for colloidal substances such as starch, since this kind of substrate could lead to clogging and compaction of the SSB. Study of the physicochemical mechanisms, the dynamics of the microbes inside SSBs, and the optimal pH is crucial for an innovative process control in future research.

6.2.4 Down plug-flow anaerobic reactor

Chen et al. [36] designed and constructed a down plug-flow anaerobic reactor (DPAR) to assess the practicability of the continuous dry digestion of swine manure. The swine manure was used as a substrate in this study and was obtained from a local pig farm situated in the central Sichuan province of China. Sludge was used as the inoculum to treat the swine waste in the laboratory and it was collected from an operating anaerobic digester. To produce biogas through the continuous DF of the swine manure without any additional stirring was the main objective of the study. In addition, the effect of the TS concentration (20%–35%) of the feedstock and the impacts of ammonia inhibition and digestate liquidity on the biogas yield were investigated as well. In this study, the DPAR was designed and constructed having a feeding inlet on the top with an internal diameter of 20 mm and a digestate outlet at the bottom with an internal diameter of 20 mm (Fig. 6.6). A double layer cylinder with a 375 mm height and a 125 mm inlet diameter made of plexiglass was used as a reactor. The effective volume of DPAR was 45 L with a 50 L of total volume.

The major findings from the study of DPAR are listed here.

- DF with swine manure used as a feedstock in a continuous process is a feasible approach for biogas production.
- The biogas yield has highly controlled by the TS concentration.
- The effect of the ammonia concentration and liquidity of digestate were also investigated, and it was observed that the production of biogas was clearly restrained at an ammonia nitrogen concentration of more than 3000 mg/L.
- The highest yield of biogas and a maximum volumetric biogas production rate were achieved with a 25% TS concentration at 25°C without inhibition.
- The TS concentration of swine manure (feedstock) should be less than 30% to attain the maximum biogas yield.

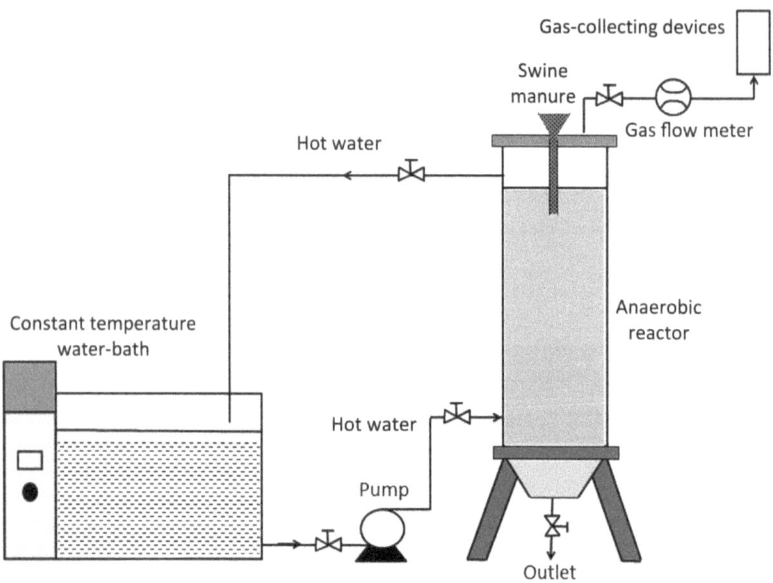

Figure 6.6 A schematic diagram of the down plug-flow anaerobic reactor. *C. Chen, D. Zheng, G.J. Liu, L.W. Deng, Y. Long, Z.H. Fan, Continuous dry fermentation of swine manure for biogas production, Waste Manag. 38 (2015) 436–442.*

The major advantages of the DPAR are listed here.

- The DPAR was operated smoothly without producing any external heat or stirring at ambient temperature, which ensure low running costs and hence, are energy saving.
- The DPAR was able to achieve a higher biogas yield at relatively high concentrations of TS in terms of the digestate discharging and without ammonia inhibition.
- The DPAR could be operated continuously and efficiently for biogas production unlike batch processes.

6.3 Conclusion

Various kinds of reactors for biogas production through DF have been studied so far toward lowering the retention time needed and making the process more cost-effective and environment-friendly. However, the appropriate design of a bioreactor depends on many factors such as the amount and type of substrate to be digested, the composition of the substrate, the process economy of the reactor, and optimization of the bioreactor conditions. This discussion showed that there is still room to increase efficiency by reducing retention time and increasing the transformation of microorganisms through heterogeneous substrates. Therefore a suitable bioreactor design is necessary to maximize biogas production utilizing DF processes as a sustainable system to meet future energy demands.

Acknowledgment

The project was supported by the University Malaysia Pahang, Malaysia (RDU 140109, FRGS RDU 160150 and PRGS 180317) and the authors are very grateful for the financial support.

References

[1] A. Yousuf, S. Sultana, M.U. Monir, A. Karim, S.R.B. Rahmaddulla, Social business models for empowering the biogas technology, Energy Sour. B: Econ. Plann. Policy 12 (2017) 99–109.

[2] J. Zhu, L. Yang, Y. Li, Comparison of premixing methods for solid-state anaerobic digestion of corn stover, Bioresour. Technol. 175 (2015) 430–435.

[3] M.E. Ersahin, H. Ozgun, R.K. Dereli, I. Ozturk, Anaerobic treatment of industrial effluents: an overview of applications, Waste Water-Treatment and Reutilization, InTech, 2011.

[4] L. Deng, Y. Li, Z. Chen, G. Liu, H. Yang, Separation of swine slurry into different concentration fractions and its influence on biogas fermentation, Appl. Energy 114 (2014) 504–511.
[5] D. Bolzonella, L. Innocenti, P. Pavan, P. Traverso, F. Cecchi, Semi-dry thermophilic anaerobic digestion of the organic fraction of municipal solid waste: focusing on the start-up phase, Bioresour. Technol. 86 (2003) 123–129.
[6] L. Dong, Y. Zhenhong, S. Yongming, Semi-dry mesophilic anaerobic digestion of water sorted organic fraction of municipal solid waste (WS-OFMSW), Bioresour. Technol. 101 (2010) 2722–2728.
[7] F. Abouelenien, Y. Nakashimada, N. Nishio, Dry mesophilic fermentation of chicken manure for production of methane by repeated batch culture, J. Biosci. Bioeng. 107 (2009) 293–295.
[8] S. Kusch, H. Oechsner, T. Jungbluth, Biogas production with horse dung in solid-phase digestion systems, Bioresour. Technol. 99 (2008) 1280–1292.
[9] R. Kothari, A. Pandey, S. Kumar, V. Tyagi, S. Tyagi, Different aspects of dry anaerobic digestion for bio-energy: an overview, Renew. Sustain. Energy Rev. 39 (2014) 174–195.
[10] J.-C. Motte, E. Trably, R. Escudié, J. Hamelin, J.-P. Steyer, N. Bernet, et al., Total solids content: a key parameter of metabolic pathways in dry anaerobic digestion, Biotechnol. Biofuels 6 (2013) 164.
[11] A. Pandey, Solid-state fermentation, Biochem. Eng. J. 13 (2003) 81–84.
[12] J. Guendouz, P. Buffiere, J. Cacho, M. Carrere, J.-P. Delgenes, High-solids anaerobic digestion: comparison of three pilot scales, Water Sci. Technol. 58 (2008) 1757–1763.
[13] L.F.- Güelfo, C. Alvarez-Gallego, D.S. Márquez, L.R. García, Start-up of thermophilic–dry anaerobic digestion of OFMSW using adapted modified SEBAC inoculum, Bioresour. Technol. 101 (2010) 9031–9039.
[14] G.K. Kafle, S.H. Kim, Anaerobic treatment of apple waste with swine manure for biogas production: batch and continuous operation, Appl. Energy 103 (2013) 61–72.
[15] H. Yabu, C. Sakai, T. Fujiwara, N. Nishio, Y. Nakashimada, Thermophilic two-stage dry anaerobic digestion of model garbage with ammonia stripping, J. Biosci. Bioeng. 111 (2011) 312–319.
[16] Y. Li, S.Y. Park, J. Zhu, Solid-state anaerobic digestion for methane production from organic waste, Renew. Sustain. Energy Rev. 15 (2011) 821–826.
[17] Y. Zhou, C. Li, I.A. Nges, J. Liu, The effects of pre-aeration and inoculation on solid-state anaerobic digestion of rice straw, Bioresour. Technol. 224 (2017) 78–86.
[18] J.-C. Motte, R. Escudié, N. Bernet, J.-P. Delgenes, J.-P. Steyer, C. Dumas, Dynamic effect of total solid content, low substrate/inoculum ratio and particle size on solid-state anaerobic digestion, Bioresour. Technol. 144 (2013) 141–148.
[19] H.K. Ahn, M. Smith, S. Kondrad, J. White, Evaluation of biogas production potential by dry anaerobic digestion of switchgrass–animal manure mixtures, Appl. Biochem. Biotechnol. 160 (2010) 965–975.
[20] M. Lantz, The economic performance of combined heat and power from biogas produced from manure in Sweden – a comparison of different CHP technologies, Appl. Energy 98 (2012) 502–511.
[21] A.H. Igoni, M. Ayotamuno, C. Eze, S. Ogaji, S. Probert, Designs of anaerobic digesters for producing biogas from municipal solid-waste, Appl. Energy 85 (2008) 430–438.
[22] P.J. Verbelen, D.P. De Schutter, F. Delvaux, K.J. Verstrepen, F.R. Delvaux, Immobilized yeast cell systems for continuous fermentation applications, Biotechnol. Lett. 28 (2006) 1515–1525.

[23] T.K. Ghose, R. Tyagi, Rapid ethanol fermentation of cellulose hydrolysate. I. Batch versus continuous systems, Biotechnol. Bioeng. 21 (1979) 1387–1400.
[24] C.R. Soccol, A. Pandey, C. Larroche, Fermentation Processes Engineering in the Food Industry, CRC Press, 2013.
[25] A. Cunningham, J. Lennox, R. Ross, Biofilms: The Hypertextbook, Biofilm Growth and Development, 2011. Retreived from: http://www.biofilmbook.com.
[26] R.J. Patinvoh, A.K. Mehrjerdi, I.S. Horváth, M.J. Taherzadeh, Dry fermentation of manure with straw in continuous plug flow reactor: reactor development and process stability at different loading rates, Bioresour. Technol. 224 (2017) 197–205.
[27] S. Lansing, J.F. Martin, R.B. Botero, T.N. Da Silva, E.D. Da Silva, Methane production in low-cost, unheated, plug-flow digesters treating swine manure and used cooking grease, Bioresour. Technol. 101 (2010) 4362–4370.
[28] P.G. Stroot, K.D. McMahon, R.I. Mackie, L. Raskin, Anaerobic codigestion of municipal solid waste and biosolids under various mixing conditions—I. Digester performance, Water Res. 35 (2001) 1804–1816.
[29] M. Adl, K. Sheng, Y. Xia, A. Gharibi, X. Chen, Examining a hybrid plug-flow pilot reactor for anaerobic digestion of farm-based biodegradable solids, Int. J. Environ. Res. 6 (2012) 335–344.
[30] M. Qian, R. Li, J. Li, H. Wedwitschka, M. Nelles, W. Stinner, et al., Industrial scale garage-type dry fermentation of municipal solid waste to biogas, Bioresour. Technol. 217 (2016) 82–89.
[31] D. Brown, Y. Li, Solid state anaerobic co-digestion of yard waste and food waste for biogas production, Bioresour. Technol. 127 (2013) 275–280.
[32] H. Benbelkacem, R. Bayard, A. Abdelhay, Y. Zhang, R. Gourdon, Effect of leachate injection modes on municipal solid waste degradation in anaerobic bioreactor, Bioresour. Technol. 101 (2010) 5206–5212.
[33] F. Shen, H. Yuan, Y. Pang, S. Chen, B. Zhu, D. Zou, et al., Performances of anaerobic co-digestion of fruit & vegetable waste (FVW) and food waste (FW): single-phase vs. two-phase, Bioresour. Technol. 144 (2013) 80–85.
[34] V. Nkemka, M. Murto, Two-stage anaerobic dry digestion of blue mussel and reed, Renew. Energy 50 (2013) 359–364.
[35] J. Mumme, B. Linke, R. Tölle, Novel upflow anaerobic solid-state (UASS) reactor, Bioresour. Technol. 101 (2010) 592–599.
[36] C. Chen, D. Zheng, G.J. Liu, L.W. Deng, Y. Long, Z.H. Fan, Continuous dry fermentation of swine manure for biogas production, Waste Manag. 38 (2015) 436–442.

CHAPTER 7

Biogas production from waste: technical overview, progress, and challenges

Pooja Ghosh[1], Goldy Shah[1], Shivali Sahota[1], Lakhveer Singh[2] and Virendra Kumar Vijay[1]
[1]Centre for Rural Development and Technology, Indian Institute of Technology, New Delhi, India
[2]Faculty of Civil and Environmental Engineering, University Malaysia Pahang, Kuantan, Malaysia

7.1 Introduction

The energy sector is currently the most important area of research due to the depleting petroleum resources, the ever-increasing world population, and the need to solve the problem of climate change by reducing the emission of greenhouse gases (GHGs). These necessitate the search for an energy source that is environmentally sustainable and can help meet the increasing global demands of the near future. Hence tremendous attention is being given to biofuel technology as an alternative source of energy [1]. Biogas has multiple advantages as it is not only an environment-friendly fuel helping to reduce GHG emissions and waste recycling, but also produces a high quality fertilizer as a by-product along with electricity/heat production [2,3].

Although, the advantages of biogas have long been known, there has been a rekindling interest in this area due to the enormous amounts of wastes available, particularly in developing countries like India where waste management is a huge problem. A wide range of wastes (agricultural, municipal, animal droppings, and food waste) can be used as feedstocks for the production of biogas.

Biogas produced from the digestion of feedstocks under anaerobic conditions majorly constitutes methane (CH_4) (40%−65% v/v) and carbon dioxide (CO_2) (35%−55% v/v) with minute amounts of hydrogen sulfide (H_2S) (0.1−3% v/v), moisture, and other trace contaminants [3]. The anaerobic digestion (AD) process involves four different stages, namely hydrolysis, acidogenesis, acetogenesis, and methanogenesis, each

Bioreactors
DOI: https://doi.org/10.1016/B978-0-12-821264-6.00007-3

step being highly regulated by microbial communities. These microbial communities are highly sensitive to variations in temperature and pH. Hence optimizing and maintaining the AD process is key to enhancing biogas yields [4]. Factors such as biomass composition, pretreatment, the addition of microbes or enzymes, operating conditions, and digester designs are known to augment the production of biogas [1].

This chapter aims to examine the current status of biogas production globally and in developing countries like India, different types of wastes available for biogas production, and the technological advancements in the field of biogas production for enhancing yields toward meeting global energy demands. The barriers in the dissemination of biogas as a source of energy are also discussed.

7.2 Current status of biogas production

Europe currently leads in global biogas production, with other regions of the world also starting to focus on this form of bioenergy to meet future energy demands in a sustainable way. Global biogas production has increased from 0.28 EJ to 1.28 EJ between 2000 and 2014, with a global volume of 59 billion m^3 biogas [5]. The renewable energy support policies in the EU have helped it reach a production of 18 billion m^3 CH_4 (654 PJ) in 2015, which accounts for 50% of the biogas produced globally. With 17,400 functioning biogas plants, the EU leads in electricity production from biogas. Along with this, it is also the world's leading producer of biomethane, which is used as a vehicular fuel [6]. The global biogas production potential is quite high as predicted by various authors. Ferreira et al. estimated the biogas production potential from the landfills of Portugal to be 1.1 TW h/year [7]. Moreda predicted that by the AD of various waste feedstocks in Uruguay, a minimum of 0.162 TW h/year of electricity can be generated [8].

Compared to developed countries, developing countries like India have still not realized their full potential when it comes to biogas production and are still struggling to make a success of waste-to-energy projects [9]. Although, there has been an increase in the number of biogas plants in India from 1.27 million to 4.54 million plants between 1990 and 2012. However, the estimated potential of digesters is as high as 12.34 million. The 2014–2015 period witnessed a production of approximately 20,700 lakh $m^3$of biogas in India. India has the potential to produce 10% of the country's energy requirements ($\sim$17,000 MW) by adopting biogas technology [10]. However, considering the huge potential that exists, biogas production potential is still untapped mainly because of the high capital cost and low revenue growth. In 2013, there were only 56 operational biogas-based power plants in India, the majority located in the states of Maharashtra, Kerala, and Karnataka [11]. Although India and other developing countries are moving at a good pace in the production of biogas, there still exists a huge scope for further growth in the sector.

7.3 Available wastes for biogas production

Every year, tons of waste are generated from industrial, agricultural, and municipal sources. Inappropriate disposal and decomposition of biodegradable waste leads to large-scale environmental pollution. The degradation of 1 Mt of organic waste has the potential to release 50−110 m^3 of CO_2 and 90−140 m^3 of CH_4 into the atmosphere. However if the same biodegradable solid waste is converted into biogas through AD, it will lessen the unfavorable environmental impacts and help reduce dependency on conventional fuels [12]. Biogas is, thus, a promising approach for the conversion of different types of waste into efficient bioenergy.

Feedstocks utilized for biogas production mainly comprise a mixture of biomasses like forest residues, agricultural wastes, municipal solid wastes (MSW), livestock and cattle dung, energy crops, conventional crops, and other lignocellulosic feedstocks [13]. Generally, every biomass has the potential to be used as a substrate for biogas production, provided they have carbohydrates, proteins, fats, cellulose, and hemicelluloses as major components. However, the biogas composition and CH_4 yield are highly dependent on the feedstock type, the digestion system, and the retention time, and therefore need to be optimized [4].

Conventionally, biogas production has mainly been coupled with the treatment of cattle manure and sewage sludge from wastewater treatment plants (WWTP). Currently, most of biogas plants digest cattle manure with other substrates to increase the organic content for enhanced biogas production. Cosubstrates commonly include harvest residues, agricultural waste, food waste, and household waste. Also, fats provide the highest biogas yields, but their retention times are high as well as having poor bioavailability. On the other hand, carbohydrates and proteins get rapidly converted into biogas, but provide relatively low gas yields (Fig. 7.1). Some of the widely used feedstocks for biogas production are discussed here.

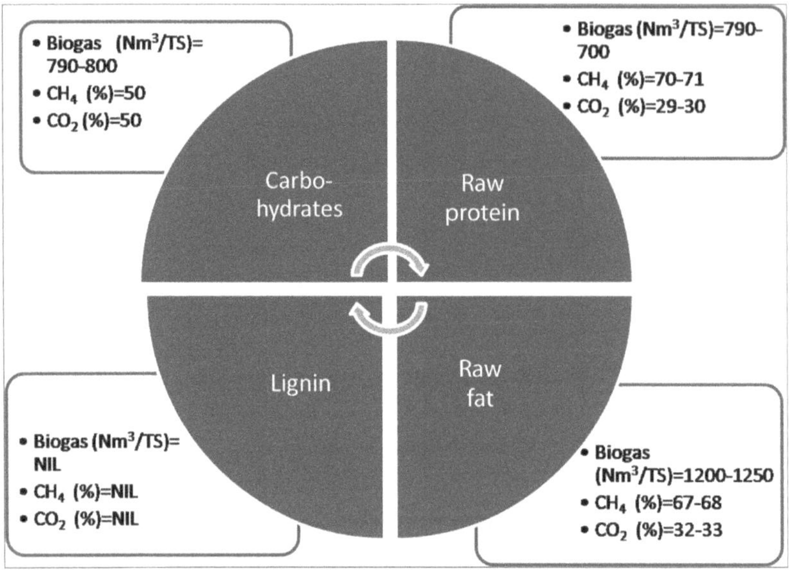

Figure 7.1 Biomethane potential and theoretical biogas contents [14].

7.3.1 Animal residues

Manures and slurries from different animals such as cows, buffaloes, horses, minks, and poultry animals have been used as feedstocks for biogas production for a long time [15]. The treatment of animal manure and slurries for biogas production is a beneficial process as it gives good quality CH_4 (50%–60%), and digested slurry can be used as a replacement to fertilizer as well as for the reduction of odors and microbial pathogens. The advantage of using them is that they contain inherent consortia of methanogens that help in methanogenesis during the AD process. However, a major limitation of animal manures and slurries is their low dry matter content (<10%) which results in low biogas production.

7.3.2 Food industry waste

Many kinds of food processing industrial waste can be digested anaerobically in a successful way such as fruit and vegetable waste (FVW), dry bread waste, dairy waste, soya flour, rice bran particles, and potato waste. The wastes can be treated separately as well as in codigestion processes [16]. The utilization of these wastes can help in food waste management. Also, the obtained digestate can act as a valuable fertilizer that is beneficial for organic farming [17].

7.3.3 Organic fraction of municipal solid waste

MSW mainly contains yard waste, office paper, corrugated printed newspaper, fruit and vegetable peel waste, leaf waste, food waste, and leaf litter, amongst which food waste accounts for the majority of the organic fraction of MSW [18]. AD is a suitable option for the treatment of MSW, provided that proper waste segregation is practiced. However, a major challenge for biogas production from MSW is unawareness among people about the segregation of the organic and inorganic fractions of MSW. Second, composition variation is also an important factor that affects biogas production. This compositional variation may be attributed to seasons, changing lifestyles, and eating habits [19] (Fig. 7.2).

7.3.4 Sewage sludge from wastewater treatment plants

Biogas production from sewage sludge has been practiced at WWTP for many decades. Mostly sewage sludge is digested alone as it contains inherent methanogens without external agitation. Anaerobic codigestion with other feedstocks can also be favorable and is reported to offer several benefits over the digestion of individual materials. The advantages of codigestion include increased cost-efficiency, augmented degradation of sewage sludge as well as increased biogas production [22]. Also, WWTPs are

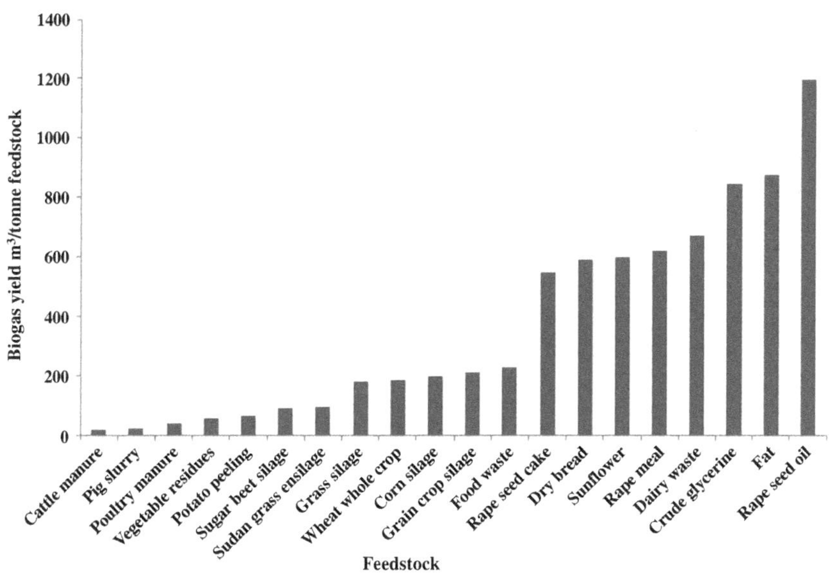

Figure 7.2 Biogas yield from different feedstocks [20,21].

potent sources of GHG emissions. Hence, biogas production from the sludge generated from WWTPs seems to be an attractive option to reduce GHG emissions [23].

7.4 Technological advancements in biogas production

There are different technological advancements happening in the biogas industry for enhancing the quantity as well as the quality of biogas produced. Since AD is a process mainly controlled by microbes at every step, from hydrolysis to methanogenesis, accelerating the microbial activity for a faster conversion of complex substrates into biogas and digestate is the aim of every technology. This is mainly possible through the use of one of four approaches, namely pretreatment of the feedstock, codigestion with a suitable substrate, seeding of microbes/enzymes, or optimizing the process parameters and digester designs (Fig. 7.3).

7.4.1 Pretreatment of wastes

A major issue with some of the feedstocks is their digestibility by microorganisms. Either they are completely recalcitrant for microbial action or the digestion is extremely slow due to the presence of inhibitory compounds. The aim of the pretreatment step is to assist the digestion step through the removal of compounds that inhibit the utilization of the feedstock by microorganisms. It not only helps in increasing biogas production, but also helps in reducing the content of volatile solids [24].

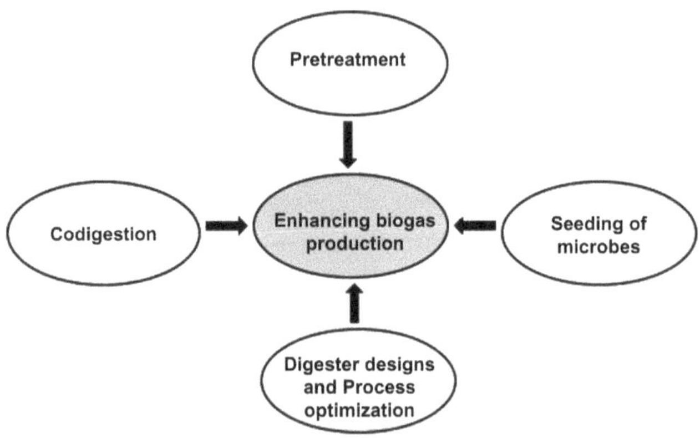

Figure 7.3 A basic outline of various approaches for enhancing biogas production.

Many researchers have reported hydrolysis to be a rate-limiting step due to the production of toxic by-products or undesirable volatile fatty acids formed during the hydrolysis process [25–28]. However, for easily biodegradable substrates, methanogenesis is reported to be a rate-limiting step [29,30]. Therefore research has been carried out extensively on various pretreatment methods for accelerating the hydrolysis step.

Types of pretreatment methods include physical, chemical, physicochemical, and biological methods. Depending on the different characteristics of different feedstocks, the effects of various pretreatment methods differ. Hence it is difficult to come to a conclusion about the most suitable method for all feedstocks as the effects will vary [31]. The most commonly used pretreatment methods are summarized in Sections 7.4.1.1–7.4.1.8.

7.4.1.1 Particle size reduction

Substrate particle size directly affects digestion as it affects the available surface area for microbial action. Mshandete et al. reported that CH_4 yields improve by reducing the size of particles to 2 mm [32].

7.4.1.2 Liquid hot water treatment

Treating feedstocks with liquid hot water helps in solubilizing hemicellulose and lignin well, thereby reducing the possibility of the production of inhibitors such as furfural. However, this process requires huge amounts of heat [33,34].

7.4.1.3 Microwave treatment

Microwave treatment of feedstocks is found to produce 4%–7% more biogas compared to untreated feedstocks [34].

7.4.1.4 Acid pretreatment

Acid pretreatment of feedstocks helps in solubilizing hemicelluloses. However, the disadvantages of acid pretreatment include high costs, risk of forming inhibitory compounds, and corrosion [35,36].

7.4.1.5 Alkali pretreatment

Taherdanak and Zilouei reported that adding alkali to the feedstock in a controlled manner enhanced the biogas yield and reduced cellulose production [37]. Clarkson and Xiao also reported that the degradation rate of paper waste in AD systems increased on the addition of optimum concentrations of sodium hydroxide solution [38]. However, a problem associated with alkali pretreatment is that alkalis frequently lead to saponification reactions [39].

7.4.1.6 Thermal/thermochemical pretreatment

Preheating of the substrate before AD has been found to reduce the volatile solids content and improve methane production. Also, thermochemical pretreatment involving the preheating of the substrate treated with chemical additives such as sodium hydroxide further increases the CH_4 production [40]. Ardic and Taner reported increases in both biomethane production as well as the biodegradation of the feedstock through pretreatment using preheated sodium hydroxide at 100°C [41].

7.4.1.7 Ultrasonic pretreatment

This involves the introduction of ultrasonic waves, which builds up mechanical shear forces helping in the disintegration of the feedstock and thereby improving biogas production by AD [24].

7.4.1.8 Enzymatic pretreatment

The addition of enzymes exogenously during AD is reported to enhance the hydrolytic step of complex substrates. Studies have shown that the addition of hydrolytic enzymes helps in solving the limitations associated with the rate limiting step of hydrolysis [42,43]. Being good sources of hydrolytic enzymes, thermophilic bacteria are increasingly being studied as good sources of hydrolytic enzymes [44]. Prabhudessai et al. reported an enhanced biogas production on the addition of enzymes (lipase and protease) to anaerobic digesters containing food processing waste [45]. In fact, they reported that mixed enzyme treatments consisting of different ratios of lipase to protease led to higher CH_4 production in comparison to single enzyme treatments.

7.4.2 Seeding of microbes

Seeding involves the addition of a material rich in microorganisms such as animal manure or municipal sludge to reduce the biogas plant start-up time [46]. Conventionally, materials such as cow dung and poultry droppings, and others have been used widely for enhancing biogas production as these are rich sources of microorganisms [47]. Compost is another material that is rich in microbes capable of producing hydrolytic enzymes and helping in the effective solubilization of organic matter that is difficult to biodegrade. Hence, these days, it is considered as a useful inoculum for enhancing the hydrolysis of feedstocks [48]. The inoculum size is an important parameter in such a case as it controls total organic matter degradation of the feedstock. Budiyono et al. reported that a concentration of 50% rumen fluid leads to better degradation of lignocellulosic feedstocks for biogas production [49]. A pilot level study by Mirdamadian et al. pointed out that with the use of a special microbial consortium, the fermentation duration can be reduced [50]. The increased biogas production can be explained by an increased production of enzymes by the microbial consortium.

7.4.3 Codigestion of wastes

A common problem faced in biogas production is low biogas yields due to the use of a single feedstock that may be either recalcitrant to digestion or rich in protein or other inhibitory compounds. These limitations can be resolved by the codigestion of different substrates with an optimum mixing ratio, while considering the C/N ratio, inhibitors, feedstock biodegradability, and total solid content. This also helps in reducing ammonia production during AD, thus reducing the chances of inhibition caused by ammonia along with enhancing the biogas production from the feedstock by increasing its digestibility [51]. Zeshan et al. found that by adjusting the C/N ratio of codigested feedstocks to 32, there occurred a reduction in ammonia production of 30% [52].

Many reports are available where different researchers have studied the effects of mixing different wastes on biogas yield. Pagés Díaz et al. utilized different agro-industrial and slaughterhouse waste mixtures and investigated their effects on biogas production [53]. Synergetic effects were observed by the authors due to the mixing of substrates, leading to a staggering 43% increase in CH_4 yields in comparison to the CH_4 yields of individual feedstocks. Callaghan et al. also reported an increase in CH_4 yields with the codigestion of FVW and cattle manure in a ratio of 1:1 [54]. In fact, codigestion has the potential to augment biogas production from 25% to 400% compared to monodigestion using the same feedstocks [55,56]. These studies clearly emphasize that codigestion improves biomethane yield, both in terms of quantity and quality. In comparison to pretreatments, codigestion is a good strategy to enhance biogas yields, while avoiding the cons of pretreatment. Nevertheless, more research and development is required for understanding the optimum mixing ratios of different substrates and how they interact to affect biogas yields.

7.4.4 Digester designs and process optimization

Considerable research has been carried out to study different reactor (e.-g., single- or multiple-stage reactors) configurations and how they affect the process of AD. Research has proven that it is difficult to enhance biogas production in a single-stage digester due to the differences in the metabolic properties, nutritional requirements, growth rates, and optimum operational factors in the various steps of AD [57,58]. Recent studies report that two-stage digestion processes, having the hydrolytic and acidogenic steps separated from the acetogenic and methanogenic steps, can increase CH_4 yields along with reducing the retention time, though the cost of such a complex system is a significant disadvantage [59]. Different types of two-stage reactors are currently being researched including reactors for thermophilic two-stage digestion, mesophilic two-stage digestion, and temperature phased anaerobic digestion [31]. Nasr et al. studied the efficiency of single-stage and two-stage digestion systems and found that an 18.5% higher energy yield was achieved through two-stage systems [60]. Fig. 7.4 depicts the advantages and disadvantages of single- and multistage systems.

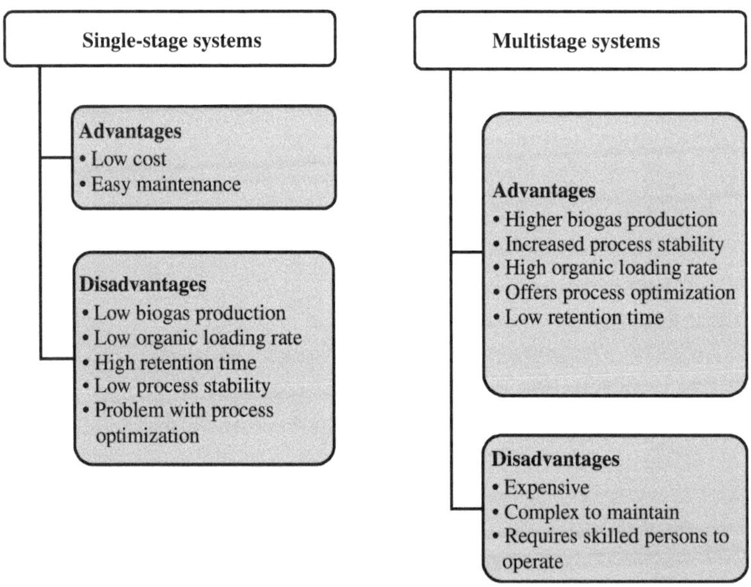

Figure 7.4 Advantages and disadvantages of single- and multistage digestion systems.

7.5 Challenges associated with biogas technology dissemination

Though biogas has emerged as a promising renewable technology to convert various types of wastes, its true production potential is still a long way from being realized. There exist both technical and nontechnical barriers preventing the dissemination of biogas technology, particularly in developing countries like India [61]. Also, these barriers vary from region to region. For example, the availability of natural resources like biomass, land, and water may be a challenge in one region, while in another region, the barrier may be related to technology preventing the expansion of biogas technology [62,63].

Fig. 7.5 shows the six major barriers associated with the dissemination of biogas technologies. The major barrier for developing countries is the economic barrier, which includes the high-investment cost for biogas plant installation along with the lack of enough financial support from governments. Another major barrier is the market competition from other low-priced energy sources such as coal and natural gas. For Brazil, it was projected that the minimum cost of energy produced from biogas will be around US$105.3/MW/h, which is much higher than that from conventional power plants [64]. Also, there exist social barriers like a lack of social acceptance for biogas from substrates like human excreta. Regulatory barriers involve the lack of coordination between national and state governments. For the diffusion of any technology in the initial stages various incentives are necessary. In India, this was the case for both solar and wind technologies, where the government have introduced investment-friendly policies for the steady growth of these renewables

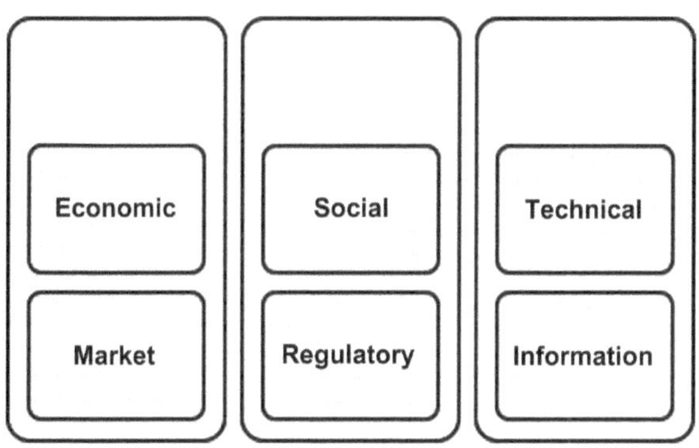

Figure 7.5 The six major barriers to biogas technologies.

over the past decade. However, for biogas technology, a lack of appropriate political framework and business models to support its dissemination can be observed [61]. Even in other countries such as Brazil, biogas has still not been listed as a primary source of energy, and there is still no specific biogas-related program for its promotion [65]. Technological barriers include the lack of proper technologies for the segregation, collection, and transportation of wastes, particularly in developing countries where waste management is generally a neglected area. Also, process standardization is quite a challenge due to huge variations in waste characteristics across different regions, which greatly hinders the extensive diffusion of biogas technology. A lack of awareness is another barrier in biogas technology dissemination.

7.6 Conclusion

Biogas is a promising renewable energy source and can be either captured from landfills or produced from a variety of wastes. If entirely utilized, biogas yields from existing organic wastes could satisfy about 20% of current natural gas use. Biogas technology should not be viewed as a competition toward already existing energy sources, but rather, it should be viewed as supplementary to what already exists and a sustainable scheme for solving growing environmental issues. With the technological advancements and political support from governments, the true potential of biogas would soon be realized and along with its application for heat and electricity generation, it could be applied for more advanced applications such as vehicular fuel throughout the world.

Acknowledgment

The authors would like to express their sincere gratitude to the Department of Science and Technology, Govt. of India for providing INSPIRE Faculty fellowship to Ghosh [DST/INSPIRE/04/2016/000362].

References

[1] D. Divya, L. Gopinath, P.M. Christy, A review on current aspects and diverse prospects for enhancing biogas production in sustainable means, Renew. Sustain. Energy Rev 42 (2015) 690–699.
[2] O. Oyewole, Biogas production from chicken droppings, Sci. World J 5 (2010).
[3] S. Sahota, G. Shah, P. Ghosh, R. Kapoor, S. Sengupta, P. Singh, et al., Review of trends in biogas upgradation technologies and future perspectives, Bioresour. Technol. Rep 1 (2018) 79–88.

[4] C. Mao, Y. Feng, X. Wang, G. Ren, Review on research achievements of biogas from anaerobic digestion, Renew. Sustain. Energy Rev 45 (2015) 540–555.
[5] W.B.A.W.W.G.B.S., World Bioenergy Association (WBA), WBA Global Bioenergy Statistics, 2017. <https://doi.org/10.1016/0165-232X(80)90063-4>.
[6] N. Scarlat, J.-F. Dallemand, F. Fahl, Biogas: developments and perspectives in Europe, Renew. Energy 129 (2018) 457–472.
[7] S. Ferreira, N.A. Moreira, E. Monteiro, Bioenergy overview for Portugal, Biomass Bioenergy 33 (2009) 1567–1576.
[8] I.L. Moreda, The potential of biogas production in Uruguay, Renew. Sustain. Energy Rev 54 (2016) 1580–1591.
[9] R. Joshi, S. Ahmed, Status and challenges of municipal solid waste management in India: a review, Cogent Environ. Sci 2 (2016) 1139434.
[10] MNRE, 2015, Annual Report, 2015–16, in: M.O.N.A.R. (Ed.), Energy, New Delhi, 2015.
[11] Consolidated Annual Review Report on Implementation of Municipal Solid Wastes (Management and Handling) Rules, 2000, in: C.P.C. BOARD (Ed.), Ministry of Environment Forests & Climate Change, New Delhi, 2013.
[12] M. Macias-Corral, Z. Samani, A. Hanson, G. Smith, P. Funk, H. Yu, et al., Anaerobic digestion of municipal solid waste and agricultural waste and the effect of co-digestion with dairy cow manure, Bioresour. Technol 99 (2008) 8288–8293.
[13] W. Qiao, X. Yan, J. Ye, Y. Sun, W. Wang, Z. Zhang, Evaluation of biogas production from different biomass wastes with/without hydrothermal pretreatment, Renew. Energy 36 (2011) 3313–3318.
[14] P. Weiland, Biogas production: current state and perspectives, Appl. Microbiol. Biotechnol 85 (2010) 849–860.
[15] I. Maile, E. Muzenda, Production of biogas from various substrates under anaerobic conditions, Proceedings of the International Conference on Innovative Engineering Technologies (ICIET) (2014) 78–80.
[16] M. Murto, L. Björnsson, B. Mattiasson, Impact of food industrial waste on anaerobic co-digestion of sewage sludge and pig manure, J. Environ. Manag 70 (2004) 101–107.
[17] C. Popescu, S. Jurcoane, Evaluation of biogas potential of some organic substrates from agriculture and food industry and co-digestion in large scale biogas plant, Roman. Biotechnol. Lett 20 (2015) 10648.
[18] Z. Yong, Y. Dong, X. Zhang, T. Tan, Anaerobic co-digestion of food waste and straw for biogas production, Renew. Energy 78 (2015) 527–530.
[19] L. Appels, J. Lauwers, J. Degrève, L. Helsen, B. Lievens, K. Willems, et al., Anaerobic digestion in global bio-energy production: potential and research challenges, Renew. Sustain. Energy Rev 15 (2011) 4295–4301.
[20] Y. Shen, J.L. Linville, M. Urgun-Demirtas, M.M. Mintz, S.W. Snyder, An overview of biogas production and utilization at full-scale wastewater treatment plants (WWTPs) in the United States: challenges and opportunities towards energy-neutral WWTPs, Renew. Sustain. Energy Rev 50 (2015) 346–362.
[21] J.L. Linville, Y. Shen, M.M. Wu, M. Urgun-Demirtas, Current state of anaerobic digestion of organic wastes in North America, Current Sustainable/Renewable Energy Reports 2 (2015) 136–144.
[22] S. Luostarinen, S. Luste, M. Sillanpää, Increased biogas production at wastewater treatment plants through co-digestion of sewage sludge with grease trap sludge from a meat processing plant, Bioresour. Technol 100 (2009) 79–85.
[23] A.T.A. Felca, R.M. Barros, G.L. Tiago Filho, I.F.S. dos Santos, E.M. Ribeiro, Analysis of biogas produced by the anaerobic digestion of sludge generated at

wastewater treatment plants in the South of Minas Gerais, Brazil as a potential energy source, Sustain. Cities Soc 41 (2018) 139–153.
[24] A. Tiehm, K. Nickel, M. Zellhorn, U. Neis, Ultrasonic waste activated sludge disintegration for improving anaerobic stabilization, Water Res 35 (2001) 2003–2009.
[25] A. Khalid, M. Arshad, M. Anjum, T. Mahmood, L. Dawson, The anaerobic digestion of solid organic waste, Waste Manag 31 (2011) 1737–1744.
[26] J. Ma, T.H. Duong, M. Smits, W. Verstraete, M. Carballa, Enhanced biomethanation of kitchen waste by different pre-treatments, Bioresour. Technol 102 (2011) 592–599.
[27] V. Vavilin, B. Fernandez, J. Palatsi, X. Flotats, Hydrolysis kinetics in anaerobic degradation of particulate organic material: an overview, Waste Manag 28 (2008) 939–951.
[28] L. Neves, R. Ribeiro, R. Oliveira, M. Alves, Enhancement of methane production from barley waste, Biomass Bioenergy 30 (2006) 599–603.
[29] H.N. Gavala, U. Yenal, I.V. Skiadas, P. Westermann, B.K. Ahring, Mesophilic and thermophilic anaerobic digestion of primary and secondary sludge. Effect of pretreatment at elevated temperature, Water Res 37 (2003) 4561–4572.
[30] A. Rozzi, E. Remigi, Methods of assessing microbial activity and inhibition under anaerobic conditions: a literature review, Rev. Environ. Sci. Bio/Technol 3 (2004) 93–115.
[31] J. Ariunbaatar, A. Panico, G. Esposito, F. Pirozzi, P.N. Lens, Pretreatment methods to enhance anaerobic digestion of organic solid waste, Appl. Energy 123 (2014) 143–156.
[32] A. Mshandete, L. Björnsson, A.K. Kivaisi, M.S. Rubindamayugi, B. Mattiasson, Effect of particle size on biogas yield from sisal fibre waste, Renew. Energy 31 (2006) 2385–2392.
[33] M. Laser, D. Schulman, S.G. Allen, J. Lichwa, M.J. Antal Jr, L.R. Lynd, A comparison of liquid hot water and steam pretreatments of sugar cane bagasse for bioconversion to ethanol, Bioresour. Technol 81 (2002) 33–44.
[34] H. Shahriari, M. Warith, M. Hamoda, K.J. Kennedy, Anaerobic digestion of organic fraction of municipal solid waste combining two pretreatment modalities, high temperature microwave and hydrogen peroxide, Waste Manag 32 (2012) 41–52.
[35] W. Xiao, W.W. Clarkson, Acid solubilization of lignin and bioconversion of treated newsprint to methane, Biodegradation 8 (1997) 61–66.
[36] J. Sumphanwanich, N. Leepipatpiboon, T. Srinorakutara, A. Akaracharanya, Evaluation of dilute-acid pretreated bagasse, corn cob and rice straw for ethanol fermentation by *Saccharomyces cerevisiae*, Ann. Microbiol 58 (2008) 219–225.
[37] M. Taherdanak, H. Zilouei, Improving biogas production from wheat plant using alkaline pretreatment, Fuel 115 (2014) 714–719.
[38] W. Clarkson, X. Xiao, Bench-scale anaerobic bioconversion of newsprint and office paper, Water Sci. Technol 41 (2000) 93–100.
[39] A. Mouneimne, H. Carrere, N. Bernet, J. Delgenes, Effect of saponification on the anaerobic digestion of solid fatty residues, Bioresour. Technol 90 (2003) 89–94.
[40] H. Carrère, B. Sialve, N. Bernet, Improving pig manure conversion into biogas by thermal and thermo-chemical pretreatments, Bioresour. Technol 100 (2009) 3690–3694.
[41] I. Ardİc, F. Taner, Effects of thermal, chemical and thermochemical pretreatments to increase biogas production yield of chicken manure, Fresenius Environ. Bull 14 (2005) 373–380.
[42] A. Hendriks, G. Zeeman, Pretreatments to enhance the digestibility of lignocellulosic biomass, Bioresour. Technol 100 (2009) 10–18.

[43] R.T. Romano, R. Zhang, S. Teter, J.A. McGarvey, The effect of enzyme addition on anaerobic digestion of JoseTall wheat grass, Bioresour. Technol 100 (2009) 4564–4571.
[44] A.S.S. Ibrahim, A.I. El-diwany, Isolation and identification of new cellulases producing thermophilic bacteria from an Egyptian hot spring and some properties of the crude enzyme, Aust. J. Basic Appl. Sci 1 (2007) 473–478.
[45] V. Prabhudessai, B. Salgaonkar, J. Braganca, S. Mutnuri, Pretreatment of cottage cheese to enhance biogas production, BioMed Res. Int 2014 (2014).
[46] A.J. Ward, P.J. Hobbs, P.J. Holliman, D.L. Jones, Optimisation of the anaerobic digestion of agricultural resources, Bioresour. Technol 99 (2008) 7928–7940.
[47] J. Eze, K. Agbo, Studies on the microbial spectrum in anaerobic biomethannization of cow dung in 10 m^3 fixed dome biogas digester, Int. J. Phys. Sci 5 (2010) 1331–1337.
[48] U. Brémond, R. de Buyer, J.-P. Steyer, N. Bernet, H. Carrere, Biological pretreatments of biomass for improving biogas production: an overview from lab scale to full-scale, Renew. Sustain. Energy Rev 90 (2018) 583–604.
[49] W.I. Nyoman, J. Seno, The kinetic of biogas production rate from cattle manure in batch mode, Int. J. Chem. Biol. Eng 3 (2010) 39–45.
[50] S. Mirdamadian, S. Khayam-Nekoui, H. Ghanavati, Reduce of fermentation time in composting process by using a special microbial consortium, World Acad. Sci. Eng. Technol 52 (2011) 475–479.
[51] J. Mata-Alvarez, J. Dosta, M. Romero-Güiza, X. Fonoll, M. Peces, S. Astals, A critical review on anaerobic co-digestion achievements between 2010 and 2013, Renew. Sustain. Energy Rev 36 (2014) 412–427.
[52] K.O. Zeshan, C. Visvanathan, Effect of C/N ratio and ammonia–N accumulation in a pilot-scale thermophilic dry anaerobic digester, Bioresour. Technol 113 (2012) 294–302.
[53] J.P. Díaz, I.P. Reyes, M. Lundin, I.S. Horváth, Co-digestion of different waste mixtures from agro-industrial activities: kinetic evaluation and synergetic effects, Bioresour. Technol 102 (2011) 10834–10840.
[54] F. Callaghan, D. Wase, K. Thayanithy, C. Forster, Continuous co-digestion of cattle slurry with fruit and vegetable wastes and chicken manure, Biomass Bioenergy 22 (2002) 71–77.
[55] C. Cavinato, F. Fatone, D. Bolzonella, P. Pavan, Thermophilic anaerobic co-digestion of cattle manure with agro-wastes and energy crops: comparison of pilot and full scale experiences, Bioresour. Technol 101 (2010) 545–550.
[56] F.A. Shah, Q. Mahmood, N. Rashid, A. Pervez, I.A. Raja, M.M. Shah, Co-digestion, pretreatment and digester design for enhanced methanogenesis, Renew. Sustain. Energy Rev 42 (2015) 627–642.
[57] V. Hubenov, S. Mihaylova, I. Simeonov, Anaerobic co-digestion of waste fruits and vegetables and swine manure in a pilot-scale bioreactor, Bulgarian Chem. Commun 47 (2015) 788–792.
[58] A. Kacprzak, L. Krzystek, S. Ledakowicz, Co-digestion of agricultural and industrial wastes, Chem. Papers 64 (2010) 127–131.
[59] L.F. Montgomery, G. Bochmann, Pretreatment of feedstock for enhanced biogas production, IEA Bioenergy Ireland (2014).
[60] N. Nasr, E. Elbeshbishy, H. Hafez, G. Nakhla, M.H. El Naggar, Comparative assessment of single-stage and two-stage anaerobic digestion for the treatment of thin stillage, Bioresour. Technol 111 (2012) 122–126.
[61] S. Mittal, E.O. Ahlgren, P. Shukla, Barriers to biogas dissemination in India: a review, Energy Policy 112 (2018) 361–370.

[62] M. Lantz, M. Svensson, L. Björnsson, P. Börjesson, The prospects for an expansion of biogas systems in Sweden—incentives, barriers and potentials, Energy Policy 35 (2007) 1830–1843.
[63] A. Shane, S.H. Gheewala, G. Kasali, Potential, barriers and prospects of biogas production in Zambia, J. Sustain. Energy Environ 6 (2015) 21–26.
[64] I.F.S. dos Santos, N.D.B. Vieira, L.G.B. de Nóbrega, R.M. Barros, G.L. Tiago Filho, Assessment of potential biogas production from multiple organic wastes in Brazil: impact on energy generation, use, and emissions abatement, Resour., Conserv. Recycl 131 (2018) 54–63.
[65] J.F.C.M. Mathias, Manure as a resource: livestock waste management from anaerobic digestion, opportunities and challenges for Brazil, Int. Food Agribus. Manag. Rev 17 (2014) 87–110.

CHAPTER 8

Life cycle assessment of waste-to-bioenergy processes: a review

Pooja Ghosh[1], Subhanjan Sengupta[2], Lakhveer Singh[3] and Arunaditya Sahay[2]
[1]Centre for Rural Development and Technology, Indian Institute of Technology, New Delhi, India
[2]Birla Institute of Management Technology, Greater Noida, India
[3]Faculty of Civil and Environmental Engineering, University Malaysia Pahang, Kuantan, Malaysia

8.1 Introduction

The major problems faced by the world today include depleting nonrenewable sources of energy, increasing global energy demand, and the need to solve the problem of climate change (CC). Based on estimations made, the world energy consumption is expected to rise by 44% from 497 EJ in 2006 to 715 EJ in 2030. Carbon dioxide (CO_2) emissions are also predicted to increase by 39% from 29 billion tons to 40.4 billion between 2006 and 2030 [1,2]. It is nearly impossible to attain sustainability based on the current fossil fuel-based energy system. Therefore there is renewed interest in alternative energy resources such as bioenergy, which looks quite prospective in solving the global problems of energy security and CC.

Along with the issue of finding a sustainable source of energy, a major challenge faced by many developing countries in particular is the problem of waste management. Waste is commonly perceived as something that has ceased having utility. However, the high organic content in waste makes it appropriate for recovering energy from it, thereby helping in solving the problems of waste management, finding an alternative energy source, and reducing environmental pollution. The Ministry of New and Renewable Energy, of the government of India, predicted that for the year 2013 there was a potential of 1700 MW of energy generation from urban organic solid waste (1500 MW from municipal solid waste and 225 MW from sewage) and 1300 MW of energy from industrial waste [3]. Hence the ever-increasing waste should be seen as a resource to be used to recover materials and energy, particularly in developing countries where waste-to-energy (WtE) projects are still in the nascent stage. With

Bioreactors
DOI: https://doi.org/10.1016/B978-0-12-821264-6.00008-5

the present global emphasis on waste management, energy security, and reducing greenhouse gas (GHG) emissions, the use of sustainable resources and life cycle thinking have become extremely important.

Globally, there are many policies that support the promotion of bioenergy. However, there are many questions being raised regarding their sustainability [4,5]. The conversion of biomass to energy involves many input and output flows that may affect the overall environmental performance of the process. To establish the sustainability of a WtE technology [e.g., incineration, gasification, pyrolysis, anaerobic digestion (AD)], life cycle assessment (LCA) is a significant instrument to measure environmental profiles and GHG emissions arising from the technological solution being used, and to compare it with the alternative waste management options such as landfilling, which is a popular method across the world. Thus LCA is an important planning tool that helps in identifying the most environmentally sustainable WtE technology.

This chapter aims to review the present global waste generation scenario, current need for waste-derived bioenergy, and the importance of LCA to establish the sustainability of a process. Also, the challenges associated with LCA studies have been dealt with.

8.2 Global waste generation scenario

Waste generation and devising ways to deal with it has increasingly been a matter of concern for modern societies. Urbanization has proliferated and waste generation is at an all-time high. As per data from the World Bank from 2012, cities across the world generated roughly 1.3 billion tons of solid waste, and this figure would probably reach 2.2 billion tons by 2025 (World Bank, 2018).[1] The most severally impacted by this is the common populace in low-and middle-income countries, where waste generation is not controlled and waste disposal and treatment is not planned and regulated with policy planning or the enforcement of law. This is out of control because of the exponential growth in the population, rapid urbanization, and rampant consumerism toward uplifting living standards, which have been instrumental in the increasing amount of solid waste generation and numerous landfills spread across the cities of the world [6]. When waste created by the daily consumption activities of urban

[1] *Source*: World Bank, 2018 <www.worldbank.org/en/topic/urbandevelopment/brief/solid-waste-management> (accessed 15.06.18).

populations is grouped with other forms of waste such as commercial, industrial, medical, agricultural, construction, and demolition waste, a staggering quantity of 7–10 billion tons of waste, generated per year across the world, is reached (United Nations Environment Program, 2015).[2]

There is no stopping this as the population in developing countries is in a continuous effort to improve their economic standard of living to reach to the kind of lifestyle standards achieved in developed countries. In addition to the waste already generated by developed nations, developing economies in Asia and Africa are soon going to surpass that in the effort of people to have the best lifestyle standards for themselves. Needless to mention that the industrial and commercial growth in these countries are at an all-time high, which leaves little room to work toward the regulation of waste generation and efficient management and treatment of what is generated. Considering the currently running sustainability scenario, it has been predicted that the world will reach "peak waste" sometime around 2075, with a staggering range of 7.3–10.9 megatons of municipal solid waste being produced every day [7].

8.3 Need for waste-derived bioenergy

Without a doubt, the elimination of current disposal practices is now a priority. Uncontrolled dumpsites full of organic contaminants, plastic-based nondegradable waste, and high heavy metal concentrations are a major concern in developing countries [8], with a major portion of the population having no access to controlled waste disposal systems and infrastructure (United Nations Environment Program, 2015).[2] Six elementary activities for the management of solid waste are reduction at the source, improving service and collection systems and infrastructure, the use of technology-enabled sorting and processing facilities for recycling, establishing large-scale composting facilitates, the use of pollution-control incinerators, and the creation of engineered, sanitary landfills with environmental controls (World Bank, 2012).[3]

However, as obviously easy these solutions seem to be, it is highly challenging to plan, implement, and sustain all of these as the challenges

[2] *Source*: United Nations Environment Program (UNEP), 2015 <https://zoinet.org/wp-content/uploads/2018/02/GWMO_report.pdf> (accessed 15.06.18).

[3] *Source*: World Bank, 2012 <https://siteresources.worldbank.org/INTURBANDEVELOPMENT/Resources/336387-1334852610766/What_a_Waste2012_Final.pdf> (accessed 15.06.18).

across countries are immense due to several demographic, economic, social, infrastructural, and policy-oriented complexities and limitations. Moreover, the way to control waste generation and deal with its management has an educational as well as infrastructural dimension to it, which is decisive in the ability, intent, and capacity to deal with waste. It is interesting to note that while developed countries have achieved technological advancement in treating food waste using incineration, AD, and other treatment alternatives determined as environment-friendly using LCA [9–11], developing countries generate far less waste per capita than developed Western countries [12]. Consequentially, developed cities have committed to developing and implementing zero waste practices [13], and corporations have started planning processes so as to achieve "zero waste" to landfill as a part of their commitment to sustainability [14].

However, achieving such transformation is unlikely to happen anytime soon due to the overconsumption that is rampant across the globe [15]. The complexity of the global waste generation scenario has invited research initiatives across the globe to explore WtE conversion alternatives with applications in different developed and developing contexts, which includes not just the popular techniques of incineration, gasification, pyrolysis, biochemical conversion, ethanol fermentation, or AD, but newer ecological technologies such as the use of microbial fuel cells (MFCs) and microbial electrolysis cells (MECs) [16]. The range of technology applications in waste-to-bioenergy processes are being continuously researched, with developments being made in bioreactor designs, optimizations in operational parameters, and the application of LCA models to identify and quantify the environmental impact of traditional and emerging technologies [10,16,17].

8.4 Different technologies for converting waste-to-energy

WtE conversion technologies can be categorized into two types, namely (1) thermal conversion technologies (incineration, pyrolysis, and gasification) and biochemical conversion (AD and fermentation); and (2) bioelectrochemical conversion technologies [16].

8.4.1 Thermal conversion technologies

8.4.1.1 Incineration

Incineration involves burning waste at high temperatures ranging between 750°C and 1100°C in the presence of oxygen to reduce the weight and volume of the waste and produce heat and energy [18]. This method is capable of reducing waste mass by almost 70% and volume by up to 90% [19]. However, a major drawback of this method is the generation of gaseous pollutants such as nitrogen oxides (NOx), sulfur oxides (SOx) and carbon oxides (COx), polyaromatic hydrocarbons, and heavy metals. This necessitates an additional treatment of the flue-gas before it can be emitted into the atmosphere [20]. The energy efficiency for the generation of heat is 80%, the cogeneration of steam and electricity varies between 20% and 30%, and pure electricity is 20% [16].

8.4.1.2 Pyrolysis

Pyrolysis involves the decomposition of waste under anaerobic conditions at temperatures ranging between 300°C and 800°C. The pyrolytic products obtained include a gaseous product (syngas), a liquid (tar), and char, with ash as an undesirable residue [21]. The yields of the products depend on the composition of the waste, the pyrolysis temperature, and the heating rate. Lower pyrolysis temperatures lead to the production of more liquid products, whereas higher temperatures yield more gaseous products. The major advantage of pyrolysis is that it is a cost-effective technology and helps curb environmental pollution. However, the production of various pollutants occurs in the exhaust gas during the pyrolysis process such as hydrogen sulfide (H_2S), ammonia (NH_3), SOx, and NOx. So strategies for treating the exhaust gas before emission into the environment are extremely important [16].

8.4.1.3 Gasification

The main aim of gasification is the production of syngas, though tar is generated concomitantly with the syngas. Compared to pyrolysis, gasification generally occurs at higher temperatures of between 550°C and 900°C in the case of air gasification and between 1000°C and 1600°C when using pure O_2 or O_2-enriched gas or steam [22]. An important advantage of gasification over combustion is that the contaminants present are easily removed as the syngas is produced at higher temperatures and pressures than those used in combustion. However, a major disadvantage includes the high capital cost involved and the production of bottom ash that needs to be removed [23].

8.4.2 Biochemical conversion technologies

Biochemical conversion involves the use of microorganisms or their enzymes to break down biomass. Widely used biochemical technologies for waste conversion include AD and fermentation. During AD, microbes convert the organic fraction of wastes into biogas [40%–65% methane (CH_4), 35%–55% CO_2, and other gases present in trace amounts like hydrogen (H_2) and H_2S]. A nutrient-rich residue is also produced that can be used as a soil conditioner [24]. The advantages of AD are that it can be applied on a wide range of substrates, even those with high moisture contents and impurities [25]. The fermentation of waste involves the conversion of sucrose into fructose and glucose via hydrolysis enzymes followed by their transformation into ethanol [16].

8.4.3 Bioelectrochemical processes

These include MFCs and MECs. MFCs use microorganisms for biocatalysis, which involves the oxidation of organic matter in a substrate and the transfer of electrons to the surface of an anode for bioelectricity production [26]. A range of organic substrates such as waste from domestic sources, sludge from wastewater treatment plants as well as animal waste can be utilized as a feedstock [27]. MECs involve the use of electrochemically energetic bacteria in order to transform waste into H_2 and other products such as CH_4, acetate, hydrogen peroxide, ethanol, and formic acid. This is similar to MFCs, the only difference is that the cathode of MECs is not exposed to air. MECs are also characterized by their higher recovery of H_2 and broader substrate diversity in comparison to MFCs [28].

8.5 Life cycle assessment for waste-derived bioenergy systems

8.5.1 Basics of life cycle assessments and its methodological framework

LCA is an environmental assessment instrument for determining the environmental burden of a product or system over its entire life cycle. Thus it often involves a "cradle-to-grave" approach starting from the acquisition of raw materials to the production, use, and finally disposal of the product [29]. The impacts at different stages of a life cycle are clustered into different categories such as global warming potential (GWP), ecotoxicity, acidification potential, eutrophication potential, effects on human health, and others. Other than this cradle-to-grave approach, it is also possible to study the single stages within the life cycle of a given process or product in order to determine the "hotspots" so that efforts can be made to reduce the impacts. It is not only important for industry or manufacturers, but also for decision makers [30].

The standardization of the procedural structure for steering an LCA has been done by the International Organization for Standardization (ISO) as depicted in Fig. 8.1. For carrying out an LCA study, the ISO 14040 series describes four stages, namely (1) defining goal and scope for the study (ISO 14041), (2) life cycle inventory (LCI) (ISO 14041), (3) life cycle impact assessment (LCIA) (ISO 14042), and (4) interpretation of the results (ISO 14043). These standards were revised in 2006 and combined to form a single standard, ISO 14044 [30]. The first step of an LCA study is defining the goal and scope of the study. This involves defining the aim, the functional unit (FU), and system boundary for the LCA study. FU is defined as "quantified performance of a product system for use as a reference unit" (ISO 2006). During a comparison of two or more product systems or processes by LCA, the role played by the FU in the LCA results is important. Also, selecting different FUs may result in different outputs and conclusions for the same study [31]. According to ISO 2006, a system boundary involves the selection of steps included within the life cycle of the process. Selecting an inappropriate system boundary that does not take into account all the phases of the life cycle, will not give a true picture of the impacts and may thus affect decision making [32]. Inventory analysis involves accounting for all the environmentally relevant flows (input mass/energy and output mass/work done) of the system lying within the system boundary. Inventory analysis is followed by the impact

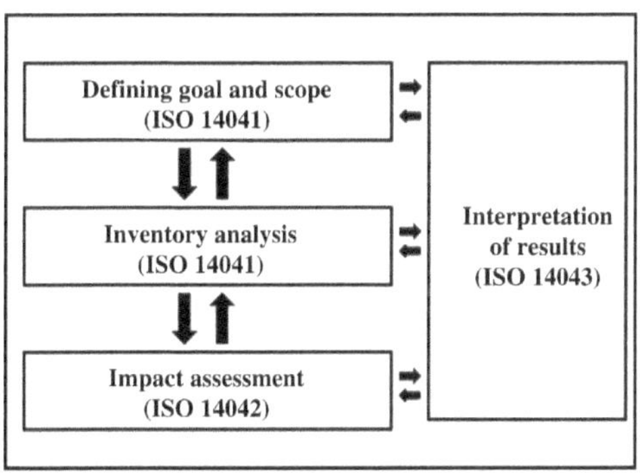

Figure 8.1 Methodological framework of a life cycle assessment study.

assessment step. This phase of LCA involves evaluating the potential environmental impacts on the basis of the inventory flow results. The final step of interpretation involves identifying major issues on the basis of the results obtained from the LCI and LCIA phases, evaluating the study by considering completeness, sensitivity, and consistency checks, and finally deriving conclusions, limitations, and recommendations.

8.5.2 Location and scope of life cycle assessment studies

Quek and Balasubramanian reviewed numerous LCA studies related to waste-to-bioenergy processes [33]. They found that the majority of these studies were from the region of Europe followed by North America, Asia and Australasia, and Africa and South America. This is an important point to be considered as the environmental and economic performance of bioenergy is much dependent on the place where these LCAs are carried out. Cherubini and Strømman reviewed LCA studies of bioenergy systems and found that half of the papers were limited to assessing GHG and energy balances only and lacked consideration for the effects of other impact categories [1]. This can be explained by the fact that mitigating CC and reducing fossil fuel consumption are major driving forces for the development of bioenergy globally.

8.5.3 Types of wastes and energy products

LCA studies conducted on waste-derived bioenergy include a number of waste residues varying from agricultural, forestry, municipal as well as from industrial activities. Used cooking oil (UCO) is the most commonly studied waste followed by lignocellulosic biomass. The most commonly studied bioenergy product has been bioethanol followed by biodiesel and biogas. Other than these bioenergy products, a large number of publications also studied the effects of the conversion of waste into heat and electricity [33].

8.5.4 Contributions of life cycle assessment research in waste-to-bioethanol processes

An LCA was carried out by de Azevedo et al. of a bioethanol production process from 1000 kg of cattle manure (CM) [34]. They studied many impact categories and the inputs and outputs that were found to have a major environmental impact included energy consumption, use of sulfuric acid (H_2SO_4) in pretreatment, buffer use in hydrolysis, and sodium phosphate use in fermentation. The effect of bioethanol production on most of the impact categories was found to be low and the LCIA showed that, in fact, the bioethanol produced from CM eradicated the requirement of manure treatment and disposal. Also, another advantage was that the residue was used efficient for the production of biofuel, thus counterbalancing the environmental effects of the bioethanol production process.

Guerrero and Muñoz performed an LCA of bioethanol production from banana agricultural wastes [35]. They considered the generation of 1 MJ of energy from the combustion of bioethanol in a car with diverse bioethanol blends as the FU. The impact categories quantified included CC, terrestrial acidification (TA), freshwater eutrophication (FE), photochemical oxidant formation (PO), particulate matter formation (PM), and fossil depletion (FD). Also, the net energy value and energy ratio (ER) were calculated to attain a positive energy balance. It was found that in comparison to the use of pure gasoline, blended gasoline reduced CC, PO, PM, and FD impacts. However, increases in FE and TA impacts were observed. The energy balance was found to be positive, with an ER value of 2.68 MJ/MJ. This study highlighted the potential use of banana agricultural waste in Ecuador for ethanol production in terms of GHG emission reduction and a net positive energy balance.

Sebastião et al. evaluated the environmental performance of an advanced bioethanol production process from pulp and paper sludge [36]. They modeled the input and output flows from an ethanol plant and studied the impacts using LCA. After analyzing the eleven impact categories of the system, the results indicated that the enzymatic hydrolysis and neutralization of calcium carbonate ($CaCO_3$) were major factors responsible for up to 85% of the overall impacts. After finding the environmental hotspots, two optimization scenarios were further evaluated, namely (1) reducing the amount of hydrogen chloride (HCl) in the neutralization stage; and (2) maximizing the ethanol yield through the cofermentation of xylose and glucose. Considerable improvements in the impact categories were observed in both the scenarios and they were able to decipher an ideal scenario based on their LCA results. According to them, a model scenario for a future sludge-to-bioethanol plant should include a combination of both the optimized scenarios studied for a higher efficiency as well as environmental performance.

Daylan and Ciliz did an LCA on the use of lignocellulosic bioethanol blends [37]. From the LCIA results, it was concluded that 1 km driven by an E10 or E85 fueled vehicle would be able to reduce GHG emissions by

12.8 g and 130.5 g CO_2 equivalent in comparison to gasoline, resulting in 4.7% and 47.1% reductions in GHG emissions, respectively. Between E85 and E10, the application of E85 fuel was found to be better in terms of CC, TA, OD, and crude oil consumption. However, the E10 fuel fared well in terms of aquatic and terrestrial eutrophication. Between gasoline and bioethanol, the use of bioethanol increased the emissions that contribute to acidification, eutrophication, and photochemical ozone depletion as bioethanol fueled vehicles release more of NOx, NH_3, and nitrous oxide (N_2O) during both the production process and the fuel combustion stage.

Wang et al. compared three different management options for wastepaper in the United Kingdom using an LCA approach [38]. The three management options were namely bioethanol production, recycling, and incineration with energy recovery. The influence of different types of pretreatment methods of wastepaper on environmental profiles was studied. The GHG emissions from the three management options were also calculated. It was found that oxidative lime pretreatment reduced GHG emissions and overall environmental burdens for a newspaper-to-bioethanol process compared to dilute acid pretreatment. Incineration involving high level technology was found to be the most favored option for managing wastepaper. However, in a number of cases, bioethanol production was also found to be environment-favorable, having equivalent emission profiles as those of recycling and incineration technologies.

These case studies clearly indicate that though generating bioethanol from wastes generally helps in reducing GHG emissions, in some scenarios there may even be an increase in some of the impacts such as acidification or eutrophication. In that case, understanding the hotspot in the production process becomes extremely important so as to make the process environmentally sustainable.

8.5.5 Contributions of life cycle assessment research in waste-to-biodiesel processes

Rajaeifar et al. carried out an LCA for expanded polystyrene (EPS) waste to be used as an additive in waste cooking oil (WCO) biodiesel [39]. The study assessed the combined life cycles of EPS waste management and biodiesel production and use. It also compared the results with the production and use of petroleum diesel. The impacts studied included CC, ecosystem quality, human health damage, and effects on resources. Petroleum diesel was found to have the highest impact in all the categories apart from the carcinogens category and was found to be the least environment-friendly option. The results obtained indicated that with increases in the concentration of EPS in the biodiesel above 50 g/L, all the associated environmental impacts decreased. Quantification of the impacts showed that using EPS in biodiesel blends (i.e., 50 and 75 g EPS) leads to a 3.3%–5.3% reduction in the human health, 3.80%–5.62% reduction in the ecosystem quality, 0.6%–0.8% reduction in the CC, and 6.3%–7% reduction in the resources damage categories, compared to biodiesel.

Lombardi et al. carried out an LCA to analyze and evaluate the environmental impacts and resource consumption of diverse pathways of energy recovery from UCO [40]. Five scenarios were studied. Scenario 1 involved the use of UCO for combined heat and power (CHP) production in a cogeneration plant. Scenarios 2–5 involved four different methods for the transesterification of UCO to yield biodiesel. Scenario 2 involved an alkali-catalyzed process using sodium hydroxide, scenario 3 involved the use of potassium hydroxide as a catalyst, scenario 4 involved an acid-catalyzed process, and scenario 5 involved a noncatalytic supercritical methanol process. The GWP and cumulative exergy consumption were studied for understanding the impacts of different scenarios. The results indicated that the recovery of UCO in a cogeneration plant resulted in lower values in terms of environmental burden than its utilization in biodiesel production.

Yano et al. also conducted an LCA of biodiesel produced from WCO to determine the environmental benefits and compare them with fossil-derived diesel fuel and fatty acid methyl ester (FAME)-type biodiesel fuel (BDF) [41]. Based on the results obtained, they predicted that in the future, a shift from FAME-type BDF to the hydrogenated biodiesel obtained from WCO would be more efficient in reducing the total environmental impacts of global warming, fossil fuel consumption, urban air pollution as well as acidification.

Dufour and Iribarren carried out an LCA on different free fatty acid (FFA)-rich wastes [42]. In terms of possible environmental impacts, biodiesel fuels obtained from FFA-rich materials were found to be better than both conventional diesel and first-generation biodiesel. Amongst the four feedstocks used, waste vegetable oils were found to be the most favorable for biodiesel production.

Mu et al. studied the potential of scum generated in wastewater treatment plants as a feedstock for biodiesel production [43]. They evaluated the sustainability of scum-to-biodiesel processing using an LCA and compared it with other scum management options, namely scum digestion and combustion. Amongst the three technologies, the scum-to-biodiesel technology was found to be the most environmentally sustainable based on several impact categories, namely fossil fuel depletion, GHG emissions, and eutrophication, whereas combustion showed the best performance in the case of acidification potential. In terms of the economic aspect as well, the scum-to-biodiesel technology was found to produce higher revenues than the other technologies. The study successfully demonstrated the viability and advantages associated with developing scum-to-biodiesel technology in wastewater treatment plants using the LCA approach.

The above case studies indicate how the sustainability of biodiesel production from different types of wastes can be assessed through the LCA approach.

8.5.6 Contributions of life cycle assessment research in waste-to-biogas processes

Edwards et al. conducted the first LCA of anaerobic codigestion (AcoD) of municipal waste and sewage sludge [44]. The studied impact categories included abiotic depletion, fossil fuel depletion, GWP, human toxicity, acidification potential, and eutrophication potential. Based on the LCA results, they concluded that AcoD had less environmental impacts compared to the presently practiced landfilling. This was true for all the categories modeled excluding the human toxicity category.

Xu et al. carried out an LCA of food waste-based biogas generation [45]. They considered three scenarios for food waste (FW) treatment, which included AcoD of FW and sludge (scenario 1), AD of FW (scenario 2), and FW to landfill (scenario 3). The results indicated that the most appropriate environmental scenario for treating FW was scenario 2, while scenario 3 had the largest environmental impact. Pérez-Camacho et al. evaluated impacts arising due to the substitution of traditional AD feedstocks (maize, grass silage, and CM slurry) with food wastes using an LCA approach [46]. The results obtained demonstrated that by this substitution GHG emissions of 163.33 CO_2-equiv. can be avoided. Along with a reduction in GHG emissions, environmental benefits such as no need for the landfilling of food wastes and the use of digestate as a substitute for synthetic fertilizers can be obtained. Edwards et al. also used an LCA approach to compare the environmental impacts of seven contemporary food waste management systems in Australia [47]. Amongst the different strategies for managing food waste, AD-based systems were found to outperform composting-based systems for GWP.

Chiu and Lo used an LCA to identify a sustainable treatment method for sewage sludge and food waste as well as to quantify the uncertainty associated with LCA studies [48]. They found that AcoD was the most advantageous in terms of environmental benefits while landfilling resulted in the greatest environmental burdens in the different waste treatment scenarios. After the AD process resulting in biogas production, the best scenario was utilizing the biogas in a combined cycle gas turbine system, which resulted in avoiding 6.75×10^4 kg CO_2 equiv. emissions, equivalent to a reduction of 44% more emissions compared to a scenario where the biogas was utilized for a CHP system.

Ayodele et al., using LCA, selected the best suitable energy generation option for municipal solid waste management in Nigeria [49]. The different WtE options that were studied included landfill gas to energy (LFGTE), a hybrid incineration and anaerobic digestion (INC/AD) system, and a hybrid incineration and landfill gas to energy (INC/LFGTE) system. The environmental impacts of these were also compared with conventional landfilling (no recovery of energy) for determining the most suitable option for all locations. The results revealed that the hybrid of incineration and anaerobic digestion (INC/AD) technology is a potentially viable option compared to other methods in terms reduction in Greenhouse gas emissions (75.7–93.3%) compared to landfilling without energy recovery. Also, the hybrid INC/AD system was found to be good in terms of effects on the ecosystem. However, LFGTE technology was found to be the best in terms of carcinogenic reduction potential. The results obtained in the paper could be of great help to decision makers regarding environmental sustainability in WtE projects in Nigeria.

8.6 Key challenges in life cycle assessment studies and future recommendations

Cherubini and Strømman reviewed the challenges associated with LCA of bioenergy systems [1]. According to them, there is a lack of uniformity regarding selection of the FU, making it difficult to compare the LCA results of different studies. For example, in many studies the FU taken is the unit of input biomass, whereas some use the number of hectares of agricultural land needed to cultivate the biomass as the FU. However, the majority of studies use the output unit (unit of heat or power produced or kilometers of transportation service) as the FU. So, there exists nonuniformity in the selection of FUs, making it difficult for decision makers to compare different studies, as even comparing different LCA of the same bioenergy by different researchers can result in different impacts, leading to the introduction of contradictory policies [50]. This type of discrepancy may also arise due to the use of different methodological approaches, namely attributional LCA (ALCA) or consequential LCA (CLCA). The ALCA methodology includes resource, material, energy, and emission flows throughout the life cycle of a product. This helps in analyzing the environmental impacts throughout the life cycle of a product. While a CLCA describes how environment-relevant flows will change in response to increases or decreases in the demand for the product system under study [51]. Thus CLCA helps to give advice to policy makers through a comparison of the environmental impacts of various alternative systems or products. ALCA is the most used LCA, but in LCA relating to bioenergy systems, the CLCA method appears to be the most commonly used as most of studies make a comparison with a fossil reference system. This is due to the fact that CLCA is more relevant for decision making and is able to address the needs of policy makers [1]. Also, Czyrnek-Delêtre et al. found that in most of the studies, the goal and scope of the study are seldom defined clearly and sensitivity analyses are most often performed [52]. They also reported that many LCA studies do not even indicate the methodological choices made, making it almost impossible to replicate the study. Another major drawback of most of LCA studies is that they consider only GHG emissions as the impact category and neglect others. In fact, Ridley et al. after assessing more than 1600 peer-reviewed papers on biofuels, found that the most discussed topics were biofuel production technologies, GHG emissions, and agricultural production of substrates [53]. However, the impact of biofuels on biodiversity and human health were very much neglected. Since the majority of LCAs take into account only GHG and energy balance as impact categories, this leads to the problem of burden shifting, where a biofuel system might achieve a high level of GHG reduction, but could affect the environment in other ways such as through acidification and eutrophication [52]. It is, therefore, highly recommended that for future LCA studies, a single FU should be fixed so as to ease the comparison between studies. Second, system boundary and methodology should be well defined. Also, impacts other than GHG emissions should be included for coming to a better conclusion regarding the effects of a given process on the environment and human health.

8.7 Conclusion

This chapter contributes to existing literature by reviewing past literature and highlighting the relevance and opportunities of research on LCA. While certain waste to bioenergy technologies have been discussed in this paper, however, it further opens up the scope for writing review papers on LCA pertaining to the individual technologies. LCA permits the quantitative measurement of environmental sustainability and delivers more data than presented by an environmental footprint assessment. If significantly researched and evolved, LCA mapping processes and findings would have significant contributions to the application and sustainability of different technologies. There would also be regional factors that may have significant impact such as the use of land and capital and differences in patterns and processes for producing biomass. In fact, LCA results may also alter with the evolution of technologies. By discussing these issues, this chapter encourages future researchers to take each of these issues and explore them for future academic and industrial contributions. This chapter facilitates an appreciation of the efficacy of LCA and the implications for policy development, and makes recommendations for future research priorities.

8.8 Acknowledgment

The authors would like to express their sincere thanks to the Department of Science and Technology, Govt. of India for providing INSPIRE Faculty fellowship to Ghosh [DST/INSPIRE/04/2016/000362].

References

[1] F. Cherubini, A.H. Strømman, Life cycle assessment of bioenergy systems: state of the art and future challenges, Bioresour. Technol 102 (2011) 437–451.
[2] A.E. Outlook, Energy information administration, Department of Energy, 92010 (2010) 1–15.
[3] H. Dhar, S. Kumar, R. Kumar, A review on organic waste to energy systems in India, Bioresour. Technol 245 (2017) 1229–1237.
[4] A. Dickie, Biofuels sustainability: a UK perspective, Renew. Energy Focus 8 (2007) 59–61.
[5] E.C. Petrou, C.P. Pappis, Biofuels: a survey on pros and cons, Energy Fuels 23 (2009) 1055–1066.
[6] T. Karak, R. Bhagat, P. Bhattacharyya, Municipal solid waste generation, composition, and management: the world scenario, Crit. Rev. Environ. Sci. Technol 42 (2012) 1509–1630.
[7] D. Hoornweg, P. Bhada-Tata, C. Kennedy, Peak waste: when is it likely to occur? J. Ind. Ecol 19 (2015) 117–128.

[8] P. Ghosh, I.S. Thakur, An integrated approach to study the risk from landfill soil of Delhi: chemical analyses, in vitro assays and human risk assessment, Ecotoxicol. Environ. Saf 143 (2017) 120–128.
[9] R.B. Tan, H.H. Khoo, Impact assessment of waste management options in Singapore, J. Air Waste Manag. Assoc 56 (2006) 244–254.
[10] H.H. Khoo, T.Z. Lim, R.B. Tan, Food waste conversion options in Singapore: environmental impacts based on an LCA perspective, Sci. Total Environ 408 (2010) 1367–1373.
[11] A.B.S. Schott, T. Andersson, Food waste minimization from a life-cycle perspective, J. Environ. Manag 147 (2015) 219–226.
[12] E. Papargyropoulou, R. Lozano, J.K. Steinberger, N. Wright, Z. bin Ujang, The food waste hierarchy as a framework for the management of food surplus and food waste, J. Clean. Prod 76 (2014) 106–115.
[13] A.U. Zaman, S. Lehmann, Urban growth and waste management optimization towards 'zero waste city', City, Cult. Soc 2 (2011) 177–187.
[14] V. Veleva, G. Bodkin, S. Todorova, The need for better measurement and employee engagement to advance a circular economy: lessons from Biogen's "zero waste" journey, J. Clean. Prod 154 (2017) 517–529.
[15] Q. Song, J. Li, X. Zeng, Minimizing the increasing solid waste through zero waste strategy, J. Clean. Prod 104 (2015) 199–210.
[16] H.D. Beyene, A.A. Werkneh, T.G. Ambaye, Current updates on waste to energy (WtE) technologies: a review, Renew. Energy Focus 24 (2018) 1–11.
[17] P. Buttol, P. Masoni, A. Bonoli, S. Goldoni, V. Belladonna, C. Cavazzuti, LCA of integrated MSW management systems: case study of the Bologna District, Waste Manag 27 (2007) 1059–1070.
[18] M.H. Kim, H.B. Song, Y. Song, I.T. Jeong, J.W. Kim, Evaluation of food waste disposal options in terms of global warming and energy recovery: Korea, Int. J. Energy Environ. Eng 4 (2013) 1.
[19] A. Tozlu, E. Özahi, A. Abuşoğlu, Waste to energy technologies for municipal solid waste management in Gaziantep, Renew. Sustain. Energy Rev 54 (2016) 809–815.
[20] M. Elsamadony, A. Tawfik, Potential of biohydrogen production from organic fraction of municipal solid waste (OFMSW) using pilot-scale dry anaerobic reactor, Bioresour. Technol 196 (2015) 9–16.
[21] J. Dong, Y. Tang, A. Nzihou, Y. Chi, E. Weiss-Hortala, M. Ni, Life cycle assessment of pyrolysis, gasification and incineration waste-to-energy technologies: theoretical analysis and case study of commercial plants, Sci. Total Environ 626 (2018) 744–753.
[22] U. Arena, F. Di Gregorio, Gasification of a solid recovered fuel in a pilot scale fluidized bed reactor, Fuel 117 (2014) 528–536.
[23] Z. Shareefdeen, A. Elkamel, S. Tse, Review of current technologies used in municipal solid waste-to-energy facilities in Canada, Clean Technol. Environ. Policy 17 (2015) 1837–1846.
[24] F. Xu, Y. Li, X. Ge, L. Yang, Y. Li, Anaerobic digestion of food waste-challenges and opportunities, Bioresour. Technol 247 (2018) 1047–1058.
[25] L. Appels, J. Lauwers, J. Degrève, L. Helsen, B. Lievens, K. Willems, et al., Anaerobic digestion in global bio-energy production: potential and research challenges, Renew. Sustain. Energy Rev 15 (2011) 4295–4301.
[26] M. Li, M. Zhou, X. Tian, C. Tan, C.T. McDaniel, D.J. Hassett, et al., Microbial fuel cell (MFC) power performance improvement through enhanced microbial electrogenicity, Biotechnol. Adv 36 (2018) 1316–1327.
[27] V.V. Pathak, S. Ahmad, A. Pandey, V.V. Tyagi, D. Buddhi, R. Kothari, Deployment of fermentative biohydrogen production for sustainable economy in

Indian scenario: practical and policy barriers with recent progresses, Curr. Sustain./ Renew. Energy Rep 3 (2016) 101–107.
[28] M. Rahimnejad, A. Adhami, S. Darvari, A. Zirepour, S.-E. Oh, Microbial fuel cell as new technology for bioelectricity generation: a review, Alex. Eng. J 54 (2015) 745–756.
[29] A. Giwa, Comparative cradle-to-grave life cycle assessment of biogas production from marine algae and cattle manure biorefineries, Bioresour. Technol 244 (2017) 1470–1479.
[30] J. Li, Y. Wang, B. Yan, The hotspots of life cycle assessment for bioenergy: a review by social network analysis, Sci. Total Environ 625 (2018) 1301–1308.
[31] B.E. Dale, S. Kim, Can the predictions of consequential life cycle assessment be tested in the real world? Comment on "using attributional life cycle assessment to estimate climate-change mitigation, J. Ind. Ecol 18 (2014) 466–467.
[32] J. Reap, F. Roman, S. Duncan, B. Bras, A survey of unresolved problems in life cycle assessment, Int. J. Life Cycle Assess 13 (2008) 374.
[33] A. Quek, R. Balasubramanian, Life cycle assessment of energy and energy carriers from waste matter–a review, J. Clean. Prod 79 (2014) 18–31.
[34] A. de Azevedo, F. Fornasier, M. da Silva Szarblewski, Rd.C. de Souza Schneider, M. Hoeltz, D. de Souza, Life cycle assessment of bioethanol production from cattle manure, J. Clean. Prod 162 (2017) 1021–1030.
[35] A.B. Guerrero, E. Muñoz, Life cycle assessment of second generation ethanol derived from banana agricultural waste: environmental impacts and energy balance, J. Clean. Prod 174 (2018) 710–717.
[36] D. Sebastião, M.S. Gonçalves, S. Marques, C. Fonseca, F. Gírio, A.C. Oliveira, et al., Life cycle assessment of advanced bioethanol production from pulp and paper sludge, Bioresour. Technol 208 (2016) 100–109.
[37] B. Daylan, N. Ciliz, Life cycle assessment and environmental life cycle costing analysis of lignocellulosic bioethanol as an alternative transportation fuel, Renew. Energy 89 (2016) 578–587.
[38] L. Wang, R. Templer, R.J. Murphy, A life cycle assessment (LCA) comparison of three management options for waste papers: Bioethanol production, recycling and incineration with energy recovery, Bioresour. Technol 120 (2012) 89–98.
[39] M.A. Rajaeifar, R. Abdi, M. Tabatabaei, Expanded polystyrene waste application for improving biodiesel environmental performance parameters from life cycle assessment point of view, Renew. Sustain. Energy Rev 74 (2017) 278–298.
[40] L. Lombardi, B. Mendecka, E. Carnevale, Comparative life cycle assessment of alternative strategies for energy recovery from used cooking oil, J. Environ. Manag 216 (2018) 235–245.
[41] J. Yano, T. Aoki, K. Nakamura, K. Yamada, S.-I. Sakai, Life cycle assessment of hydrogenated biodiesel production from waste cooking oil using the catalytic cracking and hydrogenation method, Waste Manag 38 (2015) 409–423.
[42] J. Dufour, D. Iribarren, Life cycle assessment of biodiesel production from free fatty acid-rich wastes, Renew. Energy 38 (2012) 155–162.
[43] D. Mu, M. Addy, E. Anderson, P. Chen, R. Ruan, A life cycle assessment and economic analysis of the scum-to-biodiesel technology in wastewater treatment plants, Bioresour. Technol 204 (2016) 89–97.
[44] J. Edwards, M. Othman, E. Crossin, S. Burn, Anaerobic co-digestion of municipal food waste and sewage sludge: a comparative life cycle assessment in the context of a waste service provision, Bioresour. Technol 223 (2017) 237–249.
[45] C. Xu, W. Shi, J. Hong, F. Zhang, W. Chen, Life cycle assessment of food waste-based biogas generation, Renew. Sustain. Energy Rev 49 (2015) 169–177.

[46] M.N. Pérez-Camacho, R. Curry, T. Cromie, Life cycle environmental impacts of substituting food wastes for traditional anaerobic digestion feedstocks, Waste Manag 73 (2018) 140–155.
[47] J. Edwards, M. Othman, E. Crossin, S. Burn, Life cycle assessment to compare the environmental impact of seven contemporary food waste management systems, Bioresour. Technol 248 (2018) 156–173.
[48] S.L. Chiu, I.M. Lo, Identifying key process parameters for uncertainty propagation in environmental life cycle assessment for sewage sludge and food waste treatment, J. Clean. Prod 174 (2018) 966–976.
[49] T. Ayodele, A. Ogunjuyigbe, M. Alao, Life cycle assessment of waste-to-energy (WtE) technologies for electricity generation using municipal solid waste in Nigeria, Appl. Energy 201 (2017) 200–218.
[50] A. Benoist, D. Dron, A. Zoughaib, Origins of the debate on the life-cycle greenhouse gas emissions and energy consumption of first-generation biofuels—a sensitivity analysis approach, Biomass Bioenergy 40 (2012) 133–142.
[51] J.M. Earles, A. Halog, Consequential life cycle assessment: a review, Int. J. Life Cycle Assess 16 (2011) 445–453.
[52] M.M. Czyrnek-Delêtre, B.M. Smyth, J.D. Murphy, Beyond carbon and energy: the challenge in setting guidelines for life cycle assessment of biofuel systems, Renew. Energy 105 (2017) 436–448.
[53] C.E. Ridley, C.M. Clark, S.D. LeDuc, B.G. Bierwagen, B.B. Lin, A. Mehl, et al., Biofuels: network analysis of the literature reveals key environmental and economic unknowns, Environ. Sci. Technol 46 (2012) 1309–1315.

CHAPTER 9

Bioethanol production from lignocellulosic biomass (water hyacinth): a biofuel alternative

Santhana Krishnan[1,2], Mohamad Faizal Ahmad[1,2], Nur Azmira Zainuddin[1,2], Mohd. Fadhil Md. Din[1,2], Shahabaldin Rezania[1,2], Shreeshivadasan Chelliapan[3], Shazwin Mat Taib[4], Mohd Nasrullah[5] and Zularisam Abdul Wahid[5]

[1]Centre for Environmental Sustainability and Water Security (IPASA), Research Institute of Sustainable Environment (RISE), School of Civil Engineering, Faculty of Engineering, Universiti Teknologi Malaysia (UTM), Skudai, Johor, Malaysia

[2]School of Civil Engineering, Faculty of Engineering, Universiti Teknologi Malaysia, Skudai, Malaysia

[3]Department of Engineering, Razak Faculty of Engineering and Informatics, Universiti Teknologi Malaysia, Kuala Lumpur, Malaysia

[4]Center for Coastal and Ocean Engineering, Research Institute for Sustainable Environment, Universiti Teknologi Malaysia, Kuala Lumpur, Malaysia

[5]Faculty of Civil Engineering Technology, Universiti Malaysia Pahang, Kuantan, Malaysia

9.1 Introduction

Due to rapidly growing industrial development there is an urgent need for environmentally sustainable energy sources [1]. Bioethanol as an alternative energy source has received special attention worldwide due to the depletion of fossil fuels. It also has an important role in the reduction of global warming by reducing carbon emissions from the burning process [2]. Bioethanol, which could be derived from lignocellulosic biomass, is certainly drawing increasing attention due to the advantages of being readily available, low-cost, and environment-friendly [3]. Bioethanol is a clean fuel that can be used in cars by blending it with gasoline such as E85 (85% ethanol and 15% gasoline), E100 (100% ethanol with or without a fuel additive), and oxy-diesel (typically a blend of 80% diesel fuel, 10% ethanol, and 10% additives and blending agents). In addition, ethanol also can be used for electricity generation if it is continuously produced in large quantities. In India, sugarcane molasses is the main raw material used for ethanol production, while in the United States, corn is mainly used for ethanol production [4].

Bioreactors
DOI: https://doi.org/10.1016/B978-0-12-821264-6.00009-7

The conventional method for lignin degradation is by using sulfuric acid under high pressure and temperature conditions [5]. This method involves potentially hazardous chemical constituents in the process. In terms of reducing chemical waste in the production of bioethanol, an alternative method was created using biological processes rather than chemical treatments. One such alternative method is lignin biodegradation such as the white rot fungi lignin degradation process [6]. This method is comparatively safer and can reduce the sources of pollutants from the bioethanol production process. Besides, this method can also reduce the cost of production because chemical processing is quite costly. The goal of this study was to investigate the growth rate of two species of macrophytes (*Eichhornia crassipes* and *Pistia stratiotes*) in identical conditions of nutrients supply. Another goal of the study was to investigate the feasibility of hemicellulose and lignin biodegradation by white rot fungi on water hyacinth and water lettuce. The degradation of both hemicellulose and lignin can affect the potential of extractable sugar from the plants. Furthermore, the study examined the effect of yeast concentration on the rate of sugar consumption and the limitation of sugar conversion to bioethanol by common *Saccharomyces cerevisiae* [7].

Water hyacinth is an important invasive lignocellulosic aquatic plant with a high breeding rate that can survive and be advantageous to sewage treatment plants such as waste stabilization ponds [8]. While enhancing the effectiveness of sewage treatment, such a plant can produce ethanol that can be used as car fuel. The characteristics of the rapid growth of this plant is important to fulfill the demand for daily usage in Malaysia. White rot fungi is also an abundance resource that has a good potential in lignin biodegradation [9]. The biodegradation of lignin is very crucial in this process in order to maximize the production of bioethanol. The production of ethanol from water hyacinth will not disturb food production because it is not a food source for humans. In addition, using ethanol as car fuel will have a positive impact on the environment by reducing carbon emissions.

9.2 Study background

The extraction of bioethanol from the abundant sources of lignocellulosic biomass such as water hyacinth can reduce food supply consumption. The ability of water hyacinth to produce ethanol is due to its sugar content. A high sugar content in plants indicates a high potential to produce bioethanol. Several pretreatment processes can be used in converting sugar to ethanol [10]. These pretreatment processes are important to breakdown hemicellulose and lignin, which can prompt a greater sugar extraction. A common pretreatment process is acid pretreatment, which is easy and effective in the breakdown of lignin [5]. After pretreatment, the pretreated substance will go through a fermentation process. The fermentation is to convert sugar to ethanol.

9.2.1 Bioethanol

Bioethanol fuel has an important role in the field of environmental conservation by mitigating global warming and conserving fossil fuel. It is an alcohol made of carbohydrates through a fermentation process. The production of bioethanol from biomass or waste is one way to reduce both the consumption of crude oil and environmental pollution [11]. Lignocellulosic biomasses (corn, sugar, molasses, etc.) derived from nonfood sources such as grasses and trees are also being developed as feedstocks for ethanol production. The physical and chemical characteristics of bioethanol are similar to ethanol; they just require different are resources for production. Bioethanol in its purest form is a colorless clear liquid with a mild characteristic odor that boils at 78°C and freezes at −112°C.

9.2.2 Lignocellulose

Lignocellulose refers to plant dry matter or biomass, which is also known as lignocellulosic biomass. These types of plants are the most abundantly available material on Earth used in the production of biofuel, mainly bioethanol. Since the middle of the 20th century, the lignocellulosic biomass, in the form of wood fuel, has a long history as a source of energy. There has been an increased trend in biomass being used as a precursor to liquid fuel, mainly in bioethanol production through the fermentation of lignocellulosic biomass. This can be used as an additive in the blending of petrol or gasoline for car fuels, which can decrease the usage of fossil fuel. Lignocellulose is composed of carbohydrate polymers such as cellulose, hemicellulose, and aromatic polymer (lignin). Lignin is an organic substance that binds the cells, fibers, and vessels that constitute wood and the lignified elements of plants such as in straw. It is also a complex polymer of aromatic alcohols and is commonly derived from woods. It forms an excellent strength and durability tissue becomes cell walls of almost all dry land plant cell walls. Lignified tissues such as wood are similar to fiber-reinforced plastics, in which lignin represents the plastic binder and cellulose the reinforcing fibers [12]. According to Brigham et al. [13] lignin are irregular phenyl–propane polymers that represent approximately 20%–25% of hardwood or softwood tree stems.

Hemicelluloses are complex, breached carbohydrate polymers that are formed from different monomeric sugars attached through different linkages. Substituents and noncarbohydrate components occur on hemicellulose on either the main chain or on the carbohydrate branches. The complex structure of hemicellulose is thought to confer a wide range of biophysical and biomechanical properties to plant tissues, which also occur in products made from these tissues [13]. These carbohydrate polymers contain different sugar monomers (six and five carbon sugar) and they are tightly bound to lignin. The structure of lignocellulose is complex; cellulose is surrounded by a monolayer of hemicellulose and embedded in a matrix of hemicellulose and lignin. Lignin specifically creates a barrier to enzymatic attack and the highly crystalline structure of cellulose is insoluble in water while the hemicellulose and lignin create a protective sheath around the cellulose [14].

9.2.3 Water hyacinth as lignocellulosic biomass

There are two types of water-floating macrophytes used in this study, namely *E. crassipes* and *P. stratiotes*, which are water plants that grow easily in Malaysia. They are suitable for bioethanol production processes because their rates of growth are very fast [15]. Each of them have advantages and disadvantages that have to be investigated in order to be able to choose the most economical species for the climate of Malaysia [16]. The chemical composition of *E. crassipes* is given in Table 9.1.

9.2.4 Pretreatment techniques

There are many types of biological, physical, and chemical technologies available for the pretreatment of lignocellulosic biomass [18].

Table 9.1 The percentage composition of *Eichhornia crassipes* [17].

Organic components	Percentage (%)
Hemicellulose	48.7
Cellulose	18.2
Lignin	3.5
Crude protein	13.3

Combination pretreatment techniques that use several techniques in one process for bioethanol production are also common [8]. Some pretreatment methods seem to be more economically feasible than others. However, certain environmental concerns and the production of inhibitors to fermenting yeasts during these pretreatment processes are major hurdles that have to be overcome in the commercial production of bioethanol [19]. So it is extremely important that an efficient, cost-effective, and environment-friendly pretreatment method is developed.

9.2.5 White rot fungi

Lignin modification and degradation have been most extensively studied in basidiomycetes, in which a number of enzymes and mechanisms involved in lignin attack have been elucidated. White-rot basidiomycetes (notably *Phanerochaete chrysosporium*) are the most common wood-rotting organisms, because of their ability to degrade lignin, hemicelluloses, and cellulose, often giving rise to a cellulose-enriched white material [20]. *P. chrysosporium* is a white-rot fungus that lives abundantly on dead wood. It has the special ability to degrade abundant aromatic polymer lignin without degrading the white cellulose in substrates. *P. chrysosporium* releases extracellular enzymes to breakdown the complex three-dimensional structure of lignin into simple components that can be utilized by its metabolism. Extracellular enzymes are nonspecific oxidizing agents (hydrogen peroxide and hydroxyl radicals) used to cleave lignin bonds [21]. The degradation of lignin and pollutants is made possible by the production of extracellular enzymes. Components such as lignin peroxidase and manganese peroxidase take part in the remediation of various pesticides, polyaromatic hydrocarbons, carbon tetrachloride, and various poisons [20]. The process of lignin breakdown is carried out by means of cleavage reactions. These extracellular enzymes release free radicals to initiate spontaneous breakdown to phenyl propane units in the secondary metabolism or stationary phase [22].

9.2.6 Fermentation

There are four major steps in producing bioethanol from water hyacinth biomass. The first step is the pretreatment of biomass to breakdown the lignin–hemicelluloses–cellulose complex to make it more susceptible to hydrolysis. The second step is hydrolysis to breakdown the cellulose and hemicellulose into monomer sugars. The third step is fermentation of these sugars to ethanol. The final step is product recovery and concentration by distillation.

9.3 Methodology

9.3.1 Growth rate

9.3.1.1 Preparation of cultivation tank

This study was conducted at an algae cultivation area located at Desa Bakti Wastewater Stabilization Pond. Two tanks (304 L tank volume) and several small containers were set up for cultivation. The tanks and containers were filled with wastewater that was pumped from the effluent of the wastewater treatment pond.

9.3.1.2 Cultivation of water hyacinth

In this study, two types of floating aquatic macrophytes were selected. The first one, *E. crassipes*, is labeled as type A. This type A macrophyte has a big spongy stem that enables it to float on water. The second type, *P. stratiotes*, is labeled as type B and has a similar characteristic appearance to lettuce or cabbage but with waterproof leaves that are filled with air so that it can float on water (Fig. 9.1). The mother plants were collected in Universiti Teknologi Malaysia (UTM) and normally exist on water that is contaminated with nutrients such as in a wastewater stabilization pond. Before propagating them in the tank, the mother plants were cleaned to remove all suspended solids, insects, and so forth.

9.3.1.3 Growth rate of plants

All plants in the tanks were taken out and dried for a while. Then each type of plant was weighed and the mass increment recorded. The weight was measured using balance. The length of the roots was also recorded. The leaves were measured and the growth of the leaves was recorded. The length of both the roots and leaves was measured using a ruler. This

Figure 9.1 (A) Water hyacinth and (B) water lettuce.

procedure was repeated every 3 days for 3 weeks and the data were recorded.

9.3.2 Biodegradation

9.3.2.1 Preparation of dry water hyacinth

The water hyacinth and water lettuce plants were harvested from ponds in UTM and brought to the laboratory. All plants were rinsed with tap water to remove dirt and any material attached to the plants. These were chopped into small pieces and placed inside a beaker. This was then put in an oven overnight at 105°C for drying. The dried samples were kept in zipped plastic bags to ensure that they remained dry.

9.3.2.2 Preparation of water hyacinth powder

The dried water hyacinth sample was blended using a dry mixer to get a powder. Then the water hyacinth powder was kept in a zipped plastic bag and ready for the preliminary treatment process. This procedure is important because the total area of water hyacinth was increased and the reaction time of the pretreatment process was shortened. In other words, it increases the efficiency of the pretreatment process and increases the rate of the process.

9.3.2.3 Preparation of white-rot fungi

White-rot fungi (*P. chrysosporium*) was isolated from decayed woods by the central lab of UTM. In the propagation process, potato dextrose agar (PDA) plates were used. The preparation of the PDA medium was started by sterilizing everything that was used in the process. The process requires the best aseptic technique as possible in the lab. A laminar flow chamber was used to avoid contamination during the propagation of the fungi. The propagation process of the fungi started with 3.12 g of dry PDA being mixed with 80 mL of distilled water until no scum was visible in the bottle. A cork borer, measuring cylinder, and four petri dishes were sealed with heat resistance plastic. All the sealed equipment and media were sterilized for 15 min at 121°C under 15 psi using an autoclave machine. All sterilized equipment and media were placed in a laminar flow chamber for cooling. 20 mL of media was poured into each petri dish. The cork borer was used to transfer each fungal disc from the stock plate to the prepared media. All petri dishes were sealed with parafilm to avoid contamination and placed in an incubator.

9.3.2.4 Pretreatment process

First, 10 g of substrate powder was weighed and put into a 100 mL glass beaker. Six beakers with water hyacinth powder and six beakers with water lettuce powder were prepared. A drop of Tween 80 was dropped into each beaker and 10 mg of glucose powder was added. Tween 80 was added into the substrate mixtures to enhance the growth of the fungi in the limited glucose condition [23]. All substrate mixtures were wetted with 20 mL of distilled water. Then, all mixtures were sterilized at 121°C for 15 min in an autoclave. After being autoclaved, the substrates were cooled to room temperature for a few hours. Then, five fungal discs (4 mm diameter) were added into three of the water hyacinth beakers and three of the water lettuce beakers using a cork borer and these were labeled 5WH and 5WL, respectively. Two fungal discs were added to the remaining beakers and labeled as 2WL and 2WH. All beakers were stored in a cabinet for biodelignification. Samples were taken every 10 d for 30 d for glucose determination, lignin degradation, and hemicellulose content. After 10 d, the mixtures were taken out of the cabinet and mixed with 100 mL distilled water then ground using a mortar and pestle. Then, these were filtered to remove all suspended solids with Whatman No. 1 filter paper with a 0.45 μm pore size. The filtrate was used for glucose determination and the solids retained on the filter paper was oven dried overnight for lignin and hemicellulose determination.

9.3.2.5 Determination of sugar content

The equipment used in these methods include test tubes, pipets, and DR6000 spectrophotometer. The reagents required in the dinitrosalicylic acid (DNS) method are 1% DNS reagent solution and 40% potassium sodium tartrate solution [24]. The DNS method procedure started with 3 mL of DNS reagent being added to 3 mL of glucose sample in a lightly capped test tube. To avoid the loss of liquid due to evaporation, the test tube was covered with a piece of paraffin film as a plain test tube was being used. The mixture was heated at 90°C for 15 min to develop a red-brown color. 1 mL 40% (w/v) potassium sodium tartrate (Rochelle salt) solution was added into the test tube to stabilize the color. After it cooled to room temperature in a cold-water bath, the absorbance was recorded with a spectrophotometer at 575 nm. A calibrated curve was made to convert the absorbance value into the desire concentration value. The same procedure was followed and the sample was substituted with a standard sugar solution with different concentrations. Then the data were plotted into a graph and the slope equation was used as the conversion formula.

9.3.2.6 Determination of hemicellulose

The method used to determine hemicellulose in this study was the gravimetric method, which is the Chesson method [25]. The oven-dried water hyacinth sample was ground into a powder. 1 g of the sample was weighed and mixed with 150 mL of distilled water. Then it was refluxed for 2 h at 100°C. After reflux, it was filtered with 0.45 μm filter paper and the solids retained were put in an oven at 105°C overnight. The weight lost was calculated in percentage (%). The weight lost from this procedure is soluble protein (P1). The same procedure was repeated with the same sample using 5% (v/v) sulfuric acid to obtain further weight loss. The weight lost from this procedure was soluble protein and hemicellulose (P2). In order to get hemicellulose fraction, the formula must be weight lost from second procedure (P2) minus weight lost from first procedure (P1).

9.3.2.7 Determination of lignin

The ground, oven-dried water hyacinth sample was weighed to 330 mg and soaked with 5 mL of concentrated acid, for which 72% (v/v) sulfuric acid was used. The soaking process took 2 h and the water hyacinth powder turned black in color because of the reaction of the concentrated acid with the plant structures. After the soaking process, the solution was diluted to 3% (v/v) acid concentration by adding distilled water to the water hyacinth solution. The diluted solution underwent reflux for 4 h at 125°C. After it cooled, it was filtered with 0.45 μm filter paper and the solids retained were put in an oven at 105°C. The next day, the weight of the dry filtered sample was weighed and put it in a furnace which was ignited for 3 h at 575°C. Then the sample was cooled and weighed and the data recorded.

$$\text{Acidinsolublelignin}(\%) = \frac{\text{weightafterovendry} - \text{weightafterignite}}{\text{initialweight}} \times 100\%$$

9.3.2.8 Fermentation process

The same procedure as that of the second stage was done in order to get the filtrate. The types of plants and number of fungal discs were chosen based on the results of sugar content in stage two. The pH of the filtrate was adjusted to 5.5 with diluted hydrochloric acid. The DNS method was done for initial sugar content in the filtrate. The sample was transferred into three 500 mL stirred tank reactors (Parr Instrument, United States) with 400 mL working volume for a batch fermentation process. Each batch reactor was labeled with a different concentration of yeast. Three variations of yeast concentration were used, namely 1 g/L, 2 g/L, and 3 g/L. Samples was taken every 2 h for 10 h for sugar reduction by DNS method.

9.3.2.9 Determination of sugar

The fermented solution was taken out and put in a centrifuge tube (about 45 mL). It was centrifuged at 6000 rpm for 10 min. After centrifuging, yeast cell was settled down at the bottom of the centrifuge tube. The supernatant was collected using a syringe and filtered using a nylon syringe filter of 0.45 μm. The filtered sample underwent the DNS method procedure and the absorbance value was recorded.

9.4 Results and discussion

9.4.1 Growth rate

Fig. 9.2A shows the increase in the growth of water hyacinth (in terms of percentage) from day 1 to day 21. From the graph it can be seen that the weight of the water hyacinth gradually increased and formed a linear regression line, that is, $y = 6.6667x$ and $R^2 = 1$. This means that the water hyacinth grew about 6.7% a day. The weight of the water lettuce also gradually increased and the linear regression line showed an estimation equation. The equation is $y = 20.016x$ and $R^2 = 0.9927$ which means that the water lettuce grew about 20% a day. The value of regression approaches and is near to 1, which means the equation is acceptable and valid to use for estimation.

Every plant will achieve maximum growth rates at a certain time until the weight increments are maintained and remain constant each day. Fig. 9.2B shows the relationship between percentage growth rate and time for both species. Based on the graph, the growth rate of the water hyacinth was maintained and remained the same from day 3 until day 21, which represents a 20% increment every 3 d. The growth rate for the water lettuce gradually increased from 40% on day 3 and 60% on day 6 and was maintained at 67% from day 9 until day 21. This phenomenal is

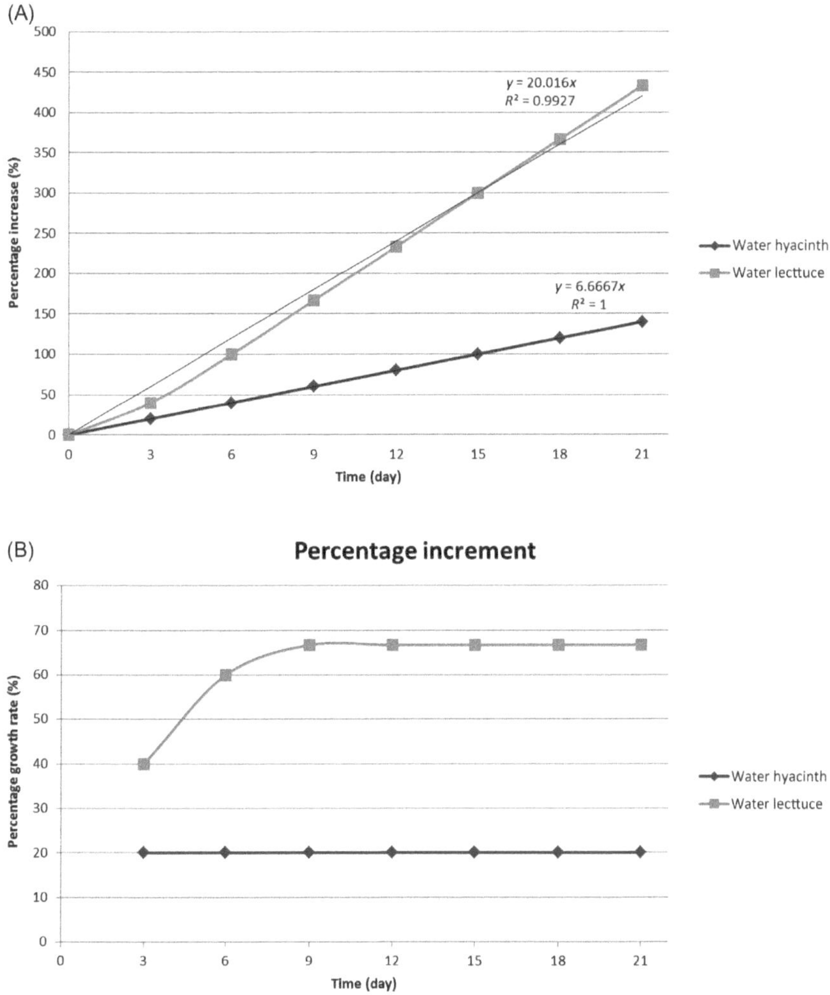

Figure 9.2 (A) Weight increments of plants and (B) growth rates of plants.

called acclimatization, which refers to requirement of plants for a period of time to adapt to a new environment and its conditions such as nutrient condition, temperature, toxicity, and so on. Every plant needs a period of time to acclimatize to its surroundings, but there was no need for this for the water hyacinth because it had already adapted to the conditions in this experiment. The analysis above shows different growth rates between water hyacinth and water lettuce. It could be seen clearly from the tanks that the water lettuce had already filled the space of the tank on day 21, whereas the tank containing the water hyacinth still had some empty space on the final day of phase one of the experiment. Therefore water lettuce is the preferable species in terms of propagation and it has the ability to provide enough supply for bioethanol production due to its rapid growth rate.

9.4.2 Biodegradation

The parameters measured in this study were sugar concentration, hemicellulose, and lignin. All parameters represent the degradation process of the water hyacinth plant structure by white-rot fungi. In this part, water hyacinth and water lettuce were used as substrate.

9.4.2.1 Hemicellulose

Fig. 9.3 depicts the hemicellulose content in water hyacinth and water lettuce plants in the different numbers of fungal discs from the start to day 30 in 10 d gaps. From the graph, hemicellulose in both species gradually decreased in ranges between 31.1% and 47.6%. All graphs show quite similar patterns of hemicellulose degradation by white-rot fungi. A graph showed that water hyacinth started with 47.6% hemicellulose content, which was degraded to 31.1% by white-rot fungi with five fungal discs at the beginning. The experiment with two fungal discs degraded to 32% hemicellulose, which means it was a lesser degradation. While water lettuce started with 45% hemicellulose content and it was degraded to 33.4% by the same species of fungi also with five fungal discs at the

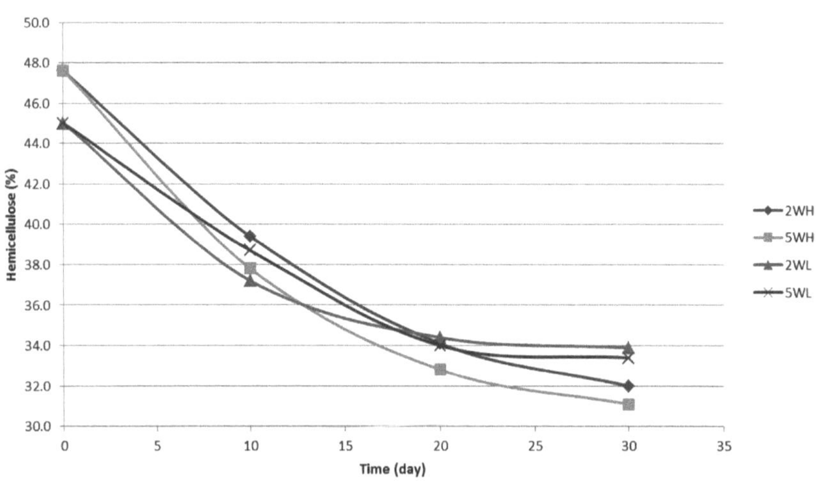

Figure 9.3 Hemicellulose degradation trends.

beginning. The experiment with two fungal discs degraded to 33.9% hemicellulose, which also shows a lesser degradation. Therefore it was concluded that the number of fungal discs affects the rate of hemicellulose degradation, however, not significantly.

Fig. 9.4 shows four sets of histograms that represent the hemicellulose reduction in different phases for water hyacinth and water lettuce with different fungal dosages. The rate of degradation was separated into three phases of time, namely days 0–10, days 10–20, and days 20–30. As can be seen, all the sets of histograms show the same pattern of gradual decrease. The histograms of hemicellulose degradation showed similar patterns to those on biological degradation that were studied by Kamra et al. [26] in which sugarcane bagasse was degraded by white-rot fungi. 5WH demonstrated the highest hemicellulose degradation in the first phase with a 20.6% degradation in 10 d. While 5WL demonstrated the lowest hemicellulose degradation in the 0–10 days phase with a 14% degradation in 10 days. 5WL should have got a higher degradation rate compared to 2WL due to the presence of fungal dosage with respect to 2WH and 5WH as the reference. This might be due to the contamination of the white-rot fungi in the experiment.

The total hemicellulose degradation by white-rot fungi in 2WH, 5WH, 2WL, and 5WL was approximately 32.8%, 34.7%, 24.7%, and 25.8%, respectively. The total degradation was calculated from day 0 until day 30. It can be seen from the total hemicellulose degradation in 30 days that the highest degradation was demonstrated by 5WH, which was water

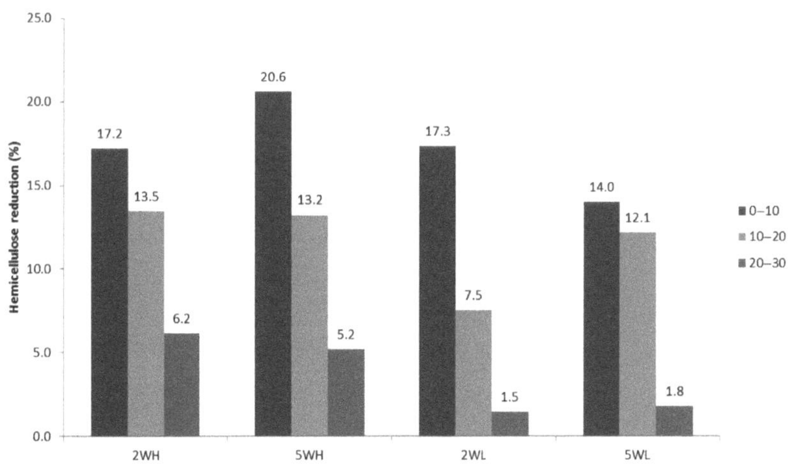

Figure 9.4 Hemicellulose degradation in three phases.

hyacinth powder with five fungal discs as the fungal dosage. Therefore it can be concluded that the water hyacinth plant is more feasible for biological pretreatment using white-rot fungi in hemicellulose degradation than water lettuce.

9.4.2.2 Sugar

Fig. 9.5 depicts the sugar concentration in the filtrate of water hyacinth and water lettuce that was biologically pretreated by white-rot fungi with different fungal dosages. The concentration of sugar in the filtrate represents the extractable sugar in water hyacinth and water lettuce plants. It can be seen that the trends of sugar concentration in the filtrate gradually increased from day 10 until day 30. All graph lines show a similar pattern of sugar increment due to the process of biological pretreatment using white-rot fungi.

All lines in the graph do not intercept each other, which mean that the lines were formed nicely and the reaction in all experiments was similar. The concentration of sugar was nicely arranged on all 3 days of sampling, which were day 10, 20, and 30. The order that can be seen in the graph is 5WH, 2WH, 5WL, and 2WL, which represents a high-to-low sugar concentration order. This shows that water hyacinth plants have more extractable sugar than water lettuce species and that the sugar increment rate was influenced by the fungal dosage.

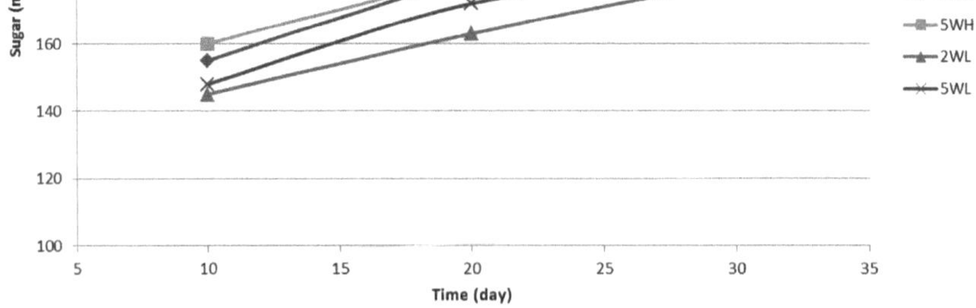

Figure 9.5 Sugar increment trends.

9.4.2.3 Lignin

Fig. 9.6 depicts the lignin content in water hyacinth and water lettuce at different fungal dosages from the start until day 30 in 10 days gaps. It can be seen from the graph that lignin in both species gradually decreased in ranges between 15.2% and 18.8% Most of the lines in the graph show a similar trend of gradually decreasing in time except for the 5WH experiment. The line for the 5WH experiment shows a very different trend because it shows an increment in lignin content from day 10 to day 20.

The graph shows that water hyacinth started with 18.8% lignin content and was then degraded to 15.5% by five fungal discs of white-rot fungi. The experiment with two fungal discs degraded to 16.1% lignin content representing a lesser degradation. While water lettuce started with 18.2% lignin content and was degraded to 15.2% also by five fungal discs of the same species of fungi. The experiment with two fungal discs degraded to 15.5% lignin content representing a lesser degradation. Therefore the number of fungal discs used affects the rate of lignin degradation.

Fig. 9.7 shows four sets of histograms that represent lignin degradation in different phases of time for water hyacinth and water lettuce with different fungal dosages. The rate of degradation was separated into three phases of time, namely days 0–10, days 10–20, and days 20–30. As can be seen, all sets of histograms show different patterns and the 5WH experiment shows a negative reduction in lignin during the 10–20 days phase,

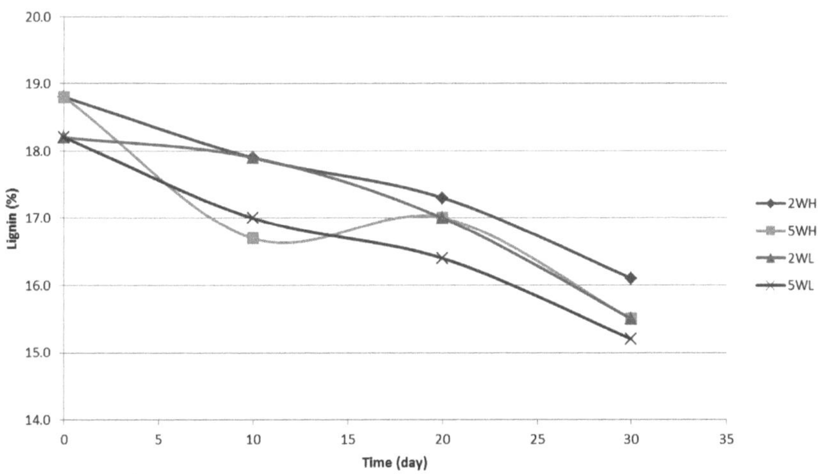

Figure 9.6 Lignin degradation trends.

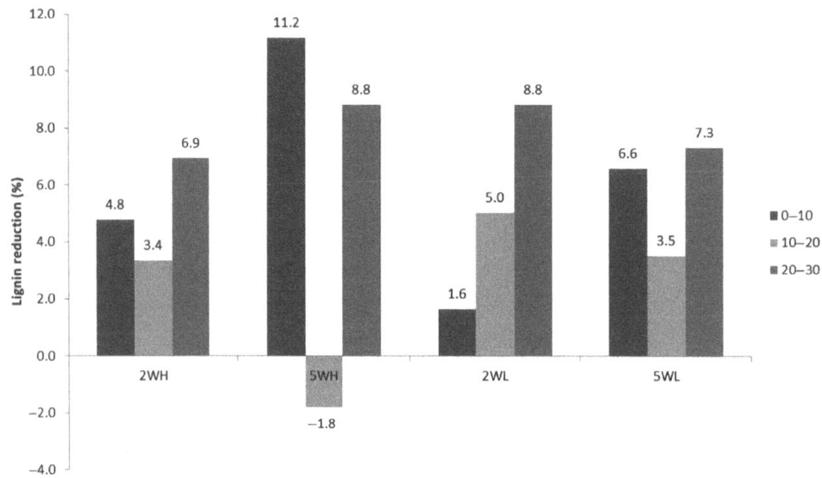

Figure 9.7 Lignin degradation in three phases.

which means that lignin increased. This happened because the 5WH lignin content on sampling day 10 indicated the lowest lignin content with 16.7% and an increase in lignin degradation rate over days 0−10 to 11.2% degradation in 10 days. According to Kamra et al. [26], lignin biodegradation by white-rot fungi should have a low biodegradation rate at the beginning and slowly increase as shown by the 2WL experiment.

The total lignin biodegradation by white-rot fungi in 2WH, 5WH, 2WL, and 5WL was approximately 14.4%, 17.6%, 14.8%, and 16.5%, respectively. The total degradation was calculated from day 0 until day 30, which was the final day of sampling in this experiment. During the operation, the highest total lignin degradation was observed in the 5WH experiment. Therefore it can be concluded that water hyacinth is more feasible for biological pretreatment using white-rot fungi in lignin biodegradation than water lettuce.

Overall, the graph pattern for both hemicellulose degradation and sugar content increments have a quite good negative correlation. Hemicellulose consists of other polysaccharides, principally xylans and mannans, which are closely associated with cellulose filaments and chemically linked with lignin [27]. According to Dhepe et al. [28], hemicellulose degrades into xylose (major) and arabinose as the hemicellulose degradation pathways. Xylose and arabinose are under the category of sugar classified as a monosaccharide of aldopentose type, which means that they contain five carbon atoms. Therefore every hemicellulose degradation process that occurs will increase the sugar concentration of the filtrate, which hemicellulose breakdowns into xylose and arabinose as additional sugar. In summary, the fungal dosage affects the rate of degradation of hemicellulose and lignin. Water hyacinth is more feasible for both lignin and hemicellulose biodegradation by white-rot fungi and produces more extractable sugar than water lettuce.

9.4.3 Fermentation

The parameter measured in this part of the study was sugar concentration. This parameter represents the sugar consumption rate of sugar conversion into ethanol by yeast. The common yeast species used in this experiment is *S. cerevisiae*, which can be found in bakers' yeast. The substrate used for the fermentation process is water hyacinth filtrate that was pretreated using white-rot fungi in 30 days.

9.4.3.1 Sugar

Fig. 9.8 shows a graph representing the sugar reduction in water hyacinth filtrate in different yeast extract concentrations from the start until 10 h in 2 h gaps of sampling. All experiments were started with 189 mg/L of sugar as shown in Table 9.2. It can be seen from the table that the sugar concentration in all reactors were reduced in ranges between 54.7 mg/L and 189 mg/L. The sugar concentration was maintained and remained

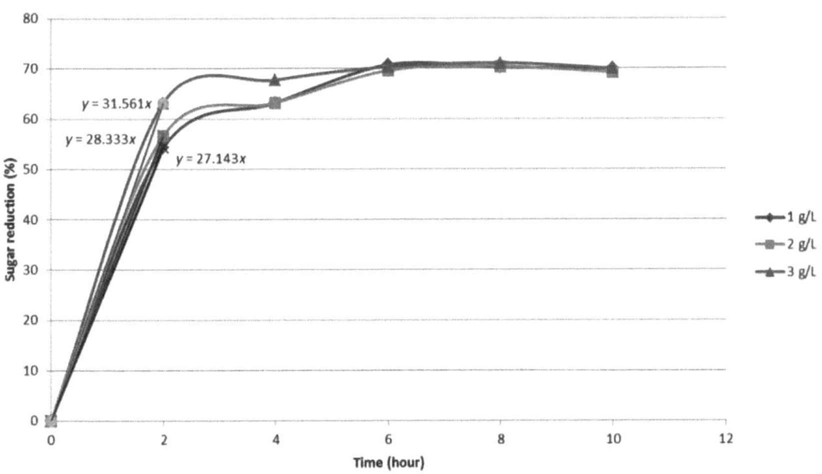

Figure 9.8 Percentage of sugar reduction versus time.

Table 9.2 Sugar concentration in fermented solution in 10 h.

Time (h)	Yeast extract concentration (g/L)		
	1	2	3
	Sugar (mg/L)		
0	189	189	189
2	86.4	81.9	69.7
4	69.7	69.7	61.0
6	55.4	57.5	56.1
8	56.4	56.1	54.7
10	56.8	58.2	56.8

constant after a certain period of time, which shows that the process of sugar conversion was stopped.

As shown in Fig. 9.8, water hyacinth filtrate with 3 g/L of yeast had the highest percentage of sugar reduction, which was an approximately 31.6% sugar reduction in 1 h followed by others with approximately 28.3% and 27.1% for 2 g/L and 1 g/L, respectively. Therefore the concentration of yeast for the fermentation process affected the rate of sugar conversion into ethanol. Increasing the yeast concentration could boost the rate of sugar consumption.

The sugar consumption was stopped at about 70% after 6 h of fermentation using *S. cerevisiae*. This means that the fermentable sugar extract from water hyacinth that underwent biological pretreatment by white-rot fungi was about 70%. 30% of the sugar extract was nonfermentable type sugar, most probably xylose. According to Schneider [29], xylose is the main product of the hydrolysis of hemicellulose from any type of lignocellulosic biomass. Xylose is classified as a nonfermentable sugar by *S. cerevisiae* [30]. According to Wohlbach and Alan [31], only genetically engineered yeast (*S. cerevisiae*) that has a xylose isomerization ability can consume xylose and convert it into bioethanol.

9.5 Conclusion

The growth rates of water hyacinth species, *E. crassipes* and *P. stratiotes*, show significantly different rates. Based on analysis, the approximate growth rates of water hyacinth and water lettuce are 6.7% and 20% a day, respectively. Furthermore, water hyacinth and water lettuce demonstrated the maximum growth rates of 20% and 67% in 3 days, respectively.

Therefore water lettuce is the preferable species in terms of propagation and the ability to provide enough supply due to its rapid growth rate. Biodegradation occurred in both species using white-rot fungi. However, water hyacinth (*E. crassipes*) was more susceptible to degradation by white-rot fungi due to the performance of hemicellulose and lignin biodegradation as compared to water lettuce. Hemicellulose breakdown affected to extractable sugar content of the biomass. Furthermore, the biodegradation rate of hemicellulose and lignin were influenced by the fungal dosage. The rate of sugar conversion into bioethanol was determined by sugar consumption. Based on analysis, the higher the concentration of yeast the higher the sugar consumption rate, which could influence the bioethanol production rate. The fermentable sugar in water hyacinth using biological pretreatment by white-rot fungi was increased to 70%.

Acknowledgments

The authors would like to thank Universiti Teknologi Malaysia for their support. This work was supported by PDRU Grant, Vot No. Q.J130000.21A2.04E53, Hitachi Scholarship program 2019, and LRGS MRUN/F2/01/2019/5.

References

[1] S. Krishnan, M.F.M.D. Din, L. Singh, S.M. Taib, Utilization of micro-algal biomass residues (MABRS) for bio-hythane production-a perspective, J. Appl. Biotechnol. Bioeng 5 (3) (2018) 166−169.
[2] B.T. Mohammad, M. Al-Shannag, M. Alnaief, L. Singh, E. Singsaas, M. Alkasrawi, Production of multiple biofuels from whole camelina material: a renewable energy crop, BioResources 13 (3) (2018) 4870−4883.
[3] C.E. Wyman, C.M. Cai, R. Kumar, Bioethanol from lignocellulosic biomass, Energy from Organic Materials (Biomass). A Volume in the Encyclopedia of Sustainability Science and Technology, second ed., Springer, 2019, pp. 997−1022.
[4] A.K. Mino, Ethanol Production From Sugarcane in India: Viability, Constraints and Implication, University of Illinois, Urbana, IL, 2010.
[5] V. Chaturvedi, P. Verma, An overview of key pretreatment processes employed for bioconversion of lignocellulosic biomass into biofuels and value added products, 3Biotech 3 (2013) 415−431.
[6] C. Wan, Y. Li, Fungal pretreatment of lignocellulosic biomass, Biotechnol. Adv 6 (2012) 1147−1157.
[7] K. Zhang, P. Wells, Y. Liang, J. Love, D.A. Parker, C. Botella, Effect of diluted hydrolysate as yeast propagation medium on ethanol production, Bioresour. Technol 271 (2019) 1−8.
[8] S. Rezania, M.F.M. Din, S.M. Taib, S.E. Mohamad, F.A. Dahalan, H. Kamyab, et al., Ethanol production from water hyacinth (*Eichhornia crassipes*) using various types of enhancers based on the consumable sugars, Waste Biomass Valoriz 9 (6) (2018) 939−946.

[9] S.B. Ummalyma, R.D. Supriya, R. Sindhu, P. Binod, R.B. Nair, A. Pandey, et al., Biological pretreatment of lignocellulosic biomass—current trends and future perspectives, Second and Third Generation of Feedstock, Elsevier, 2019, pp. 197–212.
[10] L. Singh, V.C. Kalia (Eds.), Waste Biomass Management—A Holistic Approach, Springer, Cham, 2017.
[11] A.K. Chandel, E. Chan, R. Rudravaram, M. Narasu, L. Rao, P. Ravindra, Economics and environmental impact of bioethanol production technologies: an appraisal, Biotechnol. Mol. Biol (2007) 14–32.
[12] T.A. Hsu, Pretreatment of biomass, in: C.E. Wyman (Ed.), Handbook of Bioethanol, Taylor & Francis, United Stated, 1996, pp. 179–212.
[13] J.S. Brigham, W.S. Adney, M.E. Himmel, Hemicellulases: diversity and applications, Handbook on Bioethanol, Routledge, 2018, pp. 119–141.
[14] P. Li, Q. Zhang, X. Zhang, X. Zhang, X. Pan, F. Xu, Subcellular dissolution of xylan and lignin for enhancing enzymatic hydrolysis of microwave assisted deep eutectic solvent pretreated *Pinus bungeana* Zucc, Bioresour. Technol 288 (2019) 121475.
[15] S. Rezania, M. Ponraj, M.F. Din, A.R. Songip, S.C. Fadzlin, The diverse applications of water hyacinth with main focus on sustainable energy and production for new era: an overview, Renew. Sustain. Energy Rev 41 (2015) 943–954.
[16] S. Rezania, H. Alizadeh, J. Cho, N. Darajeh, J. Park, B. Hashemi, et al., Changes in composition and structure of water hyacinth based on various pretreatment methods, BioResources 14 (3) (2019) 6088–6099.
[17] J. Nigam, Bioconversion of water hyacinth hemicellulose acid hydrolysate to motor fuel ethanol by xylose-fermenting yeast, J. Biotechnol 97 (2002) 107–116.
[18] A. Singh, N. Bishnoi, Comparative study of various pretreatment techniques for ethanol production from water hyacinth, Ind. Crop. Prod 44 (2013) 283–289.
[19] L. Khuong, R. Kondo, R. De Leon, T. Anh, S. Meguro, K. Shimizu, Effect of chemical factors on integrated fungal fermentation of sugarcane bagasse for ethanol production by a white-rot fungus, *Phlebia* sp. MG-60, Bioresour. Technol 167 (2014) 33–40.
[20] T. Bugg, M. Ahmad, E. Hardiman, R. Rahmanpour, Pathways for degradation of lignin in bacteria and fungi, Nat. Prod. Rep 12 (2011) 1883–1896.
[21] D. Martinez, L. Larrondo, N. Putnam, M. Gelpke, K. Huang, J. Chapman, Genome sequence of the lignocellulose degrading fungus *Phanerochaete chrysosporium* strain RP78, Nat. Biotechnol 22 (6) (2004) 695–700.
[22] K.E. Hammel, Mechanisms for polycyclic aromatic hydrocarbon degradation by ligninolytic fungi, Environ. Health Perspect 103 (2008) 41.
[23] N. Jacques, L. Hardy, K. Knox, A. Wicken, Effect of Tween 80 on the morphology and physiology of *Lactobacillus salivarius* strain IV CL-37 grown in a chemostat under glucose limitation, J. Gen. Microbiol 119 (1980) 195–201.
[24] G.L. Miller, Use of dinitrosalicylic acid reagent for determination of reducing sugar, Anal. Chem 31 (1959) 426–428.
[25] A. Chesson, The maceration of linen flax under anaerobic conditions, J. Appl. Bacteriol 45 (1978) 219–230.
[26] D. Kamra, N. Kewalramani, D. Lall, N. Pathak, Biolignification and changes in in vitro digestibility of sugarcane bagasse treated with white rot fungi, J. Appl. Anim. Res (1993) 133–140.
[27] E. Sjostrom, Wood Chemistry: Fundamentals and Applications, Academic Press, London, 1993.
[28] P.L. Dhepe, R. Sahu, A solid-acid-based process for the conversion of hemicellulose, Green Chem 12 (2012) 2153–2156.

[29] H. Schneider, Conversion of pentoses to ethanol by yeasts and fungi, CRC Crit. Rev. Biochem. Mol. Biol, 9, 1889, pp. 1–40.
[30] T. Paulino, M. Cardoso, T.G. Bruschi, P. Ciancaglini, G. Thedei, Fermentable and non-fermentable sugars, Biochem. Mol. Biol. Educ (2003) 180–184.
[31] D.J. Wohlbach, T.K. Alan, Comparative genomics of xylose-fermenting fungi for enhanced biofeul production, PNAS 108 (2011) 13212–13217.

CHAPTER 10

Working principle of typical bioreactors

P. Jaibiba, S. Naga Vignesh and S. Hariharan
Department of Biotechnology, Sri Venkateswara College of Engineering, Sriperumbudur, India

10.1 Introduction

A bioreactor is a vessel-like device that provides a uniform background for microorganisms to grow and maintains an uninterrupted balance in the biochemical reactions carried out by these microorganisms to produce desired metabolites [1]. The applications of bioreactors may be extended for biomass production such as single cell protein, baker's yeast, animal cells, and microalgae as well as for metabolite formation like organic acids, ethanol, antibiotics, aromatic compounds, and pigments and to transform substrates like steroids and even to produce both intra- and extracellular enzymes [2].

The bioreactor is the heart of any biochemical process as it provides a meticulous environment for microorganisms to achieve optimal growth and produce metabolites [3]. In other words, for the biotransformation and bioconversion of substrates into desirable products, reactors can be engineered or manufactured based on the growth requirements of the organisms used. They could be used for all types of biocatalysis including the production of enzymes and the growth of tissues, cells, and cellular organelles. Bioreactors are commonly designed as a cylindrical tank with an agitator and integral heating or cooling system, ranging in size from less than 1 L to more than 50,000 L, often made of steel, stainless steel, glass lined steel, or glass.

These reactors have been designed to maintain certain parameters like flow rates, aeration, temperature, pH, foam control, and agitation rate. Reactors can provide an output to specified process parameter control elements to rectify any deviation in the value of these parameters from the user-defined set point. The number of parameters that can be monitored and controlled is limited by the number of sensors and control elements incorporated into a given bioreactor [4].

Bioreactors
DOI: https://doi.org/10.1016/B978-0-12-821264-6.00010-3

Bioreactors are generally classified into two broad categories:

- Suspended growth bioreactors, which use microbial metabolism under aerobic, anaerobic, or sequential anaerobic/aerobic conditions to utilize substrates and degrade them into residuals. Examples include batch reactors, continuous stirred tank reactors (CSTRs), and plug flow reactors, etc.
- Biofilm bioreactors in which microorganisms mostly get attached to a surface and adhered within the reactor, which could be used for the treatment of wastewater, and the organisms present in the biofilm, absorb and break down toxic substances in the water. Examples include membrane, fluidized bed, packed bed, airlift, and upflow anaerobic sludge blanket (UASB) reactors.

Bioreactors can be further classified into three types based on the mode of operation used, mainly the feeding mechanism of the culture and the medium into the reactor, that is, batch, continuous and semicontinuous, or fed-batch [5]. The application of batch processes has been increased in fields relating to the production of biomolecules and pharmaceuticals, etc.

Industries should choose bioreactors based on the efficient production of bioenergy through biomass and biofuel and to limit the pollutants generated from the processes. Based on industrial requirements, the various types of bioreactors are discussed along with their working principles and applications.

10.2 Aerobic and anaerobic bioreactors

Aerobic and anaerobic bioreactors are two different types of bioreactor where in microorganisms uses carbon as a source from waste and grows either with oxygen or without oxygen and convert them into biomass and carbon dioxide. The overall reaction that happens in an aerobic reactor can be stated as [6]:

$$C_xH_y + O_2 + (\text{microorganisms/nutrients}) \rightarrow H_2O + CO_2 + \text{biomass}$$

whereas an anaerobic bioreactor is operated with a mechanism of breaking complex organic molecules into mixtures of volatile fatty acids such as acetic acid, propionic acid, and butyric acid, which can be attained by anaerobes, mostly hydrolytic and acidogenic bacterial groups [7]. The volatile fatty acids that are produced are converted to CO_2 and methane (CH_4) by acetogenic and methanogenic bacteria respectively. The overall mechanism of the anaerobic process can be expressed in equation form as [6]:

$$C_xH_y + (\text{microorganisms/nutrients}) \rightarrow CO_2 + CH_4 + \text{biomass}$$

10.2.1 Aerobic reactors

In an aerobic reactor, the product conversion rate or the degradation rate is mainly based on the bubble size and gas-to-liquid mass transfer rate. Oxygen solubility is a main parameter that should be optimally maintained because the presence of salt may hinder the solubilizing property of oxygen. Simple aerobic bioreactors can be constructed with aerated lagoons or oxidation ponds for the storage of the waste in the open environment and with a rotating disk that contains the microbe as a biofilm for regular churning.

Examples of commonly used aerobic bioreactors at an industrial scale include stirred tank bioreactors, airlift bioreactors, and inverse fluidized bed bioreactors. The most common aerobic reactor is the stirred tank reactor where air is sparged from the bottom of the reactor. In the airlift bioreactor mixing is provided by the turbulence created by gas. The oxygen transfer coefficient is high for airlift reactors compared to that of stirred tank reactors. The air is introduced from the bottom of the reactor resulting in a circulatory motion of contents within the reactor that helps obtain a maximum transfer of gas as shown in Fig. 10.1A.

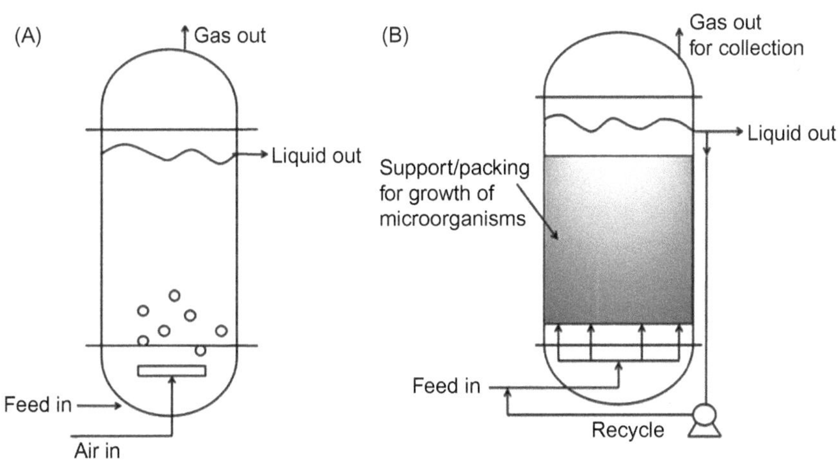

Figure 10.1 (A) Aerobic reactors and (B) anaerobic reactors.

The inverse fluidized bed reactor (FBR) is particularly used for wastewater treatment where inert particles coated with biofilm represent the solid phase, while the oxygen or air supply is the gaseous phase and the liquid phase is the wastewater. The gas flows in the countercurrent direction of the liquid flow, which improves the mass transfer rate and makes the bed easy to refluidize.

10.2.2 Anaerobic reactors

Anaerobic reactors are mostly similar to aerobic reactors but a few conditions have to be considered for maintaining an anaerobic environment. This reactor is simple in construction with a high loading rate and can withstand high toxic and organic substances. Methanogens are specialized bacteria used for supporting anaerobic reactions that have the tendency to form immobilized granules that settle at the bottom as a sludge, which is the basic principle used in anaerobic bioreactors. Based on this working principle is the UASB reactor. An anaerobic fluidized bed can be designed where the mixer of microbes should be in the form of a biofilm that is grown in carrier particles and fluidized using energy from the feed stream (Fig. 10.1B). The substrate diffuses through the biofilm then gets converted to volatile fatty acids and CH_4, which finally diffuses out into the bulk liquid. As the biofilm grows, the particles within the reactor increase in size and leave the reactor leading to a decrease in the density and concentration of the particles.

Membrane reactors can also be engineered as a type of anaerobic reactors. They are designed in different forms. Enzymes are suspended in the reactor and the mixture is withdrawn along with the enzymes and is passed through a membrane where the enzymes are retained and the products are collected or the membrane filter can be designed in such a way that it is immersed inside the reactor, but the permeate flows through the membrane will be devoid of enzymes, which are retained within the reactor. This type of reactor produces little sludge when compared to the conventional bioreactor, which is the main advantage of membrane bioreactors (MBRs). The disadvantages of these reactors include fouling of the membrane and high operational costs.

10.3 Plug flow bioreactor

A plug flow reactor is a cylindrical shaped tank where chemical reactions occur between a catalyst coated in the sides of the wall of the reactor and

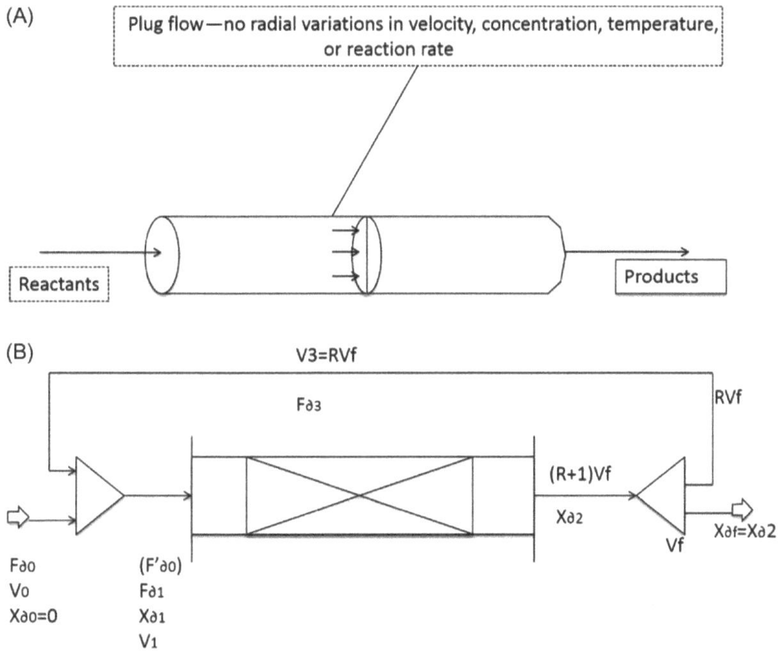

Figure 10.2 (A) Plug flow reactor design and (B) performance of plug flow reactor.

an inlet reactant (Fig. 10.2A). Plug flow reactors are mostly used for their unique property of not requiring any mixing. The plug flow reactor is based on specifications that describe the flow regime, while reactor design and variables for the design of the reactor are also to be estimated. The plug flow reactor has a continuous input and output of materials, which could be operated either as a closed or open system.

In a plug flow reactor, the nutrient (reactants) that enters the reactor will be in the form of "plug," which passes in an axial direction of the reactor with different compositions. As the plug flows through the reactor the fluid gets mixed in the radial direction not in the axial direction and the flow in the reactor will be highly turbulent.

10.3.1 Design parameters and process

For designing a reactor, several parameters should be considered such as space time, concentration, volumetric flow rate, and volume. The concentration of the feed is directly related to reactor volume or reactor length. The principle of the plug flow reactor is similar to chemical reactions where the substrate as a plug reacts with catalyst and forms product continuously without any hindrance. In a few cases, this reactor can also be used in the absence of a catalyst.

10.3.2 Performance of the reactor

A reactor should be designed to produce the desired final product with an expected production rate and a low performance cost. Generally, to attain minimal cost reactors are designed with a recycling process (Fig. 10.2B).

Levenspiel [8] described the performance of a plug flow bioreactor with a recycled stream as:

$$V = F_{AO}\left(1 + R\right)\int_{RX_f/(1+R)}^{X_f} dX_A/(-r_A)$$

where V is the volume of the reactor (m^3), F_{AO} is the mass flow rate of the feeding stream (kg/h), R is the recycle ratio, X_f is the conversion of products, and $(-r_A)$ is the kinetic expression of the chemical reaction [9].

10.3.2.1 Applications

The plug flow reactor is the second-most primary ideal reactor and is similar to the continuous stirred tank bioreactor. Here polymerization and conversion reactions are performed in noncatalytic mode. The polymerization of ethylene and the conversion of naphtha to ethylene are examples. A catalytic process is used for the synthesis of ammonia and SO_2 combustion [10]. Currently, plug flow reactors are used in the production of biodiesel and other biofuels with a recycle system [9]. The plug flow reactor is mostly preferred for bioenergy production because of its steady-state operation. In addition, the plug flow reactor does not require any agitation or baffling. When high viscous reactants are used, a high-pressure drop is developed, which may be considered as a limiting factor in the usage of this reactor, and this reactor is more complex than the continuous stirred tank bioreactor.

10.4 Upflow anaerobic sludge blanket bioreactor

UASB reactor technology was developed for wastewater treatment using anaerobic digesters. UASBs consist of a digester for the production of CH_4, which undergoes an anaerobic process, and a blanket of granular sludge is formed, through which the wastewater flows upward and the same is digested by anaerobes. Microbes in the sludge break down the organic matter and transform it into biogas. The UASB bioreactor consists of a three-phase separator system (as shown in Fig. 10.3A), which helps the reactor to separate gas, water, and sludge mixtures under high turbulence. The reactor has multiple gas chambers for the separation of biogas.

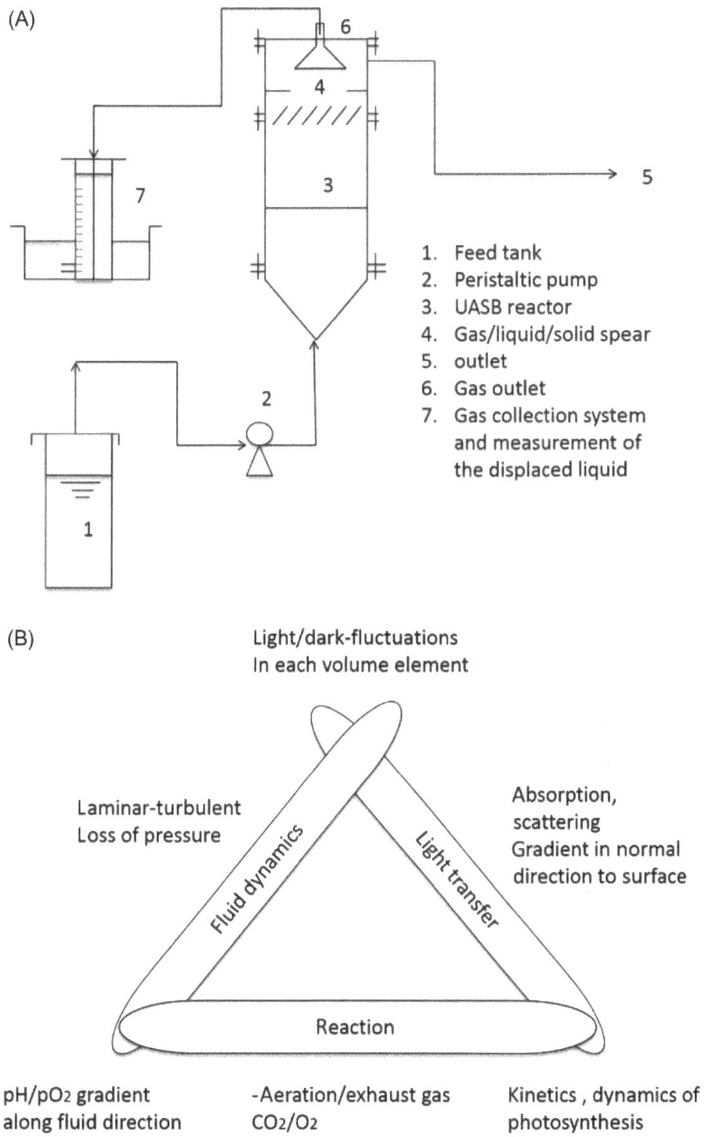

Figure 10.3 (A) Upflow anaerobic sludge blanket reactor and (B) interactions of parameters in photobioreactor.

The gas and water interfaces greatly reduce the turbulence, which creates relatively high loading rates of 10–15 kg/m^3.

10.4.1 Design of upflow anaerobic sludge blanket reactor

Generally, the temperature of a USAB reactor will be above 20°C. At equilibrium condition, the sludge withdrawn should be equal to the sludge produced daily. The sludge produced daily depends on several parameters, namely (1) the new volatile suspended solids (VSS) produced because of biological oxygen demand (BOD) removal, the yield coefficient being assumed as 0.1 g VSS/g BOD removed, (2) the nondegradable residue of the VSS coming in the inflow, and (3) ash received in the inflow, namely TSS–VSS mg/L. Thus at steady-state conditions [11]:

$$\text{SRT(solid retention time)} = \frac{\text{Total sludge present in the reactor, kg}}{\text{Sludge withdrawn per day, kg/day}}$$

Another parameter is hydraulic retention time, which is given by:

$$\text{HRT(hydraulic retention time)} = \frac{\text{Reactor volume, m}^3}{\text{Flow rate, m}^3/\text{h}}$$

The reactor volume should be so chosen depending on (1) the depth of the bioreactor, (2) the effective depth of the sludge blanket, and (3) the average concentration of sludge in the blanket.

The upflow velocity can be determined by fixing the size of the reactor. This could be expressed as:

$$\text{Upflow velocity} = \frac{\text{Reactor height, m}}{\text{HRT, h}}$$

The upflow velocity should not be more than 0.5 m/h at average flow and not more than 1.2 m/h at peak flow in order to retain the flocculent sludge in the reactor. At higher velocities, the carry-over of solids might occur and waste quality may be deteriorated. The settling compartment is formed by hoods for the collection of gas. The depth of the compartment is 2.0–2.5 m and the surface overflow rate should be kept at 20–28 m^3/m^2/day (1–1.2 m/h) at peak flow. The capacity, geometry, and hydraulics of the unit should be maintained regularly to ensure the proper working of the "gas–liquid–solid–separator" (GLSS), the gas collection hood, which separates the gas from the treated wastewater and the sludge [12].

10.4.2 Working process of upflow anaerobic sludge blanket reactor

In a UASB reactor, the waste is fed into the reactor in an upflow direction with a hydraulic retention time (HRT) of about 8–10 h.

No prior sedimentation is required for this type of reactor. This reactor does not require any specific medium, whereas the up-flowing sewage itself forms millions of small "granules" or particles of sludge that are held in suspension and provide a large surface area (sludge blanket) on which organic matter can attach and undergo biodegradation.

A high solid retention time (SRT) of 30–50 days or more occurs within the unit. No mixers or aerators are required. The gas produced can be collected and could be used as a biofuel. Excess sludge is removed regularly by a separate pipe and the drying process is carried out.

UASB reactors are best suited for treating bottling wastewaters, but low rate installations do exist, which is one of the advantages of using this type of reactor. Biomass from sugar wastewaters form granules readily so the processing of such wastewaters using UASB reactors is best. There are limitations in that this reactor cannot be used for treating pharmaceutical waste since it contains antibiotics and toxic chemicals that anaerobic treatment is not suitable for treating, and using this type of reactor is not cost efficient and consumes a large amount of power [13].

10.5 Photobioreactor

Photobioreactors are defined as closed (or mostly closed) culture systems for prototroph cultivation in which photons do not illuminate directly on the surface of the reactors, but need to pass through the transparent walls of these reactors before reaching the cultivated cells [14]. A typical photobioreactor consists of three phases, namely the liquid phase, which is the growth medium, microbial cells as the solid phase, and CO_2 and O_2 as the gaseous phase, with light radiation fields being superimposed (Fig. 10.3B).

Based on the intensity of the light, photobioreactors are classified as flat-plate, tubular, or column and are further categorized as stirred-type, bubble column, or airlift depending on the liquid flow.

10.5.1 Flat-plate photobioreactors

Flat-plate photobioreactors (Fig. 10.4A) are mostly used for the cultivation of photosynthetic microorganisms due to their large illumination surface area. Generally, flat-plate photobioreactors are made of transparent materials for enhanced utilization of light energy. Cylindrical eddies that are produced during the process are removed using horizontal baffles located in the interior of the panels at right angles to the flow, which transport the algae from dark to light at high frequency. When compared with

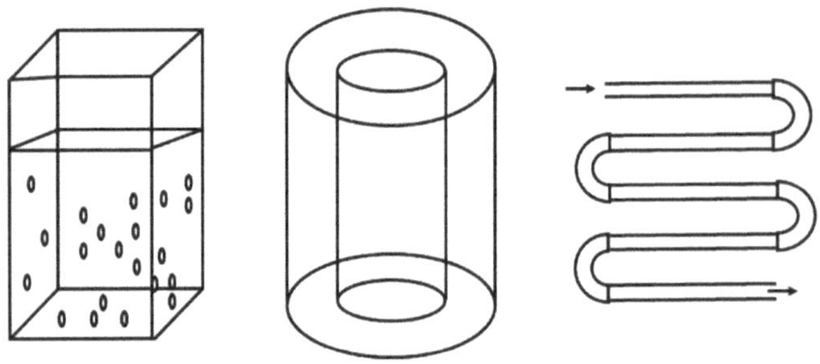

Figure 10.4 (A) Flat-plate reactor, (B) annular reactor, and (C) tubular reactor.

horizontal tubular photobioreactors, the accumulation of dissolved oxygen concentrations in flat-plate photobioreactors is relatively low. Flat-plate photobioreactors are highly suitable for mass cultures of algae.

10.5.2 Annular photobioreactor

Annular photobioreactors are constructed with acrylic cylinders 2 m in height and with diameters of around 40–50 cm of which one is kept inside and the other forms an annular chamber. The inner side of the reactor does not contribute much to radiation, but for dark periods, additional lamps can be fitted as shown in Fig. 10.4B. The airlift principle could be incorporated to increase axial transport. A downcomer is usually arranged as a section of the cross-section or as a coaxial inner cylinder. As this part of the reactor is dark, the cells flow through the riser and downcomer regularly, inducting an additional light/dark cycle in the range of 1–100 s [15].

10.5.3 Tubular photobioreactors

Tubular reactors are configured with parallelly arranged transparent tubes as shown in Fig. 10.4C. The tubes are straight and have diameters of 10 mm to maximum 60 mm, and lengths of up to several hundred meters. Incident light is reduced and is in the radial direction focused onto the axis of the tube. The aeration and mixing of the culture used are usually done by air pump or airlift systems. Since tubular photobioreactor have large illumination surface areas, they are highly suitable for outdoor mass cultures of algae. On the other hand, the major limitation of tubular photobioreactor is their poor mass transfer.

10.5.4 Photobioreactor configurations

To design a photobioreactor an ideal interaction between biomass production and environmental parameters within the reactor is necessary. The working volume of the reactor consists of a liquid and a gas phase. The amount of light that could enter and hit the surface of the reactor is determined by the total surface area of the transparent layer of the reactor. The ground area of the reactor measures the area from which light energy is collected. The irradiation at the surface is given as photon flux density [$mE/(m^2\ s)$], where E (Einstein) stands for mole of photons [16]. Irrespective of the reactor configuration, several essential parameters need to be considered.

1. *Effective and efficient provision of light.* The intensity of the light should be maintained at low to avoid excess fluorescence and heat. The distribution of light inside the reactor is influenced by the attenuation of light, which is due to the adsorption of cells that contain pigment.
2. *Supply of CO_2.* The reactor is to be fed with CO_2 for the survival of the organisms and a stoichiometric calculation should be done to predict the demand of CO_2. For instance, the carbon fraction varies from 0.45 for algae with a high carbohydrate content to 0.8 for oil-rich cells.
3. Removal of oxygen from the reactor, which may inhibit the metabolism or otherwise in extreme cases damage the culture.
4. Sensible scalability of the photobioreactor technology.

10.5.5 Working of photobioreactor

The flow of algae or photosynthetic bacteria depends on the flow of feed from the vessel to the diaphragm pump as shown in Fig. 10.5A. The pump is provided with a CO_2 inlet valve. The photobioreactor will promote biological growth by controlling the parameters.

After the algae medium has completed the flow it passes back into the feeding vessel. An oxygen sensor present within the reactor will determine the oxygen level and oxygen is also released into the feeding vessel. The harvesting rate is determined and when the algae are ready for harvesting, they are passed through the connected filtration system where they are filtered and processed.

Photobioreactors are used at lab scale to initialize the growth of microalgae. Currently, photobioreactors are used in the production of biodiesel

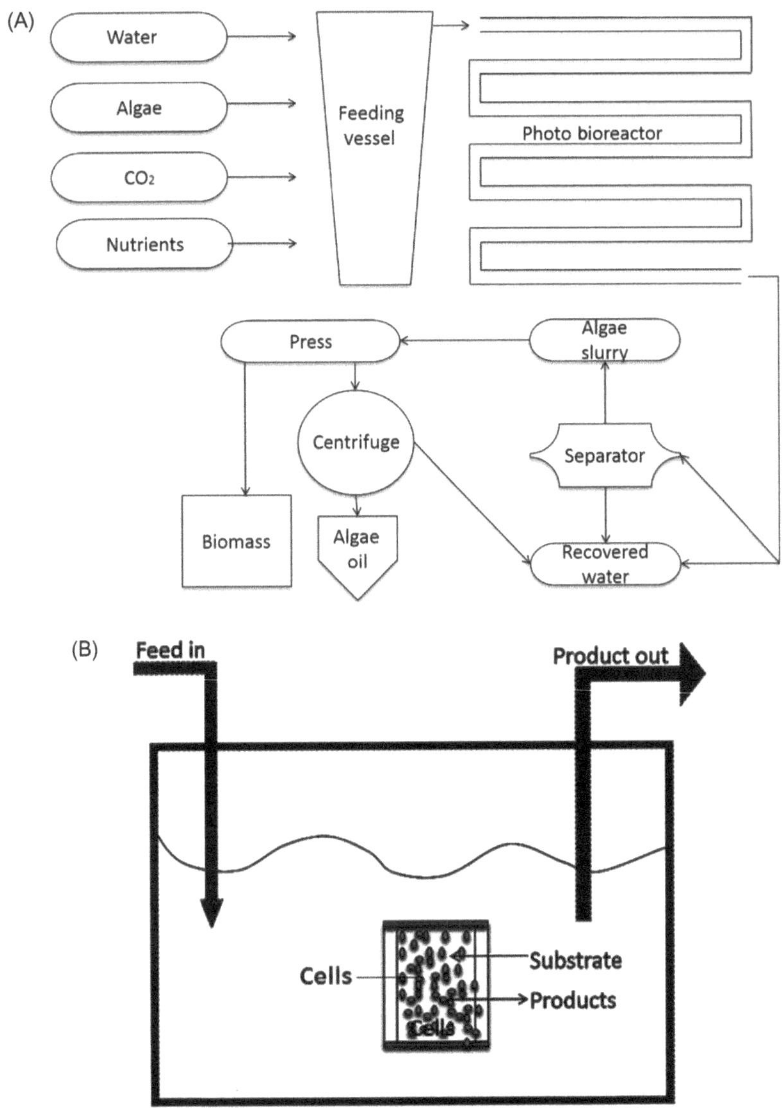

Figure 10.5 (A) Photobioreactor and (B) reverse membrane reactor.

from oils or hydrocarbons using *Botryococcusbraunii* [17], Astaxanthin from *Haemotococcuspluvialis* [18], and an analogue of fossil petrol by growing microalgae. Even though the photobioreactor is advantageous in producing many products there is a limitation in that the production costs and the capital costs are high.

10.6 Reverse membrane bioreactor

Reverse membrane bioreactors (rMBRs) represent a novel combinational technique combining the conventional membrane bioreactor and cell encapsulation techniques where cells are separated from the feed and immobilized in a membrane that is fixed within the reactor as shown in Fig. 10.5B. The principle of this reactor is similar to that of the immersed membrane reactor where the membrane is submerged in the reactor, however, in the rMBR, microorganisms are encased within membrane layers that form a sachet. Integrated permeate channel (IPC) and packed column are other forms of membrane settings [19].

The selection of membrane configuration depends on the desired product and its byproducts. For instance, a multilayer membrane column is used to produce biogas, while for ethanol production, an IPC membrane configuration-based reactor is used. Mostly synthetic membranes are used, which allow only specific nutrients to pass through such as in plant cells where the cytoplasm can be separated from the cellular components [20].

The purpose of encasing cells in the membrane layer is to increase the cell density and tolerance level. For instance, the yeast cell concentration can be increased to up to 309 g/L [21]. Because of this high density, the cell may experience more stress due to the lack of nutrition, which in turn produces a counter-stress response through the expression of stress related genes [22,23].

10.6.1 Diffusion phenomenon of reverse membrane bioreactor

Generally, in rMBRs the diffusion mechanism takes place at three different phases. (1) The diffusion of the substrate on the feed side to the membrane surface and in reverse for the product, (2) the diffusion of compounds (substrate or metabolites) through the membrane, and (3) the diffusion of the feed and the products on the cell side through a biofilm layer [20]. The diffusion of compounds through the membrane depends on various parameters, namely porosity, tortuosity, hydrophilicity, and concentration gradient, etc. The porosity of the membrane is considered as an important parameter as the pores are the connecting channels for the two media. During this condition the flux does not vary. According to Fick's first law, the flux (J) of the membrane is inversely proportional to the membrane thickness (h).

$$J \propto \frac{1}{h}$$

Greater concentration gradient will influence high flux of the compounds through the membrane.

10.6.2 Difference between conventional membrane bioreactor and reverse membrane bioreactor

The configuration of the conventional membrane bioreactor is internal or external; the pressure gradient acts as a driving force. The mechanism of mass transfer happens through convection and diffusion. The microbial or other forms of living cells are fed along with the feed and they are freely suspended in the medium. In the case of the rMBR, it is a kind of immersed bioreactor where the concentration gradient acts as a driving force and the mass transfer mechanism happens through diffusion. The living cells cannot be fed along with the feed, they need to be separated and immobilized within the membrane layers.

10.7 Immersed membrane bioreactor

Immersed membrane bioreactors (IMBRs) are a type of membrane bioreactor in which two principles, the suspended growth bioreactor and separation, take place simultaneously to produce a synergistic effect. The IMBR is based on a filtration system with membranes that are immersed in the biomass; filtration takes place by applying a vacuum to the inside of the membrane. The membranes can be placed either inside the bioreactor or in a separate tank. The membranes can be flat sheet, hollow or a combination of both (as shown in Fig. 10.6A) and an online backwash system is incorporated that reduces membrane surface fouling. Additional aeration is also required to provide air scour to reduce fouling. Hollow-fiber immersed membrane reactors are generally used for medium- to large-scale plants.

10.7.1 Configuration and design of the immersed membrane bioreactor

In the case of the immersed configuration, the membrane acts as a filter in the aerated basin and the biological treatment of the waste is improved. In the immersed configuration, the permeate flux is removed by suction, which limits transmembrane pressure (TMP) at about 0.5 bar. Based on the energy consumption, the configuration of the IMBR is more economical for two reasons; firstly, no recycle pump is needed since aeration generates a tangential liquid flow, and secondly, the operating conditions are more feasible. This type of reactor has a high biomass concentration, low operating interval, and low excess sludge production.

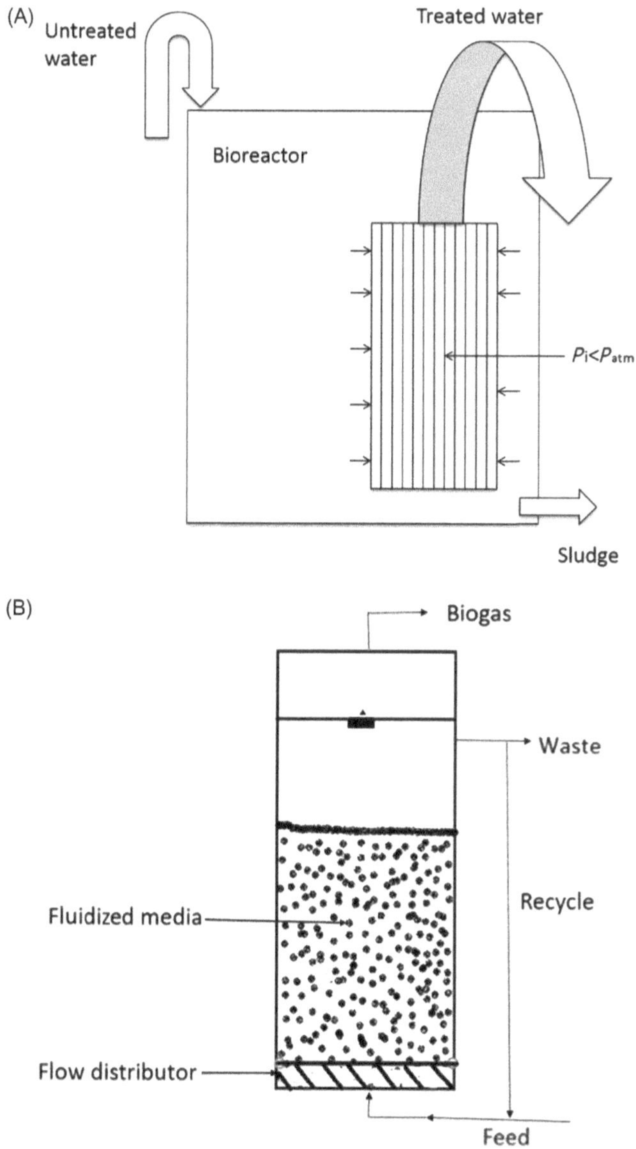

Figure 10.6 (A) Immersed membrane reactor, (B) fluidized bed reactor, and (C) packed bed reactor.

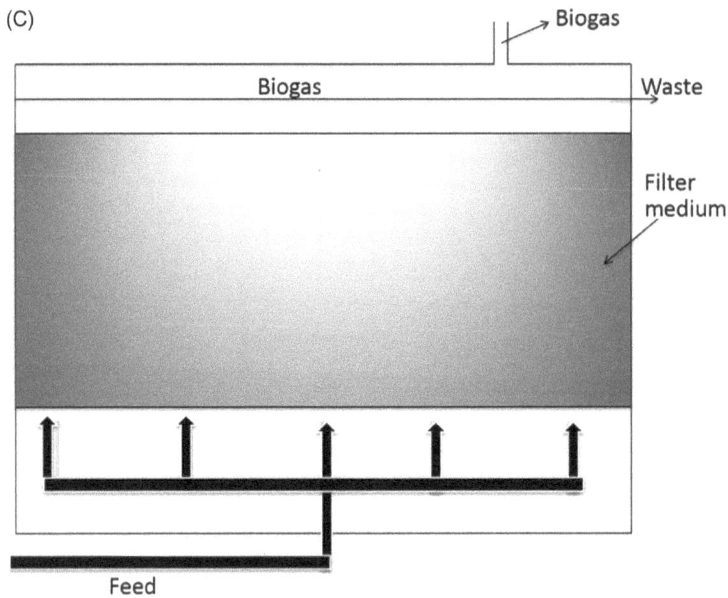

Figure 10.6 (Continued)

10.7.2 Process of immersed membrane bioreactor

The process of the IMBR depends on five key elements. These are:

- the membrane, its design, and the sustaining of permeability;
- feed water, its characteristics, and its pretreatment;
- aeration of both the membrane and bulk biomass;
- sludge withdrawal and residence time; and
- bioactivity and nature of the biomass.

These elements are largely interrelated, in particular the latter three relate to operation. The rate at which sludge is withdrawn determines the concentration of the biomass. The mixed liquor suspended solid concentration then impacts the biological properties and microbial activity and the viscosity and oxygen transfer. The wastewater to be treated enters the reactor where the pollutants are transformed by the activated sludge. The permeate is sucked through the membrane for the given filtration period until the cake thickness becomes too high, and then it is stored in the permeate tank. As the backwashing and cleaning period starts the wasted sludge is removed with a constant and continuous flow from the bottom of the reactor and is stored in the waste sludge tank. The air is supplied through three spargers, two at the bottom of the reactor for microbial consumption and one at the bottom of the membrane to scour the membrane [24].

This type of bioreactor is mostly preferred for the treatment of tannery and textile wastes as well as wastewaters and in reusing aquaculture wastes because of its low energy consumption and low capital costs. This type is more advantageous than the traditional method where the clarifier is replaced by the membrane module.

10.8 Fluidized bed bioreactor

The FBR is a truly fixed biological reactor that accumulates a maximum active attached biomass and also handles suspended solids without blocking as shown in Fig. 10.6B. The FBR is similar to the expanded bed reactor (EBR) except for the fact that a high velocity of liquid upflow is maintained in the FBR [25]. The rate of the liquid flow and the bed expansion degree determine the reactor type, that is, fluidized bed or expansion bed. EBRs have a bed expansion of 10%–20%, while FBRs have 30%–90% of expansion.

10.8.1 Mechanism and working of fluidized bed reactor

In order to achieve a maximum specific activity of biomass, the surface area for the microbial attachment is maximized using small-sized particles and the volume of the media is also minimized. The mixing regime of the FBR is between that of the packed bed reactor (PBR) and that of the stirred tank reactor.

10.8.2 Process flow in fluidized bed reactor

In the FBR, biomass is attached to the small particles such as anthracite, high density plastic beads, sand, etc., which are in suspension by means of upward flow. The large surface area of the particles with the high degree of mixing enhances the biomass development and substrate uptake.

FBR has an advantage in that a high density of particles can be used with good mixing between the phases and requiring low energy. The heat and mass transfer characteristics are superior in the FBR [26]. Moreover, the FBR has a low tendency for clogging and short circuiting, etc. The main drawback with this reactor is the loss of particles during the process, which leads to sudden changes in the density of the particles, the flow rate, and the production of gas. Even though there is this limitation, FBRs are used to produce ligninolytic enzymes, for wastewater treatment from refineries, and to reduce the chemical oxygen demand (COD) level in the wastewater.

10.9 Packed bed bioreactor

In chemical processing, the PBR, which consists of a cylindrical vessel packed with catalyst materials or adsorbents such as zeolite pellet, granular activated carbon, etc., is most commonly used. This PBR is also known as a fixed bed reactor, which is most commonly used to catalyze gas reactions. The packed bed is used to create and improve the contact between the two phases used in the process.

10.9.1 Design and configurations of packed bed reactor

The PBR consists of a cylindrical vessel, composed of a water-jacketed stainless steel column with an external diameter of 60 mm and a height of 345 mm [27] packed with pellets ranging in size from 1 to 10 mm. Ceramic balls (inert) are placed on top of the pellets to distribute the feed evenly while a metal grid and screen are placed at the bottom to support the catalyst as shown in Fig. 10.6C. The catalyst should be porous in nature to attain a high surface area. Due to their high surface area, the reactants are easily transported inside the pores by diffusion and get adsorbed to the active site. Finally, the products get desorbed and diffuse back to the bulk. The ratio of column-to-particle diameter should be 10–20 to minimize channeling and the bed height-to-diameter ratio should be around 0.5.

The performance of the reaction can be improved by the optimization of the reactor design. There are three systems that are used based on the desired reaction, namely single-bed, multibed, and multitube units. The single-bed reactor consists of a simple vessel with a large diameter that is best suited for adiabatic processes and not for exothermic or endothermic reactions, which lead to increases in temperature and affect catalyst and product stability. In order to overcome the above-mentioned issue, the reactor is divided into multistages provided with heat exchangers.

The multitube reactor is designed for heterogeneous catalytic processes that require greater heat transfer between the reactant, catalyst bed, and the heating and cooling media. Pressure drop is the main aspect in the PBR that increases when there is friction between the gas and the particle phases. This pressure drop depends mainly on the reactor length, particle diameter, and the velocity of the gas.

10.9.2 Applications of packed bed reactor

PBRs work on a principle similar to that of the plug flow reactor and are used for the synthesis of methanol, ammonia, diesters, and biofuels, etc. For instance, biodiesel is produced from waste oil using *Jatropha curcas* and KOH as a catalyst, which converts 80% of the reactant used into the desired product [27]. Moreover, this reactor is used to degrade mixtures of dyes [28]. The PBR is mostly preferred for its low maintenance and cost, but there is some limitation in that the catalyst gets deactivated when the active sites get contaminated. Therefore the regeneration and reuse of the catalyst is difficult. Poor heat transfer may lead to the generation of hotspots and the degradation of the catalyst.

10.10 Activated sludge bioreactor

In an activated sludge bioreactor, the percentage of microbes, amount of oxygen and substrate are identical as this type of reactor is provided with homogenous tank where the feed gets dispersed throughout the reactor. In a plug flow activated sludge system, the reactor is provided with a long channeled inlet tank that limits the growth of microorganisms and enhances the sludge settling ability as shown in Fig. 10.7A. The step feed reactor is a development of the plug flow reactor in which sewage is injected at several points into the aeration tank. The food-to-microbes (F/M) ratio is far better when compared to the plug flow reactor. An oxidation ditch is another type of activated sludge bioreactor that follows a modified activated sludge treatment process using long SRTs to remove biodegradable organics [29]. This is equipped with an aeration rotor or brushes for proper circulation and aeration. This reactor is similar to the complete mix reactor.

10.10.1 Mixing regime of the activated sludge bioreactor

There are two types of mixing processes that are followed in activated sludge reactors, namely plug flow and complete mixing. The former regime is characterized by the flow of the mixed liquor via the aeration tank with no mixing, whereas in the later the contents of the aeration tank are mixed uniformly. The type of regime is most important as it affects oxygen transfer, kinetics, and environmental conditions.

10.10.2 Process of activated sludge reactor

The activated sludge system involves:

- wastewater being fed into the aeration tank along with a microbial suspension,
- solid–liquid separation,

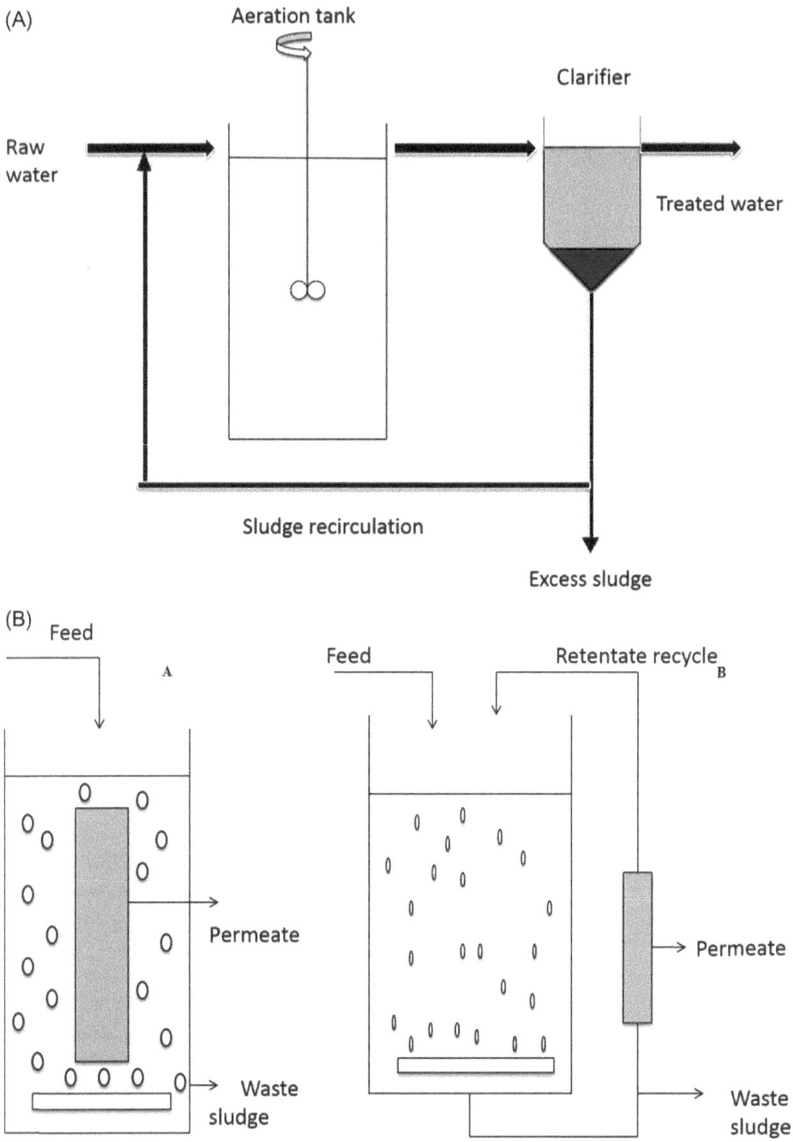

Figure 10.7 (A) Activated sludge reactor, (B) membrane bioreactors (a) internal and (b) external, and (C) working of membrane bioreactor.

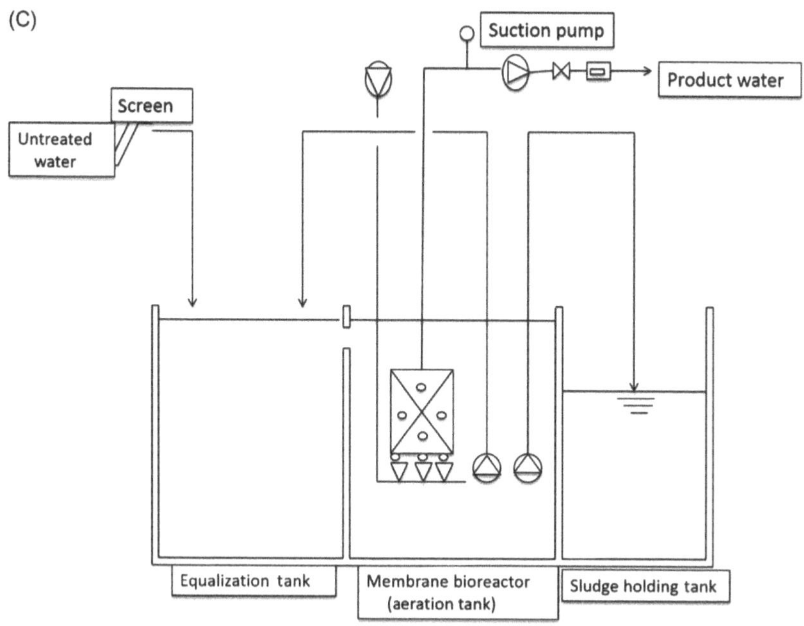

Figure 10.7 (Continued)

- the discharge of processed waste, and
- the retained biomass being returned to the aeration tank.

In the activated sludge process, sewage water with organic matter is fed into the aeration tank, which gets metabolized as it is loaded with microorganisms. This metabolized organic matter is oxidized to CO_2 and water to derive energy. Part of the newly formed cells in the process is removed from the reactor as sludge and the remaining sludge is returned to the aeration tank where the process continues.

This reactor is generally used in the treatment of sewage and wastewater. This reactor is more specifically used in the production of biofuels like bioethanol, biogas, etc., as for example, biofuel that is produced from milk-derived waste. Moreover, this reactor can be operated at high organic loading rates, but with the limitation that this reactor consumes relatively high energy and capital and operating costs.

10.11 Membrane bioreactor

A membrane bioreactor (MBR) is a type of flow reactor that is used to separate biomass and metabolites from the final product using membranes. These membranes are specially furnished materials used to separate biomolecules inside reactors based on their charge and size. An ideal feature of MBRs is that cells or enzymes are retained within the reactors, so these reactors can be continuously perfused, that is, the feed streams are delivered continuously and the products may also be removed continuously, but in some cases, the product must be removed intermittently or at the end of the process. Usually though, the feed streams are delivered continuously. The membranes are submerged in an aerated biological reactor, and the most commonly used membranes are microfiltration (MF) and ultrafiltration (UF) membranes. MF membranes have pore sizes between 0.1 and 0.5 μm and can be used to restrict cells within a reactor, whereas UF membranes typically have pore sizes between 20Å and 1000Å and are used to retain or exclude macromolecules. Membranes are made from a variety of materials including cellulose, acetate and nitrate, polyvinylidene difluoride, polysulfone, polypropylene, polytetrafluoroethylene (PTFE), and polyacrylonitrile. Other types of membranes including ceramic, silicone rubber, and ion exchange membranes are also being used. In general, membranes are packed in different modules namely plate-sheet modules, tubular modules, spiral-bound modules, and hollow-fiber modules. The most commonly used geometry for MBRs is that of the hollow-fiber module, which has been used for the growth of mammalian and plant cells and for the immobilization of bacteria, yeast, and enzymes.

10.11.1 Membrane fouling in membrane bioreactor

Membrane fouling happens due to the deposition of suspended solids or dissolved substances over the pore openings present in the surface of membranes [30]. Cake formation, pore narrowing, and pore clogging are the different forms of fouling in MBRs [31].

10.11.2 Configuration of membrane bioreactor

Membrane filtration may occur within the reactor or through recirculation across the membrane, which is subjected to a pressure drop by means of hydraulic head or a pump [32]. The UF or MF membranes utilized by MBRs have pore sizes suitable for water and solute molecules to pass through while other larger species like microorganisms and solids are retained (Fig. 10.7B). The required aeration should be provided properly for the growth of the biomass and for proper mixing within the reactor. The mixed liquid is usually kept in a solid condition ranging from 1% to 1.2% in order to provide maximum aeration and scour around the membranes.

Up-flowing bubbles cause efficient oxygen transfer in the bubble diffuser used in a submerged configuration by providing a cross-flow velocity cause on the membrane surface. This helps to maintain the flux by reducing the buildup material on the membrane surface and subsequently increases the operation cycle of the system.

10.11.3 Working of membrane bioreactor

Mostly MBRs are applied for the treatment of wastewater and other wastes. Effluent or wastewater is passed through a screen to remove the solids as shown in Fig. 10.7C. An equalization tank is included to handle the changes in the flow so that the MBR can be sized to treat the average flow rather than the peak flow. As discussed previously, the membranes are submerged in the aeration tank. Filtrate is drawn through the membranes using a suction pump. Air blowers are used to scour the membranes and to uniformly distribute suspended solids throughout the aeration tank. The continual agitation caused by the flow of air and water over the membrane surface cleans the surface of the membrane and prevents fouling.

There are two normal modes of operation, namely filtration mode and resting mode. In the filtration mode, a suction pump pulls water through the membranes to produce treated waste, whereas in the resting mode, the membrane is relaxed with an air scouring effect. Even though there is a continuous cleaning, the organic matter still gets deposited over the membrane surface resulting in an increase of TMP across the membrane. In order minimize this, chemical cleaning follows using sodium hypochlorite. For ideal MBR performance parameters like flow rate, temperature, dissolved oxygen, feed and waste BOD, TSS, and TMP, etc., are to be monitored strictly [33].

On comparing with the conventional treatment, MBRs are many times more advantageous as the secondary and tertiary clarifiers are eliminated and they have high loading rates, and so MBRs are used in municipal and industrial wastewater treatment processes. The major disadvantage of using MBRs is the high capital cost involved.

10.12 Immobilized cell bioreactor

The immobilized cell reactor (ICR) basically works on the principle of immobilization. Immobilization is the process of restricting cell mobility within a defined space [34]. Hydrophobic interaction, hydrogen bonding, and salt bridge formation between the adsorbent and the cells are driving forces for immobilization [35]. Generally, immobilization is classified into two types, namely passive and active. In the passive type, the cells get immobilized naturally within the solid matrix resulting in the formation of a biofilm. In active methods, immobilization is induced either by a chemical or physical method, which can be done in different aspects like attachment, entrapment, aggregation, and containment. The maintenance of immobilized cells within the reactor is quite difficult as these cells are independent of liquid and gas phases (Fig. 10.8). The activity of these immobilized cells could be enhanced by means of proper reactor design. To design a good reactor various criteria must be considered.

- The level of the shear forces should be low.
- The reactor should hold maximum particles.
- Mass and heat transfer should be maintained efficiently.

There are many different types of Immobilized cell bioreactor that could be operated in batch or continous mode namely packed bed reactor, air lift reactor, bubble column recator and hollow fiber membrane reactor.

The type and mode of operation depend on product and metabolites production. For instance, ethanol is fermented using immobilized *Saccharomyces cerevisiae*. The conventional production of ethanol in a

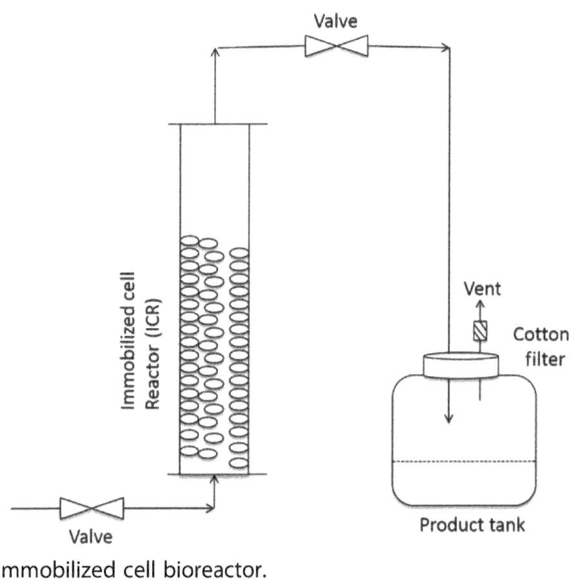

Figure 10.8 Immobilized cell bioreactor.

continuous process will be more expensive and has a high operational cost. Moreover, the concentration of the ethanol should be below the threshold range of 50–100 g/L to avoid cell death. To overcome these difficulties, immobilized cells can be used, which enhances the economics of ethanol fermentation [36].

The process of immobilization is often done using sodium alginate (2%) where immobilized cell beads are formed. Nutrient or substrate is fed into a reactor that contains the immobilized cells, these get interacted and in turn produce a product and byproduct. In most cases, ICRs are designed with two stages, namely an enricher stage and a stripper stage, which is used to treat and remove byproducts if present in large quantities [37]. Once the fermentation process is completed, the inner and the outer surfaces of the beads are compared, as the cells were initially present in the inner mass, but as the process proceeds the cells migrate and reside in the outer surface of the beads [3].

ICRs are used in the fermentation industry where the growth and production phases can be separated easily. Harvesting of the desired product can be done easily without any effort provided metabolites are released in the medium. There are some drawbacks within the ICR such as a mass transfer limitation due to intraparticle diffusion resistance, which limits the access of the substrate to the cells. This kind of problem is encountered in aerobic reactions where there is an oxygen deficiency to the cells resulting in reduced reactor performance. Another drawback is product inhibition where the product concentration is inhibited in the inner core and correspondingly the reaction rate is also reduced.

The application perspective of different bioreactors related to the production of biofuel in the industrial sector is summarized in Table 10.1.

10.13 Future perspective

The principle of bioreactors is often associated with bioreactor design and process optimization, which are mandatory for the conversion of techniques from lab scale to industrial level with maximum production that benefits society. It is mandatory to put effort into the engineering or modification of already existing bioreactors based on the societal needs of this mechanized world. The challenges to be addressed in the construction of a working principle for a bioreactor are that it should be easily handled, cost-effective, and applicable to industrial as well as human demands.

Table 10.1 Production of biofuel using bioreactors.

Product	Reactor	Substrate	Reference
Biodiesel	Tubular photobioreactor	Microalgae	[38]
	Packed bed reactor	Canola oil	[39]
	Plug flow reactor	Palm kernel oil	[40]
Bioethanol	Membrane bioreactor	Glucose	[41]
	Membrane reactor	Corn, rice, agroforestry, and industrial waste	[42]
Biogas	Membrane reactor	Organic waste	[42]
	Membrane reactor	Citrus waste	[43]
	Anaerobic reactor	Food waste (rice)	[44]
Biofuel	Fluidized bed reactor	*Prosopis juliflora*	[45]
	Hybrid anaerobic reactor, anaerobic fixed film reactor, UASB reactor	Tapioca starch wastewater	[46]
Biohydrogen	Anaerobic reactor	Glucose and other nutrients	[47]
Biomethane	UASB	Vinasse	[48]
Biobutanol	Immobilized cell reactor	Starch, lignocellulose, and algal biomass	[49]
Syngas	Fluidized bed reactor	Mixture of biomass, wood, and olive husks	[50]

UASB, Upflow anaerobic sludge blanket.

10.14 Conclusion

We are living in the world with highly advanced technologies. To satisfy the needs of the growing population and make the available living environment fit for future generations, there is need for waste management and bioenergy production. In this chapter, the principles of several types of bioreactors for bioenergy production and waste abatement were discussed.

Acknowledgment

The authors thank SVCE Management and Head of the Biotechnology Department for providing the necessary facilities.

References

[1] J. Singh, N. kaushik, S. Biswas, Bioreactors – Technology and Design Analysis, Scitech J. 1 (2014) 28–36. ISSN: 2347-7318 ISSN 2348-2311.
[2] M.R. Spier, L.P. de Souza Vandenberghe, A.B. Pedroni Medeiros, C.R. Socol, Application of Different Types of Bioreactors in Bioprocess, Nova Science Publishers, 2017. ISBN: 978-1-62100-164-5.
[3] G.D. Najafpour, Bioreactor design, Biochem. Eng. Biotechnol. (2017) 142–169.
[4] E. Abraham, S. Gupta, S. Jung, E. McAfee, Bioreactor for scale up: process control, Mesenchyal Stromal Cell (2017) 139–178.
[5] B. McNeil, L.M. Harvey, Practical Fermentation Technology, Wiley Online Library, 2008. ISBN: 9780470725306.
[6] A. Kumar, Aerobic and anaerobic reactors, Biotreat. Ind. Effl. (2005) 19–38.
[7] G. Gottschalk, Bacterial Metabolism., Springer-Verlag, New York, 1979.
[8] O. Levenspiel, Chemical Reaction Engineering. third ed., John Wiley & Sons, New York, 1999.
[9] R.J.M.C.L. Silva, T.P.C. Souza, J.P. Silva, J.G.A. Pacheco, J.M.F. Silva, Ethanolic biodiesel production: a comparative study between a plug flow reactor and reactive distillation, Braz. J. Chem. Eng. 34 (2017) 811–819.
[10] S. Rana, L. Singh, Z.A. Wahid, H. Liu, A recent overview of palm oil mill effluent management via bioreactor configurations, Curr. Pollut. Rep. 3 (2017) 254–267.
[11] S.J. Arceivala, S.R. Asolekar, Wastewater Treatment for Pollution Control and Reuse, third ed., Tata McGraw Hill Publishing Company Limited, New Delhi, 2007.
[12] Sanimas, Informed Choice Catalogue, PPT-Presentation, BORDA and USAID, 2005.
[13] E. Tilley, L. Ulrich, C. Lüthi, Ph. Reymond, C. Zurbrügg, Compendium of Sanitation Systems and Technologies (second revised ed.), Swiss Federal Institute of Aquatic Science and Technology (Eawag), Duebendorf, Switzerland, 2014. ISBN: 978-3-906484-57-0.
[14] G.C. Zittelli, N. Biondi, L. Rodolfi, M.R. Tredici, Photobioreactors for mass production of microalgae, Handbook of Microalgal Culture: Applied Phycology and Biotechnology, Wiley, 2013, pp. 227–266.
[15] M.J. Barbosa, M. Janssen, N. Ham, J. Tramper, R.H. Wijffels, Microalgae cultivation in air-lift reactors: modeling biomass yield and growth rate as a function of mixing frequency, Biotechnol. Bioeng. 82 (2003) 170–179.
[16] C. Posten, Design principles of photo-bioreactors for cultivation of microalgae, Eng. life Sci. 9 (2009) 165–177.
[17] Y.P. Nagaraja, C. Biradar, K.S. Manasa, H.S. Venkatesh, Production of biofuel by using micro algae (*Botryococcus braunii*), Int. J. Curr. Microbiol. Appl. Sci. 3 (2014) 851–860. ISSN: 2319-7706.
[18] M.M.R. Shah, Y. Liang, J.J. Cheng, M. Daroch, Astaxanthin-producing green microalga haematococcus pluvialis: from single cell to high value commercial products, Front. Plant. Sci. 7 (2016) 531.
[19] W. Doyen, W. Mues, B. Molenberghs, B. Cobben, Spacer fabric supported flat-sheet membranes: a new era of flat-sheet membrane technology, Desalination 250 (2010) 1078–1082.
[20] A. Mahboubi, P. Ylitervo, W. Doyen, H.D. Weever, M.J. Taherzadeh, Reverse membrane bioreactor: introduction to a new technology for biofuel production, Biotechnol. Adv. 34 (2016) 954–975.
[21] S.H. Cheong, J.K. Park, B.S. Kim, H.N. Chang, Microencapsulation of yeast cells in the calcium alginate membrane, Biotechnol. Tech. 7 (1993) 879–884.
[22] H.B. Klinke, A.B. Thomsen, B.K. Ahring, Inhibition of ethanol-producing yeast and bacteria by degradation products produced during pre-treatment of biomass, Appl. Microbiol. Biotechnol. 66 (2004) 10–26.

[23] M.J. Taherzadeh, K. Karimi, Pretreatment of lignocellulosic wastes to improve ethanol and biogas production: a review, Int. J. Mol. Sci. 9 (2008) 1621–1651.
[24] P. Buzatu, V. Lavric, Submerged membrane bioreactors for wastewater treatment: multi-objective optimization, Chem. Eng. Trans. 25 (2011) 267–272.
[25] S.K. Khanal, B. Giri, S. Nitayavardhana, V. Gadhamshetty, Anaerobic bioreactors/digesters: design and development, in: Current Developments in Biotechnology and Bioengineering, Elsevier, 2017, pp. 261–279.
[26] B. Fernandes, A. Mota, A. Vicente, Fundamentals of bio-reaction engineering, Curr. Dev. Biotechnol. Bioeng. Found. Biotechnol. Bioeng. (2017) 153–180.
[27] A. Buasri, B. Ksapabutr, M. Panapoy, N. Chiyut, Biodiesel production from waste cooking palm oil using calcium oxide supported on activated carbon as catalyst in a fixed bed reactor, Korean J. Chem. Eng. 29 (2012) 1708–1712.
[28] R.C. Senan, T.E. Abraham, Bioremediation of textile azo dyes by aerobic bacterial consortium aerobic degradation of selected azo dyes by bacterial consortium, Biotechnol. Prog. 19 (2004) 647–651.
[29] N.K. Shammas, L.K. Wang, Oxidation ditch, in: L.K. Wang, N.C. Pereira, Y.T. Hung (Eds.), Biological Treatment Processes, Handbook of Environmental Engineering, Humana Press, Totowa, NJ, 2009.
[30] W.J. Koros, Y.H. Ma, T. Shimidzyu, Terminology for membranes and membrane processes, IUPAC recommendations 1996, J. Membr. Sci. 120 (1996) 149–159.
[31] O.T. Iorhemen, R.A. Hamza, J.H. Tay, Membrane bioreactor (MBR) technology for wastewater treatment and reclamation: membrane fouling, Membranes 6 (2016) 1–29.
[32] H. Mallia, S. Till, Membrane bioreactors: wastewater treatment applications to achieve high quality effluent, in: (Presented) 64th Annual Water Industry Engineers and Operators Conference, 2001, pp. 57–65.
[33] R. Bernal, A.V. Gottberg, B. Mack, Using membrane bioreactors for wastewater treatment in small communities, in: SUEZ-Water Technologies and Solutions Technical Paper.
[34] M.L. Shuler, F. Kargi, Bioprocess engineering basic concepts, ed., NR Amundson, 2002.
[35] H.M. Jena, G.K. Roy, B.C. Meikap, Comparative study of immobilized cell bioreactors for industrial wastewater treatment, 2005.
[36] L. Ma, Design of a Continuous-Flow, Immobilized-Cell Fermentor System for Production of Bioethanol, Texas Tech University, 2014 (Doctoral dissertation).
[37] C.A. Plitt, W.J. Harris, D.B. Harris, Immobilized cell bioreactor, US patent 5585266, 1996.
[38] N. Mulumba, I.H. Farag, Tubular photo bioreactor for microalgae biodiesel production, Int. J. Eng. Sci. Technol. 4 (2012) 703–709.
[39] J.P. Garcaa-Sandoval, D. Dochain, E. Aguilar-Garnica, Biodiesel production in a continuous packed bed reactor with recycle: a modeling approach for an esterification system, Renew. Energy 2017.
[40] R. Sawangkeaw, K. Bunyakiat, S. Ngamprasertsith, Continuous production of biodiesel with supercritical methanol: optimization of a scale-up plug flow reactor by response surface methodology, Fuel Process. Technol. 92 (2011) 2285–2292.
[41] A. Jain, S.P. Chaurasia, Bioethanol production in membrane bioreactor (MBR) system: a review, Int. J. Environ. Res. Dev. 4 (2014) 387–394.
[42] S.M. Badenes, F.C. Ferreira, J.M.S. Cabral, Membrane bioreactors for biofuel production, Sep. Purif. Technol. Biorefineries, John Wiley & Sons, Ltd., 2012, pp. 377–407.
[43] R. Wikandari, R. Millati, M.N. Cahyanto, M.J. Taherzadeh, Biogas production from citrus waste by membrane bioreactor, Membranes 4 (2014) (2014) 596–607.

[44] M.A.H. Baky, M.N.H. Khan,M.F. Kader, H.A. Chowdhury, Production of biogas by anaerobic digestion of food waste and process simulation, in: 8th International Conference on Energy Sustainability, 2016, ES2014–6756.

[45] R. Mythili, P. Subramanian, D. Uma, Biofuel production from *Prosopis juliflora* in fluidized bed reactor, Energy Sour. A: Recov. Utiliz. Environ. Eff. 39 (2017) 741–746.

[46] A. Kamaraj, N.O. Gopal, P. Venkatachalam, A. Sampathrajan, Biofuel production from tapioca starch industry wastewater using a hybrid anaerobic reactor, Energy Sustain. Dev. 3 (2006) 73–77.

[47] L. Singh, Z.A. Wahid, Methods for enhancing bio-hydrogen production from biological process: a review, J. Ind. Eng. Chem. 21 (2015) 70–80.

[48] V.G.D. Barros, R.M. Duda, R.A.D. Oliveira, Biomethane production from Vinasse in upflow anaerobic sludge blanket reactors inoculated with granular sludge, Braz. J. Microbiol. 47 (2016) 628–639.

[49] M.F. Ibrahim, S.W. Kim, S.A. Aziz, Advanced bioprocessing strategies for biobutanol production from biomass, Renew. Sustain. Energy Rev. 91 (2018) 1192–1204.

[50] G. Maschio, A. Lucchesi, G. Stoppato, Production of syngas from biomass, Bioresour. Technol. 48 (1994) 119–126.

CHAPTER 11

Anaerobic treatment of municipal solid waste landfill leachate

Shreeshivadasan Chelliapan[1], Nithiya Arumugam[1], Mohd. Fadhil Md. Din[2], Hesam Kamyab[1] and Shirin Shafiei Ebrahimi[3]
[1]Department of Engineering & Technology, Razak Faculty of Technology and Informatics, Universiti Teknologi Malaysia, Kuala Lumpur, Malaysia
[2]Centre for Environmental Sustainability and Water Security (IPASA), Research Institute for Sustainable Environment (RISE), School of Civil Engineering, Faculty of Engineering, Universiti Teknologi Malaysia (UTM), Skudai, Johor, Malaysia
[3]School of Education, Universiti Teknologi Malaysia, Johor Bahru, Malaysia

11.1 Introduction

As the world's population increases daily, which is projected to hit 8.5 billion by 2030, the urge toward urbanization becomes a necessity [1]. The rate of urban waste being produced is rising faster than the rate of urbanization, which is a sign of mankind's growing affluence. The generation of municipal solid waste (MSW) has acquired alarming dimensions globally as the population size increases. Approximately 1.3 billion metric tons of MSW were generated in 2016 globally and it is expected to rise to about 2.2 billion by 2025 [2]. Economic development, the rate of industrialization, the various public practices of different regions, urbanization, and changing consumption patterns are some factors influencing MSW generation.

The expeditious extension of industrial and commercial sectors and enormous consumption of packaged products as a consequence of rising living standards has boosted the generation of solid waste in recent years. The increase in waste volume poses a significant challenge in disposing of the waste in a controlled and sustainable way. A big problem that many governments are trying to solve today is the fast growth of solid and hazardous wastes [3]. Researchers are trying to find the most efficient and sustainable solution for waste management. Despite the effort to divert waste from landfills, landfilling is still the primary method of waste disposal in both developed and developing countries. In spite of enjoying various advantages, one of the disadvantages of this method is the

Bioreactors
DOI: https://doi.org/10.1016/B978-0-12-821264-6.00011-5

drawback of leachate production. If a suitable treatment is not used, landfill leachate can pollute water drastically and it can even penetrate soil and subsoil. As a result, the amount of hazardous components in leachate should be compatible with standard discharges before going to landfills so that water would not be polluted and other toxics do not enter water and soil [4].

11.2 Municipal solid waste management

The first step in waste management is quantifying and qualifying wastes. Having a system for the collection and analysis of basic information about wastes is vital for environmental sustainability. Waste sources, quantities, compositions, characteristics, seasonal variations, and future generation are important factors to be analyzed before choosing the most efficient waste management strategy [5]. In fact, in order to monitor purposes and thereby to make the best effective decisions, data collection and management are key factors. There are generally four main sources of MSW, namely residential, commercial, institutional, and municipal services and it does not include wastes from sources like municipal sludge, combustion ash, and industrial nonhazardous process wastes that are probably disposed in municipal waste landfills or incinerators [6].

MSW management (MSWM) is one of the most important aspects of waste management. MSWM is literally neglected in most developing countries. Extensive negative impacts on the environment and health and safety issues have arisen due to improper MSWM. Solid waste disposal consists of three main methods, namely composting, incineration, and sanitary landfilling, but landfilling is the oldest method of waste disposal. In developing countries, MSWM consists of primary and secondary collection and open dumping [7]. Vast adverse effects on the environment and public health are caused by open dump landfilling. Therefore the development of sustainable MSWM systems has become a necessity around the world.

11.3 Landfill

There are five steps in MSWM. Table 11.1 illustrates an overview of MSWM. The MSW treatment process starts with solid waste generation in households and commercial sites. In this phase, wastes are also sorted into recyclable and nonrecyclable items. This is followed by the collection

Table 11.1 An overview of municipal solid waste management.

Process	Purpose	Characteristics
Generation	Generation of solid waste at households or commercial and waste sorting	High amount of food, organic waste, and household goods
Collection	Collection of wastes from households or commercial	High amount of food and organic waste, household goods, fallen trees
Transport	Transportation of local wastes from household or commercial to regional disposal sites	High amount of food and organic waste, household goods, fallen trees
Recovery	Facilities, equipment, and techniques employed to recover reusable or recyclable materials	Recovery of various recyclable waste
Disposal	Systematic disposal of waste materials in locations such as landfills, waste-to-energy, and incineration facilities	Treatment system and leachate generation

and transportation of wastes from households to recovery centers by the appointed contractors. In the recovery stage, wastes are further sorted out and recyclable wastes are recovered from waste piles while nonrecyclable wastes are sent for disposal in landfills. As a final dumping area for MSW and the best way to manage the huge amount of waste is landfills [8]. Nevertheless, inappropriate landfill management leads to many problems such as social and ecological problems like water, groundwater, soil, air, and noise pollution. Other problems are increased floods and diseases.

Out of 183 landfill sites in Malaysia, 171 sites are classified as level 0 or level II, while the remaining 12 sites are categorized as either level III or level IV sanitary landfills [9]. Level II refers to semisanitary landfills since they do not have any control on leachate. The classes of landfill sites in Malaysia are illustrated in Table 11.2. Level 0 and level II landfills cause the most pollution due to their lack of control on leachate leakage to underground water, which in turn pollutes soil and surface water. Leachate enters the environment from the depths of the landfill through unsaturated soil and enters the groundwater and then the surface water via hydraulic connections. In total, leachate may taint the environment and affect public health.

Table 11.2 Classes of landfill sites in Malaysia.

Level	Landfill type
0	Open dumping
I	Controlled tipping
II	Sanitary landfill with bund and daily cover
III	Sanitary landfill with leachate recirculation system
IV	Sanitary landfill with leachate treatment facilities

11.4 Overview on landfill processing

Each landfill has different processing designs. The design process consists of several major components, namely landfill development, cell preparation and development, leachate management system, landfill gas management system, capping system, surface water management, and final closure of the landfills [10]. Each properly designed and processed stage will obey the mentioned engineering plan to comply with all rules and requirements. Solid wastes generated in households and commercial sites are sorted into recyclable and nonrecyclable wastes followed by the collection and transportation of wastes from households to transfer stations and catchment areas by the appointed contractors [11]. There the wastes are further sorted from piles in which recyclable wastes are recovered while nonrecyclable wastes are sent to landfills for disposal by lorries and trucks.

At landfill sites, lorries and trucks will enter a waste reception area, be weighed on a weighing bridge, go to dump the waste in landfill cells, and then they return to the weighing bridge to be weighed again [12]. When leaving, lorries and trucks will be cleaned/have their wheels washed off before exiting the waste reception area. The dumped wastes, which contains odour, is then spread and compacted with heavy machinery. These wastes are then covered daily with a soil cover. As a matter of time, these covered wastes discharge leachate, which is transferred to leachate collection ponds via pipelines for treatment before being discharged into water bodies. Further discussion of the leachate being generated is discussed in Section 11.5. The covered wastes also release biogas, which is then collected and transferred for processing.

Landfill leachate treatment is a major engineering challenge due to the high and variable concentrations of dissolved solids, dissolved and colloidal organics, heavy metals, and xenobiotic organics [13]. A solution to the huge costs of transferring leachate to a wastewater treatment plant is on-site leachate treatment. Such a treatment is effective because it can help in cheaply discharging leachate to a water or sanitary sewer with no hauling. Technologies for leachate treatment include biological, physical, or chemical treatments. Some others are evaporation and reverse osmosis (RO). Technologies are selected based on the leachate characteristics, site constraints, and discharge limitations such as direct or indirect discharge [13].

11.5 Landfill leachate

In landfills, leachate is a liquid that enters the landfill from external sources such as rainfall, surface drainage, underground springs and groundwater, and from the liquid produced from the putrefaction of waste [14]. The generation of leachate from landfills relies on many aspects including landfill surface conditions, the availability of water, refuse conditions, the prevailing climate, underlying soil conditions, and the physical structure of the waste. Some of the most relevant factors that influence leachate generation are annual rainfall, runoff, water permeation, evaporation, transpiration, waste composition, mean ambient temperature, waste density, initial moisture content, and depth of the landfill [15].

Rainfall is a major contributor to the generation of leachate. The sources of percolating water are primarily rainwater runoff, which permeates through the landfill cover, groundwater seepage, and, to a lesser degree, the initial refuse moisture content. The major factors affecting surface runoff are surface topography (size, shape, slope, and elevation), cover material, soil permeability, soil moisture, and vegetation. Surface topography controls the flow of water at the surface [16]. Meanwhile, in a nonvegetated landfill, the amount of infiltration of water through the refuse increases due to the lower uptake by vegetation. Refuse decomposition due to microbial activity may also contribute.

11.6 Leachate characterization

Leachate is a sticky, dark liquid full of organic matter. Typical leachate has 4000 to 20,000 mg/L of chemical oxygen demand (COD). This amount varies based on the leachate maturity [17]. The standard level of COD in wastewater, according to the Malaysian standard A and B, is from 50 to 100 mg/L [6]. Treating leachate is difficult due to various limitations. The most disturbing limitation is the high level of COD created by different chemicals, and although most are organic, there are also inorganic matters such as plastic and metals. Therefore leachate is a combination of different chemicals and the most successful leachate treatment is one that applies several treatment methods together. Another limitation is that leachate can contain heavy metals, which are toxic to the environment. Table 11.3 illustrates the categories of leachate in Malaysia.

Many factors affect the quality of such leachate (i.e., age, precipitation, seasonal weather variation, and waste type and composition) depending on the standard of living of the surrounding population and the structure of the landfill [18]. Leachate composition varies greatly based on the landfill age. A young landfill has more biodegradable organic matter and anaerobic fermentation is faster so more volatile acids (VA) will be produced. Therefore acid fermentation is greater with a higher level of moisture in solid waste.

Water cycling in a sanitary landfill in a premature phase of a landfill's lifetime is called the acidogenic phase and leads to the liberation of large quantities of free VA (as much as 95% of the organic substance) [19]. Over time, the methanogenic stage takes place for a landfill. Microorganisms in this phase that grow in the waste or the VA are transformed into biogas, methane (CH_4), and carbon dioxide (CO_2). The organic portion of the leachate turns into refractory or nonbiodegradable compounds such as humic elements. Parameters such as pH, COD, biochemical oxygen demand (BOD), the ratio of BOD/COD, heavy metals, ammonium nitrogen (NH_3-N), total Kjeldahl nitrogen (TKN), and suspended solids (SS) define the leachate characteristics. With few exceptions, leachate pH lies in the range of 5.8–8.5 because of the biological activity in landfills. The BOD/COD ratio is 0.50–0.10, which decreases as a landfill ages [20]. This happens because of the huge amounts of organic molecules in solid waste. Therefore old landfill leachate is defined by its

Table 11.3 Categories of leachate.

Leachate types	Young	Intermediate	Stabilized
Age of landfill (year)	<1	1–5	>5
pH	<6.5	6.5–7.5	>7.5
BOD/COD	0.5–1.0	0.1–0.5	<0.1
COD (mg/L)	>15000	3000–15000	<3000
NH_3-N (mg/L)	<400	NA	>400
TOC/COD	<0.3	0.3–0.5	>0.5
Kjeldahl nitrogen (mg/L)	100–2000	NA	NA
Heavy metal (mg/L)	>2	<2	<2

BOD, Biochemical oxygen demand; *COD*, chemical oxygen demand; *TOC*, total organic carbon.

relatively low BOD/COD ratio and high NH_3-N. Moreover, most of the TKN is ammonia ranging from 100 to 2000 mg/L of N [20]. Overall, the relationship between the landfill age and the organic matter composition may provide useful criteria to choose a suitable treatment.

11.7 Treatment of landfill leachate

Generally, leachate treatment can be classified into two major groups, namely conventional treatments and new treatments (nonconventional). Conventional treatment comprises of leachate transfer, leachate recycling, biological treatment (aerobic or anaerobic), physical–chemical treatment involving flotation, coagulation–flocculation, chemical precipitation, adsorption, chemical oxidation, and sir stripping, whereas advanced filtration such as ultrafiltration (UF), microfiltration (MF), nanofiltration (NF), and RO is the most important part of the membrane process of new treatments [21]. MF is an effective method to eliminate colloids and SS and is more like a pretreatment to other membrane processes such as RO, NF, or UF, or to be used in association with chemical treatments [22]. UF effectively removes macromolecules, but it depends on the material type of the membrane. It fractionates organic matter and helps in evaluating the predominant molecular mass in organic pollutants in leachate. UF may be effective as a pretreatment process for RO and can be used in the biological posttreatment of leachate. NF is a flexible approach to control organic, inorganic, and microbial pollutants. However, for successful membrane technology application impressive control over membrane fouling is needed. In leachates nanofiltration, many constituents play a role in membrane fouling.

Current common treatment methods that have been used for leachate treatment are either single or combined methods and are classified into three main categories, namely biological processes (aerobic or anaerobic), chemical and physical processes, and a combination of physical–chemical and biological processes [23]. Independently, each treatment process has its own advantages as well as disadvantages. A strong knowledge of the chemistry and processes involved is necessary in order to choose the appropriate treatment methods prior to the characteristics of the wastewater as well as the treatment objectives.

Aerobic and anaerobic treatment consists of two types of growth systems, namely suspended and attached. Regarding chemical treatment, flotation involves air oxidation to breakdown pollutants whereas coagulation involves the use of chemical coagulants to sink pollutants. Adsorption as a subset of physical treatment and involves the use of an adsorbent to adsorb pollutants while membrane filtration involves the passage of wastewater through a filter membrane [24]. Electrochemical treatment is one of the most common methods in leachate treatment. It efficiently reduces two of the most concerning pollutants in leachate, namely organic matter and NH_3-N. However, its high-energy consumption and relatively expensive operation confine its usage. The coagulation–flocculation method is also favored by the industry due to its efficiency in removing nonbiodegradable matter from leachate; however, the formation of high-volume concentrated sludge restricts this method [25]. Aside from these methods, there are other emerging removal processes such as ion exchange, chemical precipitation, UF, NF, and RO, which are effective, but have relatively high operation and frequent maintenance costs. The need for careful management and disposal of chemical sludge and residue are among the drawbacks for several of these processes. Current trends in leachate treatment are the combination of different treatment methods to boost the treatment performances by means of overcoming the drawbacks and limitations of a single conventional treatment system [25]. In short, the physical–chemical process is commonly used when biological treatment is hindered due to the presence of excessive refractory compounds in leachate. Usually, these physical–chemical processes are carried out as pretreatments or as the final stage of leachate treatment.

11.8 Anaerobic treatment of leachate

Currently the main leachate treatment method used is the aerobic process. This treatment mechanism involves oxidation by air particles, sunlight, and aerobic microorganisms, which degrade the pollutants in leachate. In addition, leachate might have to be aerated in order to optimize the action of aerobic microorganisms [26]. However, this method is still insufficient to remove some toxic pollutants in leachate. On the other hand, anaerobic treatment is a biological process involving the decomposition of organic matters, which not only removes most pollutants, but also generates valuable byproducts, namely biogas in the form of CH_4. The most important reason for the popularity of anaerobic wastewater treatment is its economic advantage. Other advantages include the negation of the need for aeration, thereby lessening energy consumption, low excess sludge and therefore lower sludge management costs, the high energy content of biogas production can act as fuel, the application of high organic loading thereby saving space, and its low nutrients requirement [27].

Anaerobic digestion is a process performed by microorganisms that are able to live in oxygen-deprived conditions. The disintegration of organic substances happens in phases of hydrolysis, acidogenesis, acetogenesis, and methanogenesis as shown in Fig. 11.1. In the first stage, hydrolysis, anaerobic bacteria transform the organic substances into basic organic substances (monomers or short polymers). Fats, carbohydrates, and proteins are, respectively, converted into fatty acids, monosaccharide, and amino acids. Eq. (11.1) shows how a hydrolysis reaction converts organic waste into glucose, a simple sugar.

$$C_6H_{10}O_4 + 2H_2O \rightarrow C_6H_{12}O_6 + 2H_2 \quad (11.1)$$

In the second phase, acidogenic bacteria convert the products from the hydrolytic reaction into alcohols, ketones, short chain VA, CO_2, and

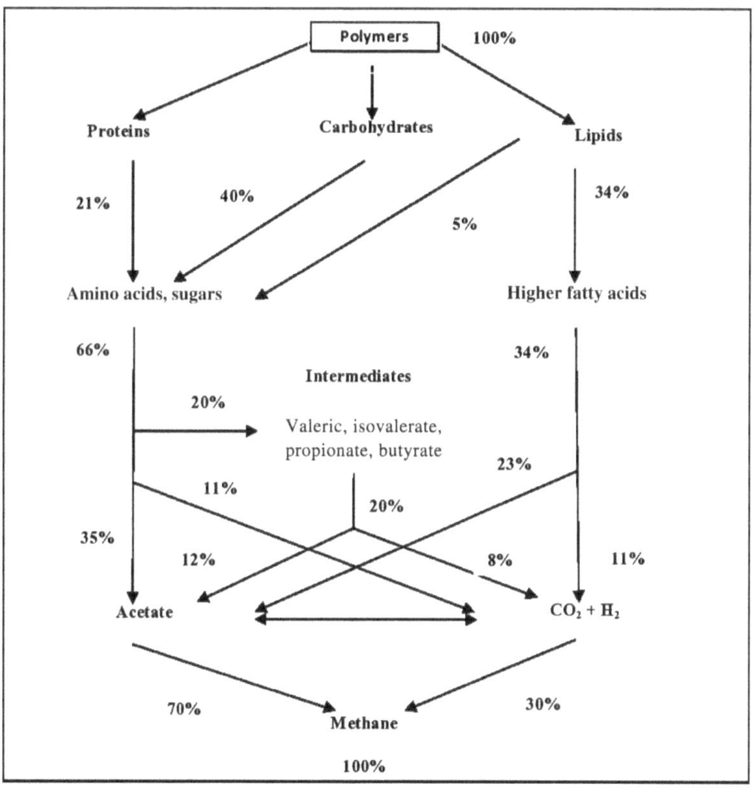

Figure 11.1 Degradation steps of anaerobic digestion process.

hydrogen (H_2). The major products of the acidogenesis phase are butyric acid ($CH_3CH_2CH_2COOH$), propionic acid (CH_3CH_2COOH), acetic acid (CH_3COOH), lactic acid ($C_3H_6O_3$), formic acid ($HCOOH$), methanol (CH_3OH), and ethanol (C_2H_5OH). From these products, acetic acid, carbon dioxide and hydrogen will omit the acetogenesis phase. Later, these products can be utilised by the methanogenic bacteria, as shown in Fig. 11.1. Eqs. (11.2)–(11.4) show three common acidogenesis reactions in which glucose is changed respectively into ethanol, propionate, and acetic acid.

$$C_6H_{12}O_6 \leftrightarrow 2CH_3CH_2OH + 2CO_2 \tag{11.2}$$

$$C_6H_{12}O_6 + 2H_2 \leftrightarrow 2CH_3CH_2COOH + 2H_2O \tag{11.3}$$

$$C_6H_{12}O_6 \rightarrow 3CH_3COOH \tag{11.4}$$

Acetogenesis is the stage where all acidogenesis products (alcohols, butyric acid, and propionic acid) are transformed into H_2, CO_2, and acetic acid by acetogenic bacteria. Eq. (11.5) shows the conversion of propionate to acetate. In addition, glucose and ethanol are changed to acetate in the third phase of anaerobic fermentation as shown in Eqs. (11.6) and (11.7).

$$CH_3CH_2COO^- + 3H_2O \leftrightarrow CH_3COO^- + H^+ + HCO_3^- + 3H_2 \tag{11.5}$$

$$C_6H_{12}O_6 + 2H_2O \leftrightarrow 2CH_3COOH + 2CO_2 + 4H_2 \tag{11.6}$$

$$CH_3CH_2OH + 2H_2O \leftrightarrow CH_3COO^- + 2H_2 + H^+ \tag{11.7}$$

Methanogenesis is the last phase of the anaerobic process where microorganisms change H_2 and acetic acid formed by acid formers into CO_2 and CH_4 gas. The bacteria that perform this conversion are anaerobes called methanogens. Waste is considered completely reduced in anaerobic treatment when CH_4 gas and CO_2 are produced.

$$CO_2 + 4H_2 \rightarrow CH_4 + 2H_2O \tag{11.8}$$

$$2C_2H_5OH + CO_2 \rightarrow CH_4 + 2CH_3COOH \tag{11.9}$$

$$CH_3COOH \rightarrow CH_4 + CO_2 \tag{11.10}$$

There are a few conceivable outcomes to plan for when designing anaerobic absorption frameworks. A framework may be straightforward like a solitary static tube-shaped digester or it may be complex and expensive. The framework that is introduced and the way it is overseen will influence plant proficiency. Simple anaerobic digesters need observation of their performances, and it is less proficient; on the other hand, an elaborate digestres will show the problems associated with the treatment system and require a regular check, and they are more efficient [27]. Despite the fact that microbial procedures are the same for all sorts of anaerobic absorption forms, each and every plant is one of a kind and ought to be planned for based on its unique information parameters and monetary plausibility. Some variables include plant limitations, waste type, the accessible zone, the climate, the demographics of the area, and the area of the plant. There are two elements that affect biogas creation, namely the characteristics of edible matter within the waste and the exchange rate of the waste portion to the digester [27]. Based on the amount of water in the digester and the method for bolstering the digester, anaerobic digestion procedures are categorized as either wet or dry ageing or ceaseless or intermittent maturation. Sections 11.8.1–11.8.5 provide a review of the treatment of leachate using anaerobic digestion.

11.8.1 Fluidized bed reactor

A fluidized anaerobic bed reactor operates by a suspended bed of particles such as polymer, plastic or sand. These elements facilitate anaerobic microorganisms on the surface of the channel. The large surface area of support particles and the high degree of mixing that result from the high vertical flows enable a high concentration of biomass to develop and efficient substrate uptake kinetics, respectively. Another advantage of the anaerobic fluidized bed system is that, due to biomass retention, the growth rate of the microorganisms and the hydraulic retention time (HRT) are uncoupled. This system permits operation with the highest organic loading rates (OLRs) among all anaerobic digesters treating wastewaters. The downside of this reactor is that the particles settle to the base when the particles aggregate too much with the biomass and become heavy. Aside from this, the loss of anaerobic microorganisms by means of effluent is a likewise noteworthy disadvantage. Gulsen and Turan [28] investigated the treatability of leachate in an anaerobic fluidized bed reactor (AFBR). The reactor was operated by gradually increasing the OLR from 2.5 to 37 kgCOD/m^3/day at a HRT of 1 day. The COD removal increased from 80% to 90% by increasing the OLR. A biogas production yield (Y_{gas}) was obtained of 0.50–0.52 L of biogas per gram of COD removed with a CH_4 content of 75%. They also used a combined AFBR and zeolite fixed bed system in sanitary landfill leachate treatment [29]. The combined anaerobic and adsorption system was effective with high COD and ammonia removal from landfill leachate.

11.8.2 Anaerobic baffled reactor

This reactor consists of a several UASB reactors. The wastewater flows over and under all baffle functioning to separate compartments, thereby neutralizing the effect of washout and holding the solids inside the reactor. In this method, the reactor does not allow any horizontal movement of the biomass and, therefore, a large amount of biomass remains in compartments. As a result, the bacteria in the reactor increases and produces more gas. The most significant advantage of the anaerobic baffled reactor (ABR) is its ability to separate acidogenesis and methanogenesis longitudinally down the reactor, allowing the different bacterial groups to develop under most favorable conditions. The application of an ABR was studied for simultaneous carbon (C) and N_2 removal from leachate [30]. The nitrified sewage was recycled in three compartments, resulting in enhanced COD removal (>95.6%) and improved N_2 removal from 12.7% to 67.4% while raising the recirculation ratio from 0.25 to 2. Burbano-Figueroa et al. [31] investigated the effect of OLR and sulfate loading rate (SLR) on landfill leachate treatment by ABR. The leachate of the landfill had a concentration of organic matter between 3966 and 5090 mgCOD/L and the amount of sulfate was not detectable. Reactors started with iron–sulfate at an SLR of 0.05 g SO_4^{2-}/L/day. The ABRs were operated at OLRs from 0.30 to 6.84 gCOD/L/day by changing the influent flows in different volumes. SO_4^{2-} was added to the influent at an SLR between 0.06 and 0.13 gSO_4^{2-}/L/day. The highest value of COD removal (66%) was at an OLR of 3.58 gCOD/L/day and SLR of 0.09 gSO_4^{2-}/L/day with a COD/SO_4^{2-} ratio of 40.

11.8.3 Anaerobic membrane bioreactor

Membrane-coupled anaerobic bioreactors have been applied as an alternative to the conventional anaerobic digestion processes because they retain all microorganisms in the reactor. Anaerobic membrane bioreactors work using an external membrane filter placed either before or after the anaerobic reactor that allows the solid washout from wastewater to be trapped and returned back to the reactor sludge. The advantages offered by an anaerobic membrane bioreactor (AMBr) over the conventional process include the high quality of the effluent and the separation of the solid retention time from the HRT. The major drawbacks of this type of reactor include organic fouling in the membrane due to the accumulation of bacteria and the loss of anaerobic microorganisms in the recirculated effluent due to the high pump flow rate. Trzcinski and Stuckey [32] explored a leachate treatment from the organic fraction of municipal solid waste (OFMSW) in a submerged anaerobic membrane bioreactor (SAMBR). The reactor was shown to be feasible at 5 days HRT, 10 L/min biogas sparging rate, and membrane fluxes in the range of 3–7 $L/m^2/h$. Under these conditions, a more than 90% COD removal was achieved. In another study by Trzcinski and Stuckey [33], leachate treatment from the OFMSW was investigated using a SAMBR followed by an aerobic membrane bioreactor. This reactor system achieved a 96% COD removal at a low HRT of 0.4 days during the anaerobic treatment of leachate. Zayen et al. [34] investigated landfill leachate treatment in an AMBr. The OLR in the AMBr was raised gently from 1 gCOD/L/day to about 6.27 gCOD/L/day. Under the same conditions, the COD reduction was 90% while the biogas removal was 0.46 L biogas gCOD. Therefore the treatment efficiency was high.

11.8.4 Anaerobic filters

An anaerobic filter is made of packed material produced using any nondegradable, shaped material or polymer that has a high surface area to volume ratio. The packed materials permit anaerobic microorganisms to join and develop like a biofilm, producing an anaerobic channel mat. The material can be made of different matter such as plastics, sand, stone, granular activated carbon, granite, quartz, or reticulated foam polymers. They can be packed in two configurations, namely loose or modular. This reactor can be utilized as part of a treatment for low- or high-quality water. This reactor has a surface that helps a biofilm to be accumulated. The advantages lead to save time on startup periods because of the more preserved inoculum. The more common inoculum is granules, but it is not necessary since ordinary municipal waste anaerobic sludge can be used if a short startup time is not important. There are some limitations with this method, which are mainly physical problems related to the deterioration of the structure of the bed by the piling up of nonbiodegradable solids. This is the reason for the channeling and shortcutting of the flow. Therefore anaerobic filters cannot be perfect for wastewaters with a large solid content. Moreover, packing materials are expensive. In order to solve some of these issues, packed bed reactors can be used in a down flow that leads the undegraded solids off the system.

Wang and Banks [35] explored the feasibility of treating an alkaline sulfate-rich leachate arising from the codisposal of MSW with cement kiln dust by means of an anaerobic filter. The results indicated that the potential for using this approach as a pretreatment could significantly reduce the COD load in a second treatment process, but problems associated with the implementation of this technology at a larger scale have been identified. Anaerobic degradation of landfill leachate and the reduction of sulfate as a function of COD/(SO_4^{2-}) ratio in an up-flow anaerobic fixed-bed reactor was investigated [36]. A most effective performance with a 91% sulfate reduction and an 87% COD removal efficiency with a 1.17 ratio of COD/(SO_4^{2-}) was reported.

11.8.5 Up-flow anaerobic sludge blanket in leachate treatment

The up-flow anaerobic sludge blanket, commonly known as a UASB reactor, works on the principle that an anaerobic sludge blanket at the bottom of the reactor acts like a filter and medium for anaerobic microorganisms to grow and utilize the organic matter in the wastewater supplied into it. Wastewater is introduced into the reactor via an inlet at the bottom of the reactor in an up-flow manner. When the wastewater passes through the sludge blanket it is filtered and treated by the microorganisms in the sludge blanket. This is the most widely used type of reactor around the globe due to its convenience as it allows for the separation of liquid, solid, and gas phases in the upper part of the reactor. Singh and Mittal [37] investigated the applicability of UASB to process leachate from a municipal landfill with an OLR of 3.00 kgCOD/m^3/day and an HRT of 12 h. The removal efficiency of soluble COD for fresh leachate was a 91%−67% reduction and for old leachate a reduction from 90% to 35%. They concluded that UASB leachate treatment leads to more toxicity of the leachate.

Abood et al. [38] studied the removal of biological nutrients from mature landfill leachate with a high N_2 load by an internal circulation up-flow sludge blanket (ICUSB) reactor. The reactor consists of a number of anaerobic–anoxic–aerobic (A2/O) bioreactors, made on an expended granular sludge blanket, granular sequencing batch reactor (SBR), and intermittent cycle extended aeration system. The best conditions for the most biological nutrients removal by the ICUSB reactor were examined in order to evaluate the influence of the parameters on its performance. The results showed that under the condition of a 12 h HRT including 4 h of pumping air into the reactor, the average removal efficiency for COD was 96.49% and for NH_3–N it was 99.39%. Liu et al. [39] treated landfill leachate by UASB plus anoxic/aerobic reactor (A/O) for greater biological N_2 removal. The concentration of biodegradable COD in the influent of the A/O reactor improved with increases in salinity, causing a drop in dissolved oxygen and thereby stimulating N_2O emissions at the aerobic zone. Lu et al. [40] studied the treatability of high-strength landfill leachate by UASB reactor, moving bed biofilm reactor, and ANITA Mox process. The COD removal efficiency of the UASB reactor was 93% from 45,000 to 3,000 mg/L at a rate of 10 kg/m^3/day. Bohdziewicz and Kwarciak [41] explored the treatment of landfill leachate using a UASB reactor and an RO process. At an HRT of 3 days and an OLR of 1.3 kgCOD/m^3/day, a COD removal efficiency of 76% was achieved in the UASB reactor. Due to the poor quality of the UASB reactor effluent, the wastewater was put into an RO posttreatment process and up to 95.4% COD removal was observed in the combined system.

Ye et al. [42] investigated the treatment of a fresh leachate with high-strength organics and calcium from an MSW incineration plant using a UASB reactor. When the reactor was fed with raw leachate (COD as high as 70,390 and 75,480 mg/L) at an OLR of 12.5 kgCOD/m^3/day, up to 82.4% of COD was removed. A two-stage UASB and SBR treatment was used by Sun et al. [43] to treat landfill leachate for COD and N_2 removal under low temperatures. The results showed that UASB enhanced COD removal and SBR improved the nutrient removal efficiency. A 96.7% COD removal efficiency was observed for the system. A combined process consisting of a short-cut nitrification reactor and an anaerobic ammonium oxidation UASB reactor was developed by Liu et al. [44] to treat diluted effluent from a UASB reactor treating high ammonium content municipal landfill leachate. The ammonium and nitrite removal efficiencies reached over 93% and 95%, respectively, after 70 days continuous operation, at a maximum total N_2-loading rate of 0.91 kg/m^3/day.

11.9 Conclusion

There are a number of studies using anaerobic reactors for landfill leachate treatment. Each anaerobic reactor is unique by itself and has different treatment performances according to the nature of the landfill leachate, reactor configuration, seed sludge, and start-up and operational conditions as well as the standard requirements of the effluents. In general, biological treatment is still being used widely for the treatment of landfill leachate, however, there are not many commercially available leachate treatment systems using the anaerobic system. Most of anaerobic treatment of landfill leachate was studied at the laboratory scale. The treatment of landfill leachate using anaerobic reactors will lead to an effective treatment solution for landfill leachate as compared to available aerobic treatments. However, there are some drawbacks such as the long HRT, low contaminant removal, and sensitivity to temperature. Another important aspect that requires consideration is the investment, operation, and management of the construction of an anaerobic leachate treatment plant. Moreover, anaerobic treatment alone may not able to achieve emission standards. Since leachate concentration is high and contains complex refractory compounds, it is vital to combine the aerobic biological treatment process with other processes. Due to social progress and technology development, leachate biotechnology will mature and develop great aspects for application and development. Since compared to sewage, the quality of leachate is different and unstable, merely biological treatment technology is not effective and should be enhanced by pre- or postprocessing technologies.

Acknowledgments

This research work was funded by Universiti Teknologi Malaysia under the Research University Grant, Vote Number: Q. K130000.2510.13H11. Hesam Kamyab is a researcher of Universiti Teknologi Malaysia (UTM) under the Post-Doctoral Fellowship Scheme (PDRU Grant) for the project: "Alternative Innovation of Enhancement Technologies for Algal Oil Extraction" (Vote No. Q. J130000.21A2.03E31) and Enhancing the Lipid Growth in Algae Cultivation for Biodiesel Production.

References

[1] Y. Peng, Perspectives on technology for landfill leachate treatment, Arab. J. Chem 10 (2017) s2567–s2574.

[2] G.K. Singh, K. Gupta, S. Chaudhary, Solid waste management: its sources, collection, transportation and recycling, Int. J. Environ. Sci. Dev 5 (2014) 347–351.

[3] S. Renou, J.G. Givaudan, S. Poulain, F. Dirassouyan, P. Moulin, Landfill leachate treatment: review and opportunity, J. Hazard. Mater 150 (2008) 468–493.
[4] P. Mandal, B.K. Dubey, A.K. Gupta, Review on landfill leachate treatment by electrochemical oxidation: drawbacks, challenges and future scope, Waste Manage 69 (2017) 250–273.
[5] S.H. Fauziah, P. Agamuthu, Trends in sustainable landfilling in Malaysia, a developing country, Waste Manag. Res 30 (2012) 656–663.
[6] A. Fazeli, F. Bakhtvar, L. Jahanshaloo, N.A. Che Sidik, A.E. Bayat, Malaysia's stand on municipal solid waste conversion to energy: a review, Renew. Sustain. Energy Rev 58 (2016) 1007–1016.
[7] K. Kawai, T. Tasaki, Revisiting estimates of municipal solid waste generation per capita and their reliability, J. Mater. Cycle Waste Manag 18 (2016) 1–13.
[8] M.A. Kamaruddin, M.S. Yusoff, H.A. Aziz, Y.T. Hung, Sustainable treatment of landfill leachate, Appl. Water Sci 5 (2014) 113–126.
[9] M.D.M. Samsudina, M.M. Dona, Municipal solid waste management in Malaysia: current practices, challenges and prospect, J. Teknologi 62 (2013) 95–101.
[10] A. Arvin, M. Peyravi, M. Jahanshahi, H. Salmani, Hydrodynamic evaluation of an anaerobic baffled reactor for landfill leachate treatment, Desalin. Water Treat 57 (2016) 19596–19608.
[11] M.J.K. Bashir, H.A. Aziz, M.S. Yusoff, New sequential treatment for mature landfill leachate by cationic/anionic and anionic/cationic processes: optimization and comparative study, J. Hazard. Mater 186 (2011) 92–102.
[12] Y. Arij, S. Fatihah, A.R. Rakmi, Performance of pilot scale anaerobic biofilm digester (ABD) for the treatment of leachate from a municipal waste transfer station, Bioresour. Technol 260 (2018) 213–220.
[13] M. El-Fadel, F. Sleem, J. Hashisho, P.E. Saikaly, S. Ghanimeh, Impact of SRT on the performance of MBRs for the treatment of high strength landfill leachate, Waste Manag 73 (2018) 165–180.
[14] K.Y. Foo, L.K. Lee, B.H. Hameed, Batch adsorption of semi-aerobic landfill leachate by granular activated carbon prepared by microwave heating, Chem. Eng. J 222 (2013) 259–264.
[15] J.L. Gao, V. Oloibiri, M. Chys, S. De Wandel, B. Decostere, W. Audenaert, et al., Integration of autotrophic nitrogen removal, ozonation and activated carbon filtration for treatment of landfill leachate, Chem. Eng. J 275 (2015) 281–287.
[16] P. Kijjanapanich, A.T. Do, A.P. Annachhatre, G. Esposito, D.H. Yeh, P.N.L. Lens, Biological sulfate removal from construction and demolition debris leachate: effect of bioreactor configuration, J. Hazard. Mater 269 (2014) 38–44.
[17] T.A. Kurniawan, W.H. Lo, G.Y.S. Chan, Physico-chemical treatments for removal of recalcitrant contaminants from landfill leachate, J. Hazard. Mater 129 (2006) 80–100.
[18] S. Elyasi, T. Amani, W. Dastyar, A comprehensive evaluation of parameters affecting treating high-strength compost leachate in anaerobic baffled reactor followed by electrocoagulation–flotation process, Water Air Soil Pollut 226 (2015) 116.
[19] S. Fudala-Ksiazek, M. Sobaszek, A. Luczkiewicz, A. Pieczynska, A. Ofiarska, A. Fiszka-Borzyszkowska, et al., Influence of the boron doping level on the electrochemical oxidation of raw landfill leachates: advanced pre-treatment prior to the biological nitrogen removal, Chem. Eng. J 334 (2018) 1074–1084.
[20] M. Kawai, M. Kishi, M.R. Hamersley, N. Nagao, J. Hermana, T. Toda, Biodegradability and methane productivity during anaerobic co-digestion of refractory leachate, Int. Biodeterior. Biodegrad 72 (2012) 46–51.
[21] J. Luo, G. Qian, J. Liu, Z.P. Xu, Anaerobic methanogenesis of fresh leachate from municipal solid waste: a brief review on current progress, Renew. Sustain. Energy Rev 49 (2015) 21–28.

[22] H. Sun, Y. Peng, X. Shi, Advanced treatment of landfill leachate using anaerobic–aerobic process: organic removal by simultaneous denitritation and methanogenesis and nitrogen removal via nitrite, Bioresour. Technol 177 (2015) 337–345.
[23] Q.Q. Zhang, B.H. Tian, X. Zhang, A. Ghulam, C.R. Fang, R. He, Investigation on characteristics of leachate and concentrated leachate in three landfill leachate treatment plants, Waste Manag 33 (2013) 2277–2286.
[24] W. Zhang, L. Zhang, A. Li, Anaerobic co-digestion of food waste with MSW incineration plant fresh leachate: process performance and synergistic effects, Chem. Eng. J 259 (2015) 795–805.
[25] W. Dastyar, T. Amani, S. Elyasi, Investigation of affecting parameters on treating high-strength compost leachate in a hybrid EGSB and fixed-bed reactor followed by electrocoagulation–flotation process, Process. Saf. Environ 95 (2015) 1–11.
[26] R. Abdallah, H. Djelal, A. Amrane, W. Sayed, F. Fourcade, T. Labasque, et al., Dark fermentative hydrogen production by anaerobic sludge growing on glucose and ammonium resulting from nitrate electro reduction, Int. J. Hydrogen Energy 41 (2016) 5445–5455.
[27] T. Abbasi, S.M. Tauseef, S.A. Abbasi, Anaerobic digestion for global warming control and energy generation – an overview, Renew. Sustain Energy Rev 16 (2012) 3228–3242.
[28] H. Gulsen, M. Turan, Anaerobic treatability of sanitary landfill leachate in a fluidized bed reactor, Turk. J. Eng. Env. Sci 28 (2004) 297–305.
[29] M. Turan, H. Gulsen, M.S. Celik, Treatment of landfill leachate by a combined anaerobic fluidized bed and zeolite column system, J. Environ. Eng 131 (5) (2005) 815.
[30] Z. Yin, L. Xie, X. Cui, Q. Zhou, Effective carbon and nitrogen removal with reduced sulfur oxidation in an anaerobic baffled reactor for fresh leachate treatment, J. Biosci. Bioeng 123 (2017) 84–90.
[31] O. Burbano-Figueroa, M. Jaramillo, M. Moreno-Moran, P. Fernández-Izquierdo, Effect of sulfate loading rate and organic loading rate on anaerobic baffled reactors used for treatment of sanitary landfill leachates, Braz. J. Chem. Eng 32 (2015) 385–395.
[32] A.P. Trzcinski, D.C. Stuckey, Effect of sparging rate on permeate quality in a submerged anaerobic membrane bioreactor (SAMBR) treating leachate from the organic fraction of municipal solid waste (OFMSW), J. Environ. Manag 168 (2016) 67–73.
[33] A.P. Trzcinski, D.C. Stuckey, Inorganic fouling of an anaerobic membrane bioreactor treating leachate from the organic fraction of municipal solid waste (OFMSW) and a polishing aerobic membrane bioreactor, Bioresour. Technol 204 (2016) 17–25.
[34] A. Zayen, S. Mnif, F. Aloui, F. Fki, S. Loukil, M. Bouaziz, et al., Anaerobic membrane bioreactor for the treatment of leachates from Jebel Chakir discharge in Tunisia, J. Hazard. Mater 177 (2010) 918–923.
[35] Z. Wang, C.J. Banks, Treatment of a high-strength sulphate-rich alkaline leachate using an anaerobic filter, Waste Manag 27 (2007) 359–366.
[36] O.B.D. Thabet, H. Bouallagui, J. Cayol, B. Ollivier, M.L. Fardeaub, M. Hamdi, Anaerobic degradation of landfill leachate using an upflow anaerobic fixed-bed reactor with microbial sulfate reduction, J. Hazard. Mater 167 (2009) 1133–1140.
[37] V. Singh, A.K. Mittal, Toxicity and treatability of leachate: application of UASB reactor for leachate treatment from OKHLA landfill, New Delhi, Water Sci. Technol 65 (2012) 1887–1894.
[38] A.R. Abood, J. Bao, Z.N. Abudi, Biological nutrient removal by internal circulation upflow sludge blanket reactor after landfill leachate pre-treatment, J. Environ. Sci 25 (2013) 2130–2137.
[39] M. Liu, Q. Yang, Y. Peng, T. Liu, H. Xiao, S. Wang, Treatment performance and N_2O emission in the UASB-A/O shortcut biological nitrogen removal system for landfill leachate at different salinity, J. Ind. Eng. Chem 32 (2015) 63–71.

[40] T. Lu, B. George, H. Zhao, W. Liu, A case study of coupling upflow anaerobic sludge blanket (UASB) and ANITA Mox process to treat high-strength landfill leachate, Water Sci. Technol 73 (2016) 662–668.
[41] J. Bohdziewicz, A. Kwarciak, The application of hybrid system UASB reactor–RO in landfill leachate treatment, Desalination 222 (2008) 128–134.
[42] J. Ye, Y. Mu, X. Cheng, D. Sun, Treatment of fresh leachate with high-strength organics and calcium from municipal solid waste incineration plant using UASB reactor, Bioresour. Technol 102 (2011) 5498–5503.
[43] H. Sun, Q. Yang, Y. Peng, X. Shi, S. Wang, S. Zhang, Advanced landfill leachate treatment using a two-stage UASB–SBR system at low temperature, J. Environ. Sci 22 (2010) 481–485.
[44] J. Liu, J. Zuo, Y. Yang, S. Zhu, S. Kuang, K. Wang, An autotrophic nitrogen removal process: short-cut nitrification combined with ANAMMOX for treating diluted effluent from an UASB reactor fed by landfill leachate, J. Environ. Sci 22 (2010) 777–783.

CHAPTER 12

Advancements in hydrothermal liquefaction reactors: overview and prospects

S.N. Sahu[1], N.K. Sahoo[1], S.N. Naik[1] and D.M. Mahapatra[2]
[1]Centre for Rural Development and Technology, Indian Institute of Technology Delhi, New Delhi, India
[2]Biological & Ecological Engineering, Oregon State University, Corvallis, OR, United States

12.1 Introduction

In the past few decades, alternative, renewable sources of fuels have gained attention to tackle the day-by-day increases in greenhouse gas (GHG) emissions and the rapid decline of conventional fossil fuels [1,2]. In this context, biomass-based biocrude has received much interest as a renewable and sustainable alternative to crude petroleum in meeting the demands of petrochemicals and fuels [3,4]. Biomass from various sources has wide potential as renewable and alternative sources of energy, and algae feedstocks are among the most promising and attractive sources. Biomass feedstocks can be further processed using different types of conversion processes such as thermochemical, chemical, and biochemical methods for the conversion of biomass to energy. However, hydrothermal liquefaction (HTL) has special advantages that make the process an attractive option for efficient conversion while deriving valuable by-products. In addition, a wide range of gaseous, liquid, and solid fuel options are possible through the conversion of biomass using HTL. Furthermore, HTL has great potential due to its flexibility in the conversion of different categories of biomass feedstocks such as aquatic wastes, lignocellulosic biomass, herbaceous matter, food-processing wastes, fungus, cyanobacteria, and biochemical wastes from macro/microorganisms [5] Thermochemical conversion has four different process options for the conversion of biomass, namely direct combustion, pyrolysis, gasification, and liquefaction. Biomass can be processed using conventional thermochemical techniques such as pyrolysis and gasification, which usually require a dry feedstock, in contrast to HTL that uses high moisture content biomass, and therefore

Bioreactors
DOI: https://doi.org/10.1016/B978-0-12-821264-6.00012-7

negates the cost of drying processes. HTL conversion of biomass is suitable for the production of biooil from biomass with varying moisture contents. HTL is an environment-friendly process of biomass conversion that requires subcritical water conditions using hot water for extraction at temperatures ranging between 100°C and 374°C under high pressure to maintain water in the liquid state. During the process of conversion, the properties of water change and the hydrogen-bonding network between the molecules is weakened by the thermal motion of the molecules, resulting in a higher concentration of hydroxide and hydrogen ions in the water [6–9]. Consequently, this also holds the possibility of solvent viscosity, surface tension, and dielectric constant being lowered. Furthermore, the polarity of water below its critical point is almost equal to that of ethanol at ambient temperature [10–13]. During the subcritical conversion process, water participates in a series of complex reactions with macromolecules including hydrolysis, fragmentation, aromatization, dehydration, and deoxygenation to produce lower molecular weight compounds, thus resulting in different fractions such as liquid biocrude, solid biochar, and aqueous and gaseous fractions [14,15]. The HTL process converts lipid and other cellular components such as protein, fiber, and carbohydrate in algae into biocrude oil [16]. Hence, both high-lipid microalgae and low-lipid macroalgae feedstocks are suitable for HTL. Many earlier studies have separately examined microalgae and macroalgae as feedstocks for HTL for biomass conversion into biofuels [1,2,17,18]. In addition to this, the HTL of biomass produces a high-energy biocrude (30–40 MJ/kg), which is lower in oxygen and moisture content compared to biocrude produced using pyrolysis, which provides a more stable product [19,20]. This biocrude can be subsequently refined to obtain a diverse array of renewable fuels ranging from petroleum to aviation fuel [21]. This chapter highlights the basics of the HTL process, biomass feedstocks used for HTL, the continuous HTL process, and some advancements made in the HTL process.

12.2 Background on hydrothermal liquefaction

HTL refers to hydrous pyrolysis, a process used for the reduction of complex organic materials (biowaste or biomass) into crude oil and other valuable chemicals. It is carried out in a closed, oxygen-free reactor (Fig. 12.1) by pressurizing inert gases (e.g., N_2 or He) or reducing gases (e.g., H_2 or CO) at a certain temperature (250°C–380°C) and pressure (5–28 MPa).

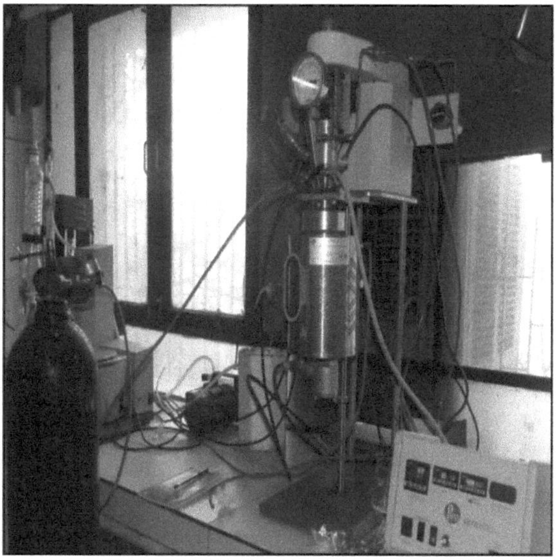

Figure 12.1 Photograph of hydrothermal liquefaction reactor (Parr Instruments Company, United States).

A phase diagram of water is shown in Fig. 12.2 [20]. In the HTL process, hot-compressed water is used as both a solvent and a reaction medium. In addition, HTL using hot-compressed water as a solvent has the advantages of it being abundant, nontoxic, nonflammable, inexpensive, and naturally stored in biomass.

The HTL of biomass generates a liquid biocrude and an aqueous phase containing soluble hydrocarbons. HTL is a thermochemical conversion process in which high temperatures and pressures are used to decompose complex organic materials including biomass. HTL is one of a family of thermochemical conversion processes (i.e., pyrolysis, gasification, and torrefaction) that uses heat to chemically decompose organic materials. The major processing differences of HTL from other thermochemical conversion processes are that it occurs at high pressures and is often performed on wet biomass using water as a critical reactant in the decomposition process.

One of the safest and most environment-friendly media for most organic reactions is water. Among all the solvents employed, the most common is water, as it is environmentally benign and the least expensive. It exists in three phases, namely solid, liquid, and gas. The vapor pressure curve separating the liquid and vapor phases ends at the critical point ($T_c = 373°C$, $p_c = 22.1$ MPa). Furthermore, beyond the critical point,

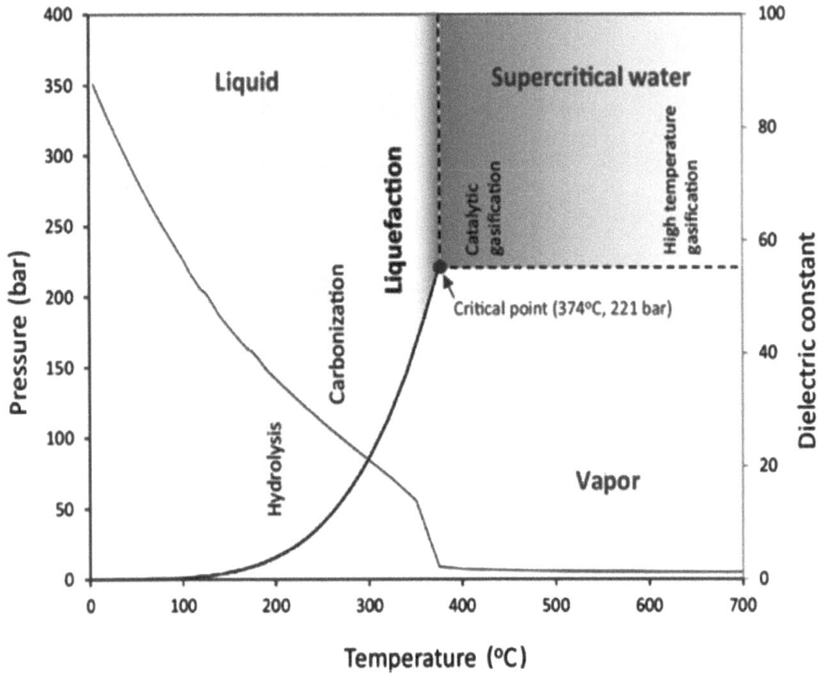

Figure 12.2 Phase diagram of water [20].

water properties can be changed without any phase transition. In addition, one important condition is the supercritical state, which refers to the zone of temperature and pressure at the critical point at which water acts as both a reactant and a catalyst. At this supercritical condition, various properties such as the ionic product, density, viscosity, and dielectric constant of water show rapid variation. Supercritical water is an excellent solvent for most homogeneous organic reactions owing to its high miscibility, the absence of any phase boundaries, and the fact that supercritical water acts as a nonpolar solvent. Supercritical water is able to react with different compounds. The changeability of the dielectric constant of water makes it a suitable medium for solvating organic molecules. HTL causes reactions to occur in a single phase leading to higher reaction rates due to improved nucleophilic substitutions and eliminations [22] and subsequent hydrolysis reactions [23]. On the other hand, the phase transition of water to its organic form causes the precipitation of salts due to its decreased solubility, which often results in blockage issues. Similarly, the viscosity of water tends to decrease with an increase in temperature, leading to a higher diffusion coefficient and mass transfer.

Water plays an essential role in HTL, and it is therefore critical to understand the fundamentals of the chemistry of water when subjected to high temperature conditions. However, when the temperature increases, the properties of water molecules change substantially, that is, the relative permittivity (dielectric constant) of water decreases quickly when the temperature increases, and when the thermal energy increases, the shared electron by oxygen and H_2 atoms tends to circulate more evenly while the electronegativity of the oxygen molecule is reduced (making it less polar). This polarity change makes water more affinitive to organic hydrocarbons, most of which are nonpolar molecules. Some of the parameters that play an important role in HTL of biomass are reaction temperature, residence time, type of catalyst, type of solvent, and biomass-to-water ratio. Moreover, temperature is one of the most dominant operational parameters in biomass liquefaction as it strongly affects the yields and properties of biocrude. According to the cited literatures, the HTL temperature ranges between 200°C and 450°C and liquefaction on biomass species highly depends on solvents used, affecting both product yields as well as content too [24]. According to the literature, there are different types of solvents that can be used for biomass HTL and most common and easy handled solvent is water. However, some organic solvents such as methanol and ethanol have also been applied for the conversion of biomass using HTL with better results than that of water in some cases. For wood biomass, HTL can also be effective with water or organic solvents when optimal conditions and catalysts are employed. However, for waste biomass in particular it is preferably to treat the biomass with alcohols as water is not an effective solvent. In addition, algal HTL is more preferable to the use of alcohols and the use of a cosolvent system such as water and alcohol (either methanol or ethanol) performs better than a single solvent. However, the choice of the appropriate solvent depends on the biomass feed and operating parameters.

A schematic diagram of the HTL process for the separation of gas, aqueous phase, and acetone soluble is depicted in Fig. 12.3. The gaseous phase mainly contains carbon dioxide (CO_2), CO, methane (CH_4), N_2, and H_2, whereas the aqueous phase contains polar that are soluble in water. The acetone soluble fraction of the HTL process is called biocrude, which is the major component, and the solid left behind is biochar. Comparative data between HTL processing and other thermochemical conversion technologies are shown in Table 12.1.

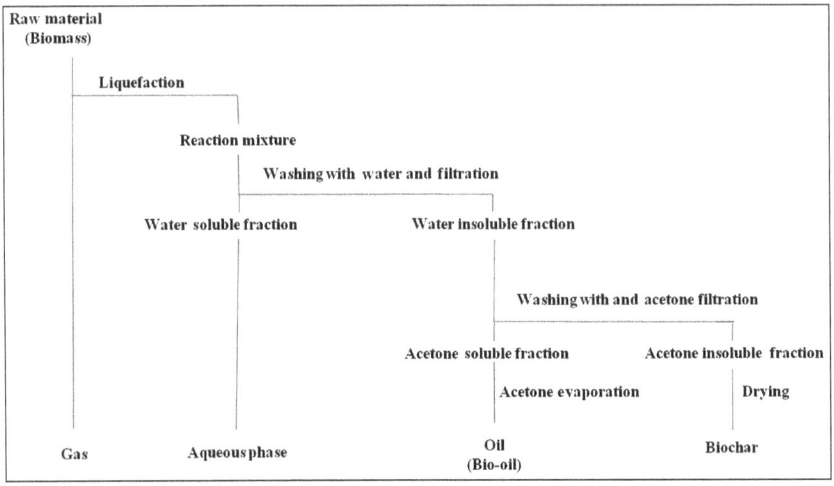

Figure 12.3 Schematic flow diagram of the continuous hydrothermal liquefaction reaction system for the conversion of various biomass feedstocks.

Table 12.1 Thermochemical conversion technologies summary [25,26].

Process	Residence time	Temperature (°F)	Heating rate	Important products
Carbonization	Hours-days	572–932	Very low	Charcoal
Pyrolysis	Hours	752–1112	Low	Solids, liquids, gases
Fast pyrolysis	5–30 min	1292–1652	Medium	Solids, gases
Flash pyrolysis	< 1 s	1202–1832	High	Liquids, gases
Gasification	10–20 s	> 930	High	Gases
Hydrothermal Liquefaction	15–120 min	< 932 + High pressure	Medium	Liquids

12.3 Hydrothermal liquefaction biomass feedstocks

The HTL process can utilize different types of biomasses from modern society like sewage sludge, manure, wood, and compost and plant material along with various wastes from households, dairy production, and similar industries. However, most biomasses processed in HTL are used because of their hydrophilic nature and the reasonable ease in forming water slurries of biomass particles at pumpable concentrations, typically 5%–35% dry solids. However, lignocellulosic biomass is comparatively low in moisture content, therefore, recovery and reuse of the water for slurry preparation is imperative. Moreover, for high-moisture biomasses like algae some dewatering process is required in order to lessen the processing costs of excessive water. HTL uses high-moisture biomass, and therefore negates some of the high costs of the dewatering/drying of biomass. Hence, HTL is suitable for biocrude production from biomass with varying moisture contents such as wood, waste, and algae biomasses. In addition, in HTL processing, wet feedstocks are particularly suited and algae biomass has received increasing attention in this research field. Moreover, woody biomass requires grinding prior to processing in HTL, whereas some macroalgae strains and certain manures/sledges are of suitable small size for their direct processing. Manure/sewage sludge has not been processed in continuous HTL systems, although results from batch systems are promising for their application in continuous systems [27,28]. Furthermore, there are advantages of using hydrothermal processing for sludges and manures as it also ensures the sterilization of bioactive contaminants [29].

12.3.1 Woody biomass and wood-processing waste

Lignocellulosic biomass such as woody biomass comes from the forestry sector in the form of forest residues from the wood-processing industry and from agricultural crop residues. These biobased wastes are good candidates for fuels production because of their availability in large quantities and at low prices compared to timber. HTL processing of woody biomass has been studied over the past few years using feedstocks like white pine sawdust, *Fraxinus mandshurica*, *Pinus massoniana*, *Populus tomentosa*, pine wood powder, cypress, bamboo shoot shell, paulownia, switchgrass (*Panicum virgatum*), corn stover, kenaf, sorghum, and wheat straw [30]. Woody biomass contains cellulose (30%–50%), hemicellulose (15%–35%), and lignin (20%–35%) [3,31]. HTL processing of woody biomass yields ranges between 17 and 68 wt.% depending on the operation parameters, type of catalysts, and solvents used [30]. The HTL processing of woody biomass has been extensively studied in the past, however, its commercial application is still awaited due to many challenges such as low biocrude yield from biomass conversion.

12.3.2 Wastes

HTL processes have been applied using a variety of residual biomasses and wastes as feedstocks such as agricultural and municipal wastes, etc. HTL processing using raw primary sewage sludge [32], sewage sludge from wastewater treatment plants [33], and waste plastics [34] as feedstocks have been carried out for biocrude production and with excellent throughput and results. The yields of biocrude from the HTL of sewage sludge were almost 15 wt.% along with a complex mixture of N_2-containing species, mainly amines (2-ethyl-6-methyl-benzenamine, etc.), pyridines (3-ethyl-pyridine, 2-ethyl-5-methyl-pyridine, etc.), indoles (methyl-1H-indole, 2,3-dimethyl-1H-indole, etc.), amides (hexadecanamide, *N,N*-diethyl-dodecanamide, (*Z*)-9-octadecenamide, etc.), imidazole, and nitrile. In addition, it contains non-N_2-containing species, mainly phenol and phenolic compounds (e.g., 2-methyl-phenol, methoxyphenol, ethyl-phenol), esters (1,2-benzenedicarboxylic acid-bis(2-methylpropyl)ester, 2,2-dimethyl-propanoic acid-2-phenylethyl ester, etc.), alcohols, hydrocarbons (naphthalene, cholest-2-ene, stigmastane, etc.) and acids [33]. Moreover, biocrude yields from waste plastics range from 2 to 95 wt.% depending on the type of plastic (polyethylene, polypropylene, polystyrene, polyvinyl chloride, or polyethylene terephthalate). The composition of gas products from waste plastic HTL conversion includes methane, ethane, ethene, propene, propane, and butane [34]. In addition to this, solid wastes and swine manure were also processed by HTL. The biocrude produced from the HTL of swine manure had a heating value of 36 MJ/kg and a viscosity of 843 cP at 50°C. The biocrude obtained from the HTL of swine manure is not comparable to transportation fuels like gasoline due to its unfavorable properties [35,36]. The HTL of reed and corn stover residues results in a water-insoluble fraction (lignin), a solvent (chloroform) extract (mainly esters, phenols, hydrocarbons, and their derivatives), and after solvent removal the remaining products are organic acids (levulinic acid, acetic acid, propionic acid, and formic acid) [37]. However, it should be noted that the high water content of the aforementioned wastes is a drawback to their economic exploitation through other thermochemical technologies [4]. Human waste is a significant and growing problem in urban conglomerates all over the world. Rampant urbanization in conjunction with unscientific waste management practices has resulted in large quantities of human wastes production leaving a challenging task for their disposal. In addition, these man-made wastes originate from households as well as from food-processing industries. Food processing wastes include field crops such as soft and hard wheat, rice, grain maize, barley, oats, sunflower and soybeans; and arborous crops such as grapevine, peach, apricot, apple, pear, cherry, plum, and hazelnut. The HTL of some of municipal and food-processing wastes has been studied in the past. A mixture of cabbage, boiled rice, boiled and dried sardine, butter, and clamshells, simulating typical municipal wastes, was studied for HTL biocrude production that could be co-fed to a refinery or alternatively used as a source of valuable chemicals [38]. Having established that there is a significant quantity of available biomass residues, thermochemical direct liquefaction seems to be an attractive technology to efficiently convert waste biomass into liquid fuels.

12.3.3 Microalgae

Microalgal feedstocks are generally collected from natural sources like lakes and water bodies [39–42], landfill leachate disposal sites [43] especially when microalgae form blooms, or from artificial cultivation reactors/tanks, and algal-based ponds and lagoons [44–46]. Cultured microalgae are separated from diluted culture media using various biological, physical, and chemical techniques [47,48]. Microalgae can be harvested in two-step separation procedures such as thickening procedures in which the microalgae slurry is concentrated to about 2%–7% of the total suspended matter and dewatering procedures where the concentration of the microalgae slurry is 15%–25% of the total suspended solids. Biomass slurry could be used for HTL conversion directly or after further drying of the slurry. Microalgae biomass is a complex mixture of various biomacromolecules such as polysaccharides, triglycerides, lipids, and proteins, etc. [49–51]. The composition of microalgae may vary from strain to strain [51]. The HTL conversion process can be carried out for all biochemical components of the microalgae biomass as this allows for maximum production. The HTL process has the potential to utilize both pure and mixed cultures in addition to high-lipid content strains [52].

HTL has advantages in biomass processing for both dry and wet biomasses and does not require pretreatment prior to the conversion process. Consequently, HTL for algae conversion has a lower energy consumption compared to other conversion processes [14]. The HTL of algae biomass provides four products, namely biocrude, solid residue, gas, and water-soluble products [53]. Additionally, the process water from the HTL process is rich in nutrients like nitrogen, phosphorous, and elements like Fe, Mg, K, Ca as well as some other minerals and polar organic compounds [54]. Other benefits of algal HTL are the simultaneous production of value-added polysaccharides as well as biocrude and the preservation of the nutritional value of solid residual algal biomass as an animal feedstock additive.

The HTL of algal biomass has been studied for the past few years, including the HTL of *Dunaliella tertiolecta* biomass [55] with high biocrude yields in the range of 25–64 wt.%, *Entermorpha prolifera* biomass [56] with biocrude yields of 16–33 wt.%, and *Chlorella pyrenoidosa* biomass [57] with biocrude yields 60 wt.%, etc. In algal HTL, the heating value of biocrude generally ranges from 28 to 37 MJ/kg and consists of unsaturated fatty acids, ketones, aldehydes, and hydrocarbons [57].

12.3.4 Catalytic and noncatalytic hydrothermal liquefaction

HTL or simply direct liquefaction is basically pyrolysis in hot liquid water. HTL does not require a catalyst as such, however significant amounts of research and development on catalyst-based HTL is in progress. Such catalytic hydrothermal conversion is one of the most promising processes for biofuels generation in future. However, to make the process cost competitive with petroleum-based fuels, further studies are warranted on the frontiers of HTL contributing valuable and inexpensive future solutions to meeting global energy demands.

Different types of biomass resources such as algal biomass, woody biomass, lignin, rice husk, and others have been used for noncatalytic HTL. In the case of algal biomass, noncatalytic HTL results in high biooil yields. Brown et al. carried out a noncatalytic HTL process using biomass from a marine microalga, *Nannochloropsis* sp., at a temperature range of between 200°C and 500°C with a residence time of 60 min. According to their reports, the highest biooil yield of 43 wt.% was obtained at 350°C with a heating value of 39 MJ/kg, which was comparable to that of a petroleum crude oil. However, the HTL of *D. tertiolecta* was carried out at temperatures between 300°C and 380°C with residence times of 10–100 min [58] and a biooil yield of 36.9 wt.% with a heating value of 26.62 MJ/kg at 360°C with 30 min holding time and a feedstock ratio of materials-to-water of 1:10 was reported.

Biomass HTL using homogenous catalysts are highly effective [59]. Furthermore, among homogenous catalysts, base and basic salts have been widely used in the HTL of biomass [1]. Heterogeneous catalysts, in the hydrothermal upgrading, were generally used for the hydrodeoxygenation of model compounds such as phenol [60]. The HTL of woody biomass at 280°C with a 15 min residence time using catalyst like K_2CO_3, KOH, Na_2CO_3, and NaOH results in biooil yields in relation to the alkaline catalyst of the order: $K_2CO_3 > KOH > Na_2CO_3 > NaOH$ [61].

The biooil yield depends on the hydrothermal conditions (i.e., temperature, residence time, the types of catalysts, the concentrations of catalysts, and the type of reactor) as well as on the type of biomass. Furthermore, increased base concentration (KOH and Na_2CO_3) or reaction temperature usually promotes biooils formation with lower molecular weight hydrocarbons. Using catalysts such as $Ba(OH)_2$ and Rb_2CO_3 in the HTL of biomass results in a significant increase in both phenolic/neutral oil and gas yields for all biomass feedstocks excluding lignin. Moreover, the degradation mechanism of biomass under hydrothermal conditions in the presence of alkaline or acidic solutions is not clear. In the case of using alkaline solutions in the HTL of biomass, it is difficult to separate liquid and solid products, but this can be solved by the addition of hydrochloric acid prior to the separation of the solid and liquid products [62]. The HTL process is revealed to convert triglycerides to fatty acids and alkanes using some heterogeneous catalysts and the initial decomposition products include free fatty acids and glycerol. The HTL of microalgae using heterogeneous catalysts has been revealed to slightly increase biocrude yields, but the higher heating value and the level of deoxygenation also simultaneously increase by up to 10% [5,63].

12.3.5 Continuous flow system of hydrothermal liquefaction

HTL technologies are focused on increasing the yields of biocrude production by the reduction of processing time as well as on the minimization of cost for the development of an environment-sustainable process. Earlier research on biomass HTL has been reported in both the batch and continuous modes for the processing of different types of biomass feedstocks. However, for an efficient and larger scale, a continuous HTL process design is required that is far better than batch systems for large-scale commercial production. In lieu of this, there are a number of advantages of continuous HTL such as cost and waste reduction, safety, and chemical and the process operability, etc. Continuous HTL has great potential for large-scale commercial conversion of various wet biomasses to energy-rich fuels and valuable chemicals [6–8]. During continuous HTL biomass is subjected to hydrothermal states and water molecules degrade the larger molecules in biomass into smaller fragments (Fig. 12.4).

HTL, also known as hydrous pyrolysis [64], is a very flexible technology as far as various types of feedstocks are concerned. HTL converts a wide variety of biobased and waste feedstocks such as woody biomass,

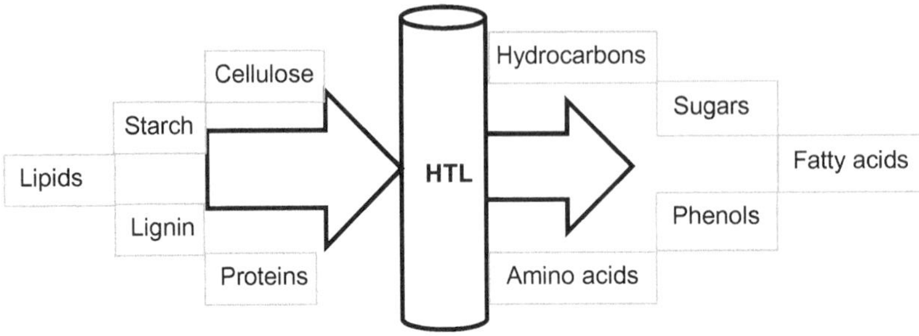

Figure 12.4 Macromolecular breakdown of biomass by hydrothermal liquefaction.

industrial wastes, food wastes, swine manure, algae, arborous crops, various wastes from the forest industry, and so on. During the HTL process, biomass is directly converted to biocrude in the presence of a solvent and in some cases catalysts at temperatures lower than 400°C. Most earlier studies in the past have been carried out in small-scale batch type-reactors with slow heating rates and long residence times. In this context, in order to become more efficient, feasible, and chemically controllable research in continuous system reactors is required [65,66]. Pedersen et al. [67] reported that a continuous flow reactor yields a significant amount of heating values. Elliot et al. also stressed the potential of introducing continuous flow reactors rather than the conventional batch process for higher scale processing of various waste biomasses and higher yields of HTL products. Table 12.2 presents a summary of HTL results on the respective feedstocks reported on by earlier researchers. The wet manure and sewage sludge feedstocks have not been processed in continuous systems, although results from batch systems are promising for their application in continuous systems [27,28]. One of the most essential advantages of HTL for sludges and manures is the effect of sterilizing bioactive contaminants [29]. However, for commercialization, further advancement in the frontiers of the continuous flow process need to be developed.

12.3.6 Hydrothermal liquefaction efficiency

It is widely known that a key factor for global warming and associated climate change is GHG emissions from the utilization of fossil fuels. In this context, the HTL procedure converts various waste biomasses into biocrude that can be utilized as an alternative to commercial fossil-derived fuels. However, the economy of the HTL process is still questionable because of its requirement for high pressure and temperature conditions. On the contrary, one of major benefits of the HTL process is that it

Table 12.2 Summary of feedstocks used for hydrothermal liquefaction (HTL) and biocrude result in continuous HTL (Elliott et al. [68]).

Feedstock (dry basis)	Ash	H/C	O%	N%	HHV (MJ/kg)	Size
Lignocellulosic Umeki et al. [69], Wang et al. [70]	3–8	1.2	35–45	0.5–3	12–20	1–100,000 mm
Macroalgae Ross et al. [71]	15–35	1.2	25–40	3–7	10–20	1–10,000 mm
Microalgae Biller and Ross [63]	7–26	1.6	25–30	5–9	25–30	1–100 μm
Manures Vardon et al. [27], Wang et al. [70]	10–20	1.5	35–45	3–6	10–20	1–10,000 μm
Sewage sludge Fonts et al. [72]	20–50	1.6	50	3–8	14	1–100,000 μm

Biocrude (continuous HTL results)

	Lignocellulosic	Macroalgae	Microalgae
Yield (%) (dry free ash)	35	27	38–64
Energy recovery (%)	64	52	60–78
N (%)	0.3	3–4	4–8
O (%)	12	6–8	5–18
	Tews et al. [73], NABC [74]	Elliott et al. [75]	Jazrawi et al. [76], Elliott et al. [77]

HHV, Higher heating value.

consumes ~10%–15% of the energy in the biomass feedstock, thus yielding an energy efficiency of 85%–90%. HTL technology can recover more than 70% of a biomass feedstock's carbon content. Another advantage of this process is that the biocrude produced does not require much treatment/upgradation for commercial utilization.

The by-products obtained through the HTL process as hydrochar can be used as a fertilizer amendment due to its high salt content and organic carbon where as biooil obtained can be used as potential fuel substitute to fossil fuels [78]. Important chemical compounds of biooil obtained by HTL as described by Gollakota et al. include fatty acids (myristic acid, palmitic acid, steric/oleic acid, tetradecanoic and octanoic acid, hexadecanoic acid, arachidic acid, and eicosapentaenoic acid); monoaromatics (phenol, benzene/toluene/styrene, cholesterol, cholestene, and vitamin E); alkane/alkene and polyaromatic compounds (naphthalene, quinoline, indene, anthracene, phenanthrene, pyrene and carbozole, and fluorine); nitrogen compounds (piperidines, pyrrols/pyrrolidines, indoles, pyridines, pyrazines, pyrrolidinones, amides, amines, and nitriles); and some oxygenated compounds such as esters, aldehydes, ketones, alcohols, acetic acid, and furans, etc.

Transitions in the operational parameters of HTL (temperature, residence time, particle size, moisture, and reactor configuration) can cause several changes to process chemistry such as hemicellulose depolymerization (oligomers, monomers), alteration/degradation of lignin (phenolic compounds), and increased availability of cellulose. The products obtained in this process are valuable sources of materials for the chemical, pharmaceutical, food, and energy industries. Moreover, the use of hydrothermal processing in aquatic biomass (macro- and microalgae) is an exciting technology for biocrude oil production as well as the extraction of polysaccharides for different applications and hydrolysis into sugars for further utilization in various useful processes such as fermentation.

12.4 Conclusion

The conversion of various wastes/biomasses/agricultural residues into valuable sources of chemicals and energy is crucial for sustainable development [78–81]. In this regard, hydrothermal processing technologies have significant potential for the conversion of various biomasses and organic substrates with high moisture contents to valuable bioproducts coupled with bioenergy. Innovations in HTL through knowledge of the fundamentals of the process and its applications in feedstocks adopted for HTL and understanding the underlying mechanisms of HTL processes and process flows are highly important to improving the overall efficiency of the process. By standardizing the conditions affecting the process such as feed quality, liquid–solid ratio in the feed, reaction time, mixing, temperature, pressure, catalyst, and solvent biocrude yields could potentially be optimized. More research emphasis on various breakthroughs that augment the process efficiency in HTL will eventually aid in the commercialization of the process. This would be a platform for further technology development toward a green and sustainable economy.

References

[1] K. Anastasakis, A.B. Ross, Hydrothermal liquefaction of the brown macroalga *Laminaria saccharina*: effect of reaction conditions on product distribution and composition, Bioresour. Technol. 102 (2011) 4876–4883.

[2] D.L. Barreiro, W. Prins, F. Ronsse, W. Brilman, Hydrothermal liquefaction (HTL) of microalgae for biofuel production: state of the art review and future prospects, Biomass Bioenergy 53 (2013) 113–127.

[3] F. Behrendt, Y. Neubauer, M. Oevermann, B. Wilmes, N. Zobel, Direct liquefaction of biomass, Chem. Eng. Technol. 31 (2008) 667–677.

[4] S. Bensaid, R. Conti, D. Fino, Direct liquefaction of lingo-cellulosic residues for liquid fuel production, Fuel 94 (2012) 324–332.

[5] P. Biller, R. Riley, A.B. Ross, Catalytic hydrothermal processing of microalgae: decomposition and upgrading of lipids, Bioresour. Technol. 102 (7) (2011) 4841–4848.

[6] M. Kumar, O.A. Olajire, A. Kumar, A review on the current status of various hydrothermal technologies on biomass feedstock, Renew. Sustain. Energy Rev. 81 (2018) 1742–1770.

[7] G. Kumar, S. Cho, P. Siva, P. Anburajan, D.M. Mahapatra, J. Park, et al., Insights into evolutionary trends in molecular biology tools in microbial screening for biohydrogen production through dark fermentation, Int. J. Hydrogen Energy (2018). Available from: https://doi.org/doi.org/10.1016/j.ijhydene.2018.09.040.

[8] R. Kumar, L. Singh, Z.A. Wahid, D.M. Mahapatra, H. Liu, Novel mesoporous $MnCo_2O_4$ nanorods as oxygen reduction catalyst at neutral pH in microbial fuel cells, Bioresour. Technol. 254 (2018) 1–16.

[9] D. Prabakar, V.T. Manimudi, K.S. Subha, S. Swetha, D.M. Mahapatra, K. Rajendran, et al., Advanced biohydrogen production using pretreated Industrial waste: outlook and prospects, Renew. Sustain. Energy Rev. 96 (2018) 306–324.

[10] S. Daneshvar, F. Salak, T. Ishii, K. Otsuka, Application of subcritical water for conversion of macroalgae to value-added materials, Ind. Eng. Chem. Res. 51 (2012) 77–84.

[11] A.A. Galkin, V.V. Lunin, Subcritical and supercritical water: a universal medium for chemical reactions, Russ. Chem. Rev. 74 (2005) 21.

[12] M. Herrero, A. Cifuentes, E. Ibanez, Sub and supercritical fluid extraction of functional ingredients from different natural sources: plants, food-by-products, algae and microalgae: a review, Food Chem. 98 (1) (2006) 136–148.

[13] P. Mishra, S. Thakur, D.M. Mahapatra, Z.A. Wahid, H. Liu, L. Singh, Impacts of nano-metal oxides on hydrogen production in anaerobic digestion of palm oil mill effluent – a novel approach, Int. J. Hydrogen Energy 43 (2017) 2666–2676.

[14] S.S. Toor, L. Rosendahl, A. Rudolf, Hydrothermal liquefaction of biomass: a review of subcritical water technologies, Energy. 36 (2011) 2328–2342.

[15] L.H. Zhang, C.B. Xu, P. Champagne, Overview of recent advances in thermo chemical conversion of biomass, Energy Convers. Manag. 51 (2010) 969–982.
[16] V. Patil, K.Q. Tran, H.R. Giselrød, Towards sustainable production of biofuels from microalgae, Int. J. Mol. Sci. 9 (2008) 1188–1195.
[17] G. Yoo, M.S. Park, J.W. Yang, M. Choi, Lipid content in microalgae determines the quality of biocrude and energy return on investment of hydrothermal liquefaction, Appl. Energy 156 (2015) 354–361.
[18] D. Zhou, L. Zhang, S. Zhang, H. Fu, J. Chen, Hydrothermal liquefaction of macroalgae enteromorpha prolifera to bio-oil, Energy Fuels 24 (2010) 4054–4061.
[19] A.V. Bridgwater, Review of fast pyrolysis of biomass and product upgrading, Biomass Bioenergy 38 (2012) 68–94.
[20] A.A. Peterson, F. Vogel, R.P. Lachance, M. Fröling, M.J. Antal Jr., J.W. Tester, Thermochemical biofuel production in hydrothermal media: a review of sub and supercritical water technologies, Energy Environ. Sci. 1 (2008) 32–65.
[21] E.D. Larson, A review of life-cycle analysis studies on liquid biofuel systems for the transport sector, Energy Sustain. Dev. 10 (2006) 109–126.
[22] X. Xu, C. De Almeida, M.J. Antal, Mechanism and kinetics of the acid-catalyzed dehydration of ethanol in supercritical water, J. Supercrit. Fluids 3 (1990) 228–232.
[23] M. Sasaki, M. Furukawa, K. Minami, T. Adschiri, K. Arai, Kinetics and mechanism of cellobiose hydrolysis and retro-aldol condensation in subcritical and supercritical water, Ind. Eng. Chem. Res. 41 (2002) 6642–6649.
[24] H. Durak, T. Aysu, Effects of catalysts and solvents on liquefaction of *Onopordum heteracanthum* for production of bio-oils, Bioresour. Technol. 166 (2014) 309–317.
[25] G.W. Huber, S. Iborra, A. Corma, Synthesis of transportation fuels from biomass: chemistry, Chem. Rev. 106 (9) (2006) 4044–4098.
[26] D.L. Klass, Biomass for Renewable Energy, Fuels, and Chemicals, Academic Press, San Diego, CA, 1998.
[27] D.R. Vardon, B.K. Sharma, J. Scott, G. Yu, Z. Wang, L. Scideman, et al., Chemical properties of biocrude oil from the hydrothermal liquefaction of *Spirulina* algae, swine manure, and digested anaerobic sludge, Bioresour. Technol. 102 (2011) 8295–8303.
[28] S. Yin, R. Dolan, M. Harris, Y. Tan, Subcritical hydrothermal liquefaction of cattle manure to bio-oil: effects of conversion parameters on bio-oil yield and characterization of bio-oil, Bioresour. Technol. 101 (2010) 3657–3664.
[29] M. Pham, L. Schideman, J. Scott, N. Rajagopalan, M.J. Plewa, Chemical and biological characterization of wastewater generated from hydrothermal liquefaction of *Spirulina*, Environ. Sci. Technol. 47 (4) (2013) 2131–2138.
[30] A. Dimitriadis, S. Bezergianni, Hydrothermal liquefaction of various biomass and waste feedstocks for biocrude production: a state of the art review, Renew. Sustain. Energy Rev. 68 (2017) 113–125.
[31] A.J. Panshin, C. DeZeeuw, Textbook of Wood Technology, fourth ed., McGraw-Hill, New York, 1980, p. 722.
[32] A. Suzuki, T. Nakamura, S. Yokoyama, T. Ogi, K. Koguchi, Conversion of sewage sludge to heavy oil by direct thermochemical liquefaction, J. Chem. Eng. Jpn. 21 (1988) 288–293.
[33] Y. Zhai, H. Chen, B. Xu, B. Xiang, Z. Chen, C. Li, et al., Influence of sewage sludge based activated carbon and temperature on the liquefaction of sewage sludge: yield and composition of bio-oil, immobilization and risk assessment of heavy metals, Bioresour. Technol. 159 (2014) 72–79.
[34] P.T. Williams, E. Slaney, Analysis of products from the pyrolysis and liquefaction of single plastics and waste plastic mixtures, Resour. Conserv. Recycl. 51 (2007) 754–769.

[35] W.T. Cheng, Y. Zhang, J. Zhang, G. Yu, L.C. Schideman, P. Zhang, et al., Hydrothermal liquefaction of mixed-culture algal biomass from wastewater treatment system into bio-crude oil, Bioresour. Technol. 152 (2014) 130−139.
[36] D. Cheng, L. Wang, A. Shahbazi, S. Xiu, B. Zhang, Characterization of the physical and chemical properties of the distillate fractions of crude bio-oil produced by the glycerol-assisted liquefaction of swine manure, Fuel 130 (2014) 251−256.
[37] J. Lu, X. Li, R. Yang, J. Zhao, Y. Liu, Y. Qu, Liquefaction of fermentation residue of reed- and corn stover-pretreated with liquid hot water in the presence of ethanol with aluminum chloride as the catalyst, Chem. Eng. J. 247 (2014) 142−151.
[38] T. Minowa, M. Murakami, Y. Dote, T. Ogi, S.Y. Yokohama, Oil production from garbage by thermochemical liquefaction, Biomass Bioenergy 8 (1995) 117−120.
[39] D.M. Mahapatra, H.N. Chanakya, T.V. Ramachandra, Assessment of treatment capabilities of Varthur Lake, Bangalore, Int. J. Environ. Technol. Manag. 14 (1−4) (2011) 84−102.
[40] D.M. Mahapatra, H.N. Chanakya, T.V. Ramachandra, Role of macrophytes in urban sewage fed lakes. I, Int. Omics Appl. Biotech. 2 (7) (2011) 1−9.
[41] D.M. Mahapatra, H.N. Chanakya, T.V. Ramachandra, C:N ratio of sediments in a sewage fed Urban Lake, Int. J. Geol. 5 (3) (2011) 86−92.
[42] D.M. Mahapatra, N.V. Joshi, T.V. Ramachandra, Insights to bioprocess and treatment competence of urban wetlands, J. Environ. Manag. 206 (2017) 1179−1191.
[43] B.P. Naveen, D.M. Mahapatra, T.G. Sitharam, P.V. Sivapullaiah, T.V. Ramachandra, Physicochemical and biological characterization of urban municipal landfill leachate, Environ. Poll. 220 (Part A) (2016) 2−12.
[44] D.M. Mahapatra, H.N. Chanakya, T.V. Ramachandra, Treatment efficacy of algae based sewage treatment plants, Environ. Monit. Assess. 185 (2013) 7145−7164.
[45] T.V. Ramachandra, D.M. Mahapatra, The science of carbon footprint assessment, The carbon Footprint Handbook, CRC Press, Taylor & Francis Group, Boca Raton, 2015, pp.)1−44.
[46] T.V. Ramachandra, D.M. Mahapatra, S.P. Bhat, N.V. Joshi, Biofuel production along with remediation of sewage water through algae, Algae and Environmental Sustainability, Developments in Applied Phycology., vol 7, Springer, New Delhi, 2015, pp.)33−51.
[47] D.M. Mahapatra, H.N. Chanakya, T.V. Ramachandra, *Euglena* sp. as a suitable source of lipids for potential use as biofuel and sustainable wastewater treatment, J. Appl. Phycol. 25 (2013) 855−865.
[48] D.M. Mahapatra, T.V. Ramachandra, Algal biofuel: bountiful lipid from *Chlorococcum* sp. proliferating in municipal wastewater, Curr. Sci. 105 (2013) 47−55.
[49] D.M. Mahapatra, H.N. Chanakya, T.V. Ramachandra, Bioremediation and lipid synthesis of myxotrophic algal consortia in municipal wastewater, Bioresour. Technol. 168 (2014) 142−150.
[50] D.M. Mahapatra, H.N. Chanakya, T.V. Ramachandra, Book chapter: algae derived single cell proteins: economic cost analysis and future prospects, Protein Byproducts: Transformation From Environmental Burden Into Value-Added Products, firstst ed., Elsevier, San Diego, 2016, pp. 275−301.
[51] D.M. Mahapatra, V.S. Varma, S. Muthusamy, K. Rajendran, Wastewater algae to value-added products, in: R. Singhania, R. Agarwal, R. Kumar, R. Sukumaran (Eds.), Waste to Wealth. Energy, Environment, and Sustainability., Springer, Singapore, 2018.
[52] A. Lababpour, Continuous hydrothermal liquefaction for biofuel and biocrude production from microalgal feedstock, Chem. Bio. Eng. Rev. 5 (2) (2018) 90−103.
[53] T.M. Brown, P. Duan, P.E. Savage, Hydrothermal liquefaction and gasification of *Nannochloropsis* sp, Energy Fuels 24 (2010) 3639−3646. 2010.

[54] A.B. Ross, P. Biller, M.L. Kubacki, H. Li, A. Lea-Langton, J.M. Jones, Hydrothermal processing of microalgae using alkali and organic acids, Fuel 89 (2010) 2234–2243.
[55] Z. Shuping, W. Yulong, Y. Mindge, I. Kaleem, L. Chun, J. Tong, Production and characterization of bio-oil from hydrothermal liquefaction of microalgae *Dunaliella tertiolecta* cake, Energy 35 (2010) 5406–5411.
[56] B. Jin, P. Duan, Y. Xu, F. Wang, Y. Fan, Co-liquefaction of micro- and macroalgae in subcritical water, Bioresour. Technol. 1491 (2013) 103–110.
[57] B. Jin, P. Duan, C. Zhang, Y. Xu, L. Zhang, F. Wang, Non-catalytic liquefaction of microalgae in sub-and supercritical acetone, Chem. Eng. J. 254 (2014) 384–392.
[58] S. Zou, Y. Wu, M. Yang, C. Li, J. Tong, Bio-oil production from sub- and supercritical water liquefaction of microalgae *Dunaliella tertiolecta* and related properties, Energy Environ. Sci. 3 (2010) 1073–1078.
[59] S. Yin, Z. Tan, Hydrothermal liquefaction of cellulose to bio-oil under acidic, neutral and alkaline conditions, Appl. Energy. 92 (2012) 234–239.
[60] B. Yoosuk, D. Tumnantong, P. Prasassarakich, Amorphous unsupported Ni–Mo sulfide prepared by one step hydrothermal method for phenol hydrodeoxygenation, Fuel 91 (2012) 246–252.
[61] S. Karagöz, T. Bhaskar, A. Muto, Y. Sakata, T. Oshiki, T. Kishimoto, Low-temperature catalytic hydrothermal treatment of wood biomass: analysis of liquid products, Chem. Eng. 108 (2005) 127–137.
[62] J.E. Miller, L. Evans, A. Littlewolf, D.E. Trudell, Batch microreactor studies of lignin and lignin model compound depolymerization by bases in alcohol solvents, Fuel 78 (11) (1999) 1363–1366.
[63] P. Biller, A.B. Ross, Potential yields and properties of oil from the hydrothermal liquefaction of microalgae with different biochemical content, Bioresour. Technol. 102 (2011) 215–225.
[64] C. Jazrawi, Hydrothermal Treatment of Algal Biomass: From Batch to Continuous Pilot Plant Operations (Thesis), University of Sydney, Sydney, Austalia, 2014.
[65] L. Garcia Alba, C. Torri, C. Samorì, J. van der Spek, D. Fabbri, S.R.A. Kersten, et al., Hydrothermal treatment (HTT) of microalgae: evaluation of the process as conversion method in an algae biorefinery concept, Energy Fuels 26 (2012) 642–657.
[66] S.K. Tanneru, P.H. Steele, Liquefaction of dried distiller's grains with solubles (DDGS) followed by hydroprocessing to produce liquid hydrocarbons, Fuel 150 (2015) 512–518.
[67] T.H. Pedersen, I.F. Grigoras, J. Hoffmann, S.S. Toor, I.M. Daraban, C.U. Jensen, Continuous hydrothermal co-liquefaction of aspen wood and glycerol with water phase recirculation, Appl. Energy 162 (2016) 1034–1041.
[68] D.C. Elliott, P. Biller, A.B. Ross, A.J. Schmidt, S.B. Jones, Hydrothermal liquefaction of biomass: developments from batch to continuous process, Bioresour. Technol. 178 (2015) 147–156.
[69] K. Umeki, K. Yamamoto, T. Namioka, K. Yoshikawa, High temperature steam-only gasification of woody biomass, Appl. Energy 87 (3) (2010) 791–798.
[70] L. Wang, A. Shahbazi, M.A. Hanna, Characterization of corn stover, distiller grains and cattle manure for thermochemical conversion, Biomass Bioenergy 35 (1) (2011) 171–178.
[71] A.B. Ross, J.M. Jones, M.L. Kubacki, T. Bridgeman, Classification of macroalgae as fuel and its thermochemical behaviour, Bioresour. Technol. 99 (14) (2008) 6494–6504.
[72] I. Fonts, G. Gea, M. Azuara, J. Ábrego, J. Arauzo, Sewage sludge pyrolysis for liquid production: a review, Renew. Sustain. Energy Rev. 16 (5) (2012) 2781–2805.

[73] I.J. Tews, Y. Zhu, C.V. Drennan, D.C. Elliott, L.J. Snowden-Swan, K. Onarheim, et al., Biomass Direct Liquefaction Options: Technoeconomic and Life Cycle Assessment, Pacific Northwest NationalLaboratory, Richland, Washington, USA, 2014. PNNL-23579.
[74] NABC, 2014. <http://www.nabcprojects.org/> (07.08.14).
[75] D.C. Elliott, T.R. Hart, G.G. Neuenschwander, L.J. Rotness, G. Roesijadi, A.H. Zacher, et al., Hydrothermal processing of macroalgal feedstocks in continuous-flow reactors, ACS Sustain. Chem. Eng. 2 (2) (2013) 207–215.
[76] C. Jazrawi, P. Biller, A.B. Ross, A. Montoya, T. Maschmeyer, B.S. Haynes, Pilot plant testing of continuous hydrothermal liquefaction of microalgae, Algal Res. 2 (3) (2013) 268–277.
[77] D.C. Elliott, T.R. Hart, A.J. Schmidt, G.G. Neuenschwander, L.J. Rotness, M.V. Olarte, et al., Process development for hydrothermal liquefaction of algae feedstocks in a continuous flow reactor, Algal Res. 2 (4) (2013) 445–454.
[78] A.R. Gollakotaa, N. Kishore, S. Gu, A review on hydrothermal liquefaction of biomass, Renew. Sustain. Energy Rev. 81 (2018) 1378–1392.
[79] H.N. Chanakya, D.M. Mahapatra, R. Sarada, R. Abitha, Algal biofuel production and mitigation potential in India, Mitig. Adapt. Strat. Glob. Change 18 (2013) 113–136.
[80] H.N. Chanakya, D.M. Mahapatra, R. Sarada, V.S. Chauhan, R. Abitha, Sustainability of large scale algal biofuel production in India, J. Indian Inst. Sci. 92 (1) (2012) 63–98.
[81] T.V. Ramachandra, D.M. Mahapatra, B. Karthick, R. Gordon, Milking diatoms for sustainable energy: biochemical engineering vs. gasoline secreting diatom solar panels, Ind. Eng. Chem. 48 (19) (2009) 8769–8788.
[82] T.V. Ramachandra, D.M. Mahapatra, S. Samantray, N.V. Joshi, Biofuel from urban wastewater: scope and challenges, Renew. Sustain. Energ Rev. 21 (2013) 767–777.

CHAPTER 13

An overview of algal photobioreactors for resource recovery from waste

Surjith Ramasamy, S. Arun and Kannan Pakshirajan
Department of Biosciences and Bioengineering, Indian Institute of Technology Guwahati, Guwahati, India

13.1 Introduction

Countries such as Germany (293 PJ) and China (295 PJ) produce maximum bioenergy from biogas plants [1]. Ammonium volatilization, nutrient pollution, and sodacity are the main drawbacks of applying anaerobic digestate to agricultural crops. The discharge of wastewater rich in ammonium, phosphate, and other nutrients leads to algal bloom in natural water bodies, which needs to be treated properly. Throughout the world, 20% of all nitrogen and phosphorus are distributed in wastewater [2].

The utilization of fossil fuels in various industrial and transportation sectors has resulted in an unsustainable economy and the emission of greenhouse gases such as CO_x, NO_x, SO_x, etc. Carbon dioxide (CO_2) levels may reach up to 45,000 MT by 2040, which would result in major air pollution problems [3], and therefore CO_2 sequestration has become of interest. Compared with physicochemical methods, microalgae-based CO_2 sequestration is more attractive as the process is low-cost, energy-saving, and devoid of waste disposal problems. The process can be made more economical when it is combined with wastewater treatment to produce biofuel or any value-added products without the use of additional fertilizers and fresh water [4].

In fact, nutrient-rich anaerobic digestate from biogas plants is a major source of water pollution. There are several biotechnology-based methods for transforming and concentrating waste into valuable products including biohydrogen, biogas, and lipids produced from wastewater. Organic acids, methane, and hydrogen are used to make bioplastic, which is a high-value product, while biomass obtained from this process can be used as a single cell protein or biofertilizer [5]. Organisms with high growth rates such as

Bioreactors
DOI: https://doi.org/10.1016/B978-0-12-821264-6.00013-9

bacteria, microalgae, and certain plants are often used for recovering nutrients and toxic metals from waste streams. Mainly heterotrophic bacteria, phototrophic anaerobic bacteria, algae, and oxygenic photosynthetic bacteria are used for producing such value-added products from different waste streams. Autotrophic organisms and heterotopic bacteria obtain energy and electrons from organics present in wastewater using external oxygen and this tends to produce a lot of sludge. On the other hand, phototrophic anaerobic bacteria use energy from light, nutrients, and electrons from wastewater, whereas phototrophic algae use CO_2 as a carbon source. Thus phototrophic algae and bacteria are capable of converting greenhouse gases, CO_2, nitrous oxide, and others into value-added compounds such as biomass, carotenoids, lipids, and hydrogen, thereby reducing the carbon footprint with the help of light [6]. Carbohydrate containing biomass is also a highly suitable substrate for producing bioethanol, hydrogen, and methane. The recovery of nutrients by these phototrophic organisms abate air, water, and soil pollution. Whereas atmospheric nitrogen is available in plenty, other micronutrients, for example, phosphorus, potassium, and magnesium from natural sources, are highly limited and, therefore, need to be recovered from waste. For instance, ammonium, nitrate, phosphate, and potassium are commonly found in different wastewaters, anaerobic digesters, and the explosive and fertilizer industries.

Compared with terrestrial plants and agricultural crops for biofuel production, microalgae has distinct advantages including a reduced requirement of land and water and a high lipid content, etc. [7]. Autotrophic organisms in water require a 106:16:1 ratio of carbon/nitrogen/phosphorus (C:N:P), which may vary based on several physicochemical parameters. Microalgae contain a three-times higher nutrient content than terrestrial plants [8] and require 0.04–0.09 kg of N and 0.003–0.015 kg of P to produce 1 kg of biomass. Microalgae are capable of growing on various inorganic and organic carbon sources such as CO_2, sodium bicarbonate, sodium carbonate, glucose, acetate, and glycerol.

Open-type wastewater treatment ponds such as facultative ponds are ineffective as compared with closed and well-designed reactor systems. This is due to the fact that the efficiency of open treatment systems depend on external environmental factors such as contamination due to bacteria and other organisms, which cannot be controlled, unlike in a closed system. The product formation, growth rate, and physiochemical characteristics of algae are highly influenced by various abiotic factors such as light intensity, temperature, pH, and mixing rheology [9]. Moreover, the treatment efficiency depends on the concentration of pollutants, light intensity, and temperature [10]. Closed and controlled reactor systems such as tubular, airlift, flat panel, and column reactors yield maximum biomass productivity by providing the optimal conditions required for algal growth; and therefore, bioreactors offer long-term, continuous cultivation of algae from waste substrates. Bioreactor-based cultivation is suitable for long-term, continuous cultivation from waste substrates. In addition, closed photobioreactors are effective in preventing waste and CO_2 loss, which tend to maintain uniform cell biomass and growth rate.

For the proper design of photobioreactors, nutrient mass transfer and maximum utilization of photons are of primary concern [11]. Media optimization and system biology approaches are followed to maximize the process efficiency and reduce production costs. Table 13.1 presents different photobioreactors studied for application in wastewater treatment with algae and under different cultivation conditions.

However, the large-scale application of these photobioreactors is restricted due to a lack in understanding of reactor performance under large-scale conditions. Some closed-type reactors have major drawbacks such as overheating, oxygen accumulation, cell damage due to poor light intensity, and insufficient mixing. Hence this chapter discusses all these aspects for the proper design and application of photobioreactors for algae cultivation on waste substrates and resource recovery.

13.2 Photobioreactors used for algal cultivation

13.2.1 Column photobioreactor

Column reactors such as bubble column and airlift reactors are widely used for algal cultivation. Column photobioreactors have a maximum radius and height of up to 20 cm and 400 cm, respectively; in order to obtain a high surface area, it is desirable to keep the ratio of the diameter to height of the reactor vessel small. However, increasing the column height beyond 400 cm leads to nutrient gradient in the reactor, which leads to starvation conditions for the organisms. Moreover, a long residence time of oxygen (O_2) can be detrimental to algae in such reactors [29]. These reactors offer high mass transfer coefficient values of nutrients in the liquid–gas system.

Table 13.1 Different algal photobioreactors studied for wastewater and resource recovery and their operation conditions.

Sl. no.	Microalgae species	Reactor configuration	Waste substrate	Cultivation conditions	Biomass/ product productivity	Operation conditions	Reference
1.	*Chlorella vulgaris*	Borosilicate tubular bioreactor	Effluent from concentrate I (N/P 0.7)	Mixotrophic	195.1 mg SS/L days	Light intensity 150 μmol/m^2 s Photoperiod 14:10 12 day incubation Temperature 20°C ± 1°C	[12]
2	*Chlorella zofingiensis*	10 L tubular column	Mixed biogas slurry and municipal wastewater	Mixotrophic	0.28 g/L day biomass, 96.3 mg/L day lipid	Light intensity 150 μmol/m^2 s Photoperiod 12:12 12 day incubation Temperature 25°C ± 1°C	[13]
3	*Cholorella sorokiniana*	Airlift photobioreactor	Bioindustrial wastewater	Mixotrophic	0.023 g dw/L day	Temperature 25°C ± 1°C 70 day incubation Sunlight	[14]
5.	*C. vulgaris*	Membrane photobioreactor	Synthetic municipal wastewater	Phototrophic	2 mg/L biomass	Light intensity 2000 lux Hydraulic retention time (HRT) 2 days	[15]
6.	*Scenedesmus obliquus*	Vertical alveolar flat panel photobioreactor	Cattle wastewater	Mixotrophic	Biomass productivity 213–358 mg/L day	HRT 12 day Photoperiod 24:0 Light intensity 58 μmol/m^2 s	[16]

7.	*C. zofingiensis*	Tubular bubble column photobioreactor	Piggery wastewater	Mixotrophic	106.28–296.16 mg/L day biomass productivity 11.85–30.14 mg/L day	Temperature 25° C ± 1°C HRT 10 days Light intensity 230 ± 20 μmol/m^2 s	[17]
8.	*Botryococcus braunii*	Submerged membrane photobioreactor	Livestock wastewater	Heterotrophic	3500 mg/L	HRT 3, 4, and 5 Days SRT 16 days Temperature 25°C ± 1°C Light intensity 220 μmol E/m^2 s	[18]
10	*Halochlorella rubescens*	Twin Layer photobioreactor	Municipal wastewater	Mixotrophic	Microalgal growth 6.3 g/m^2 day	HRT 8 days Light intensity 22–220 μmol/m^2 s	[19]
11	*Euglena* sp.	Sequencing batch membrane photobioreactor	Real secondary effluent wastewater	Phototrophic	550 mg/L biomass concentrationLipid content 10.09%	Biomass retention time (BRT) 60 days HRT 2, 4, and 8 days Light intensity 10,000 lux	[20]
12	*S. obliquus*	Airlift tubular photobioreactor	Domestic wastewater	Mixotrophic	Maximum areal productivity 21.76.26 ± 0.3 g SS/m^2 day 20.80 ± 0.22 wt.%	HRT 5 days	[21]

(Continued)

Table 13.1 (Continued)

Sl. no.	Microalgae species	Reactor configuration	Waste substrate	Cultivation conditions	Biomass/ product productivity	Operation conditions	Reference
14	*Chlorella pyrenoidosa*	Airlift circulation photobioreactor	Anaerobic digested starch processing wastewater	Photoautotroph	630 mg/L d biomass productivity 69 mg/L day lipid productivity	Two-phase strategy cultivation Temperature 15°C Light intensity 60 μmol/m^2 s HRT 3–4 days Temperature 35°C Light intensity 220 μmol/m^2 s HRT 6–8 days	[22]
15	*C. vulgaris*	Membrane photobioreactor	Secondary effluent from domestic wastewater	Mixotrophic	1.035–1.524 g/L biomass concentration	HRT 2 days BRT 21.1 days Light intensity 10.5–112.3 μmol/ m^2 s	[23]
16	*Scnedesmus* sp.	Bubble column photobioreactor	Olive mill wastewater	Photoautotrophic Heterotrophic	Biomass productivity 86 mg/L day Biomass production 0.86-1.4 g/L	HRT 21 days	[24]
17	*C. vulgaris*	Photobioreactor	Saline wastewater	Photoautotrophic	Biomass concentration stage 3 1.0380 mg/L Lipid productivity stage 3 54.25 mg/L day 40% lipid content	Batch cultivation 20 days	[25]

18	*C. vulgaris*	Membrane photobioreactor	Sewage	Photoautotrophic	39.93 mg/L day	Light intensity 8000 lux Temperature 25° C ± 2°C HRT 2.5 days	[26]
19	*C. vulgaris*	Biofilm membrane photobioreactor	Secondary effluent	Phototrophic	Volumetric microalgal production 0.072 g/L day	HRT 2 days Temperature 25° C–28°C Light intensity 8000 lux	[27]
20	*C. sorokiniana*	Photobioreactor	Distillery wastewater	Photoautotrophic	Biomass concentration 12 g/L	HRT 4 days Temperature 27°C Light intensity 180 μmol photons/ m^2s	[28]

13.2.2 Bubble column reactor

A bubble column reactor is tubular with a sparger at the bottom, which supplies gas for mixing the contents without the need for any internal mixing units (e.g., impeller). A high mass transfer rate of CO_2 and nutrients can be achieved with an increase in retention time and number of small-sized bubbles. The flow rheology of the fluid and the geometry of the reactor can affect the light availability. In this type of reactor, a diameter exceeding 20 cm reduces the permeability of light, whereas the height is limited to 400 cm to avoid shade in a sequential reactor system. Air bubbles can cause a cloud effect in the middle of the reactor that increases the path length of the light inside. The cloud effect is caused by a group of bubbles that reduces the intensity of light reaching the microalgae. Placing perforated plates along the reactor height increases bubble dispersion for an enhanced mixing efficiency. Bubble column reactors are most suitable for high-density cultures. Schematics of different photobioreactors are shown in Fig. 13.1. Quang and Richmond cultivated *Isochrysis galbana* in outdoor conditions during summer and achieved 4.6 g/L of biomass and a maximum productivity of 1.6 g/L/day [30]. Cell density has a direct influence on productivity; a slight variation in optimal cell density can affect biomass productivity. During photoadaptation, productivity reduces by between 5% and 35%, which signifies that the higher the cell density the faster the adaption. The effects of cell density are measured based on productivity and dissolved oxygen in the reactor. Light inhibition is prominent in bubble column reactors due to short path length as compared with that in raceway ponds. Photoinhibition is reported for an initial cell density below that of the optimal level in the range 2.8–3.8 g/L as cells undergo photooxidative death within a short time span under natural light of a strong intensity. Hulatt and Thomas varied the power input for sparging, and 10, 20, and 50 W/m^3 energy was used to study its effect on biomass production. The maximum biomass production was high at 50 W/m^3 but net energy gain through biodiesel of the biomass obtained was negative, whereas 10 W/m^3 energy had 39% net energy gain.

13.2.3 Airlift photobioreactor

Airlift reactors consist of a tubular vessel with more than one interconnecting zone comprising of a riser, where the gas mixture is sparged, and a downcomer without any sparger. Blenke introduced the airlift loop reactor (ALR), which is classified into three types based on the driving

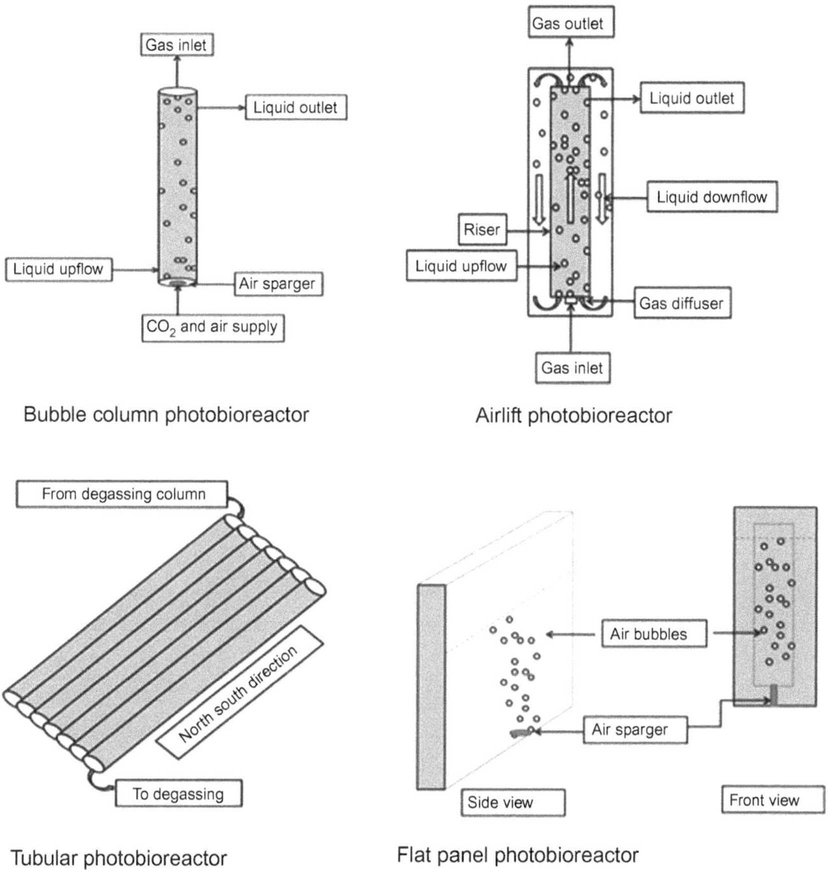

Figure 13.1 Schematic showing different photobioreactors used for algal cultivation.

force applied, namely hydrostatic, hydromechanics propeller (or pump), and hydrodynamic jet flow driven ALR [31]. A maximum algal biomass yield (1.20 g/mol) was achieved under a photon flux density (PFD) of 58 $\mu mol/m^2s$, whereas at more than 1200 $\mu mol/m^2s$ of PFD the biomass yield was reduced to 0.272 g/mol [32]. Rectangular airlift photobioreactors (ALPBRs) are used for algal growth to improve mixing, thereby enhancing the photosynthesis efficiency [32]. A maximum specific growth rate of 0.023 h^{-1} was achieved by cultivating *Phaeodactylum tricornutum* in a 12 L concentric-tube ALPBR with a superficial gas velocity of 0.055 m/s [33]. An external loop ALPBR with swirling motion was designed to enhance mixing [34]. In comparison with a bubble column reactor, an ALPBR with the same volume yielded a better value of maximum specific growth rate of $7.41 \times 10^{-2}\, h^{-1}$ and a cell concentration of 8.88×10^{6} cells/mL; these values in the bubble column reactor were of $3.80 \times 10^{-2}\, h^{-1}$and 5.8×10^{6} cells/mL, respectively. The diameter of an ALPBR determines the photon loss or even photoinhibition in narrow ALPBRs. A large diameter with high turbulent flow inside the solar receiver lowers the photoinhibition effects, which tends to improve biomass production. Overheating, photoinhibition or photon loss, biofouling, and high initial-investment costs are some of the drawbacks of ALPBRs. A multiphysics model has been established to predict the effects of shear stress, fluid mixing, radiant transport of light and heat, and algal growth kinetics following the Eulerian approach in ALPBRs [35].

13.2.4 Flat panel photobioreactor

Flat panel photobioreactors have short light paths and offer good surface area-to-volume ratios for utilizing maximum light energy. Introduced in 1953 by Burlew, the flat-plate photobioreactor (FPPBR), can be divided into two types based on the use of a recirculation pump (pump-driven) or compressed air (airlift). Issarapayup et al. used an ALPBR for cultivating *Haematococcus pluvialis* NIES-144 by varying parameters such as area of downcomer and riser cross section, size, and superficial gas velocity [36]. A maximum cell density of 4.1×10^{5} cell/mL and a specific growth rate of 0.52 day^{-1} were achieved using a 17 L flat panel airlift photobioreactor (FP-ALPBR), whereas a 90 L reactor yielded a low cell density (40×10^{4} cell/mL) and specific growth rate (0.39 day^{-1}). The performances of a 3 L and a 17 L cylindrical ALPBR were compared in a study, which revealed a high growth rate and cell density with the 3 L reactor [36]. Between the 17 L and 90 L flat panel airlift reactors, the 90 L FP-ALPBR produced 1 g of biomass for an investment cost of USD1.16, which is more cost effective than other PBR systems. In order to overcome construction costs and problems due to flow control, Tredici et al. used an alveolar panel reactor unit with 16 mm-thick plexiglass sheets of 95% light transparency, a surface-to-volume ratio of 80 m^2/m^3, and a culture thickness of 12.5 mm [37]. The outdoor cultivation of *Anabaena azollae* in a flat panel reactor with a surface area of 5 m^2 yielded 28 g/m^2 of biomass with a productivity of about 16 g/m^2/day. The thickness of an FBPBR decides the surface-area-to-volume ratio and the path length of light; the smaller the thickness, the better the light transfer is, which enhances biomass production [38,39]. Sudden variations in temperature, inhibition due to light, biofilm formation, and expensive materials for construction are considered the main drawbacks of this reactor for its large-scale application. Tilt angle, that is, the angle at which a reactor is placed on the ground is important in attaining maximum light exposure in order to achieve high biomass production. The position of the sun plays an important role in deciding the tilt angle; during summer a small tilt angle (10–30 degrees) is suitable for biomass, whereas in winter, a large tilt angle around 60 degrees is required. Geographic latitude also influences the tilt angle, and the direction the reactor faces, for example, west-east or south-north is very important as it can affect the biomass production in such PBRs [40].

13.2.5 Tubular photobioreactor

Tubular reactors are either straight, bent, or spiral in shape with different arrangements such as horizontal, inclined, or vertical. An array of tubes placed in a tubular reactor offer a large surface-area-to-volume ratio for effective utilization of light energy. This type of reactor has an external unit integrated into the main unit for the exchange of gas and heat, which maintains the CO_2 and O_2 concentration and the temperature for optimal growth. A recirculation pump is used for mixing and circulation and the cells are harvested using an external unit.

Horizontal tubular photobioreactors provide a higher surface-area-to-volume ratio than their vertical counterparts because of the possibility of decreasing the diameter of the tubes without affecting the structural integrity of such reactors. Vertical columns require a minimum diameter to support the weight above, whereas horizontal columns are free from this requirement. Horizontal reactors can also overcome the disadvantages that vertical reactors have of the requirement for a strong material with a minimum diameter to support the media and internal pressure. The angle of the incident of light is better in horizontal reactors than in vertical reactors, which can cause the temperature to rise very quickly in horizontal units and this is otherwise difficult to control at a large-scale without an external cooling system [41].

Near-horizontal tubular bioreactors are placed at 45 degrees, beyond which the construction costs would increase. An array of transparent pipes are connected with a bottom pipe in order to supply gas, which increases the fluid velocity, gas hold up, and nutrient mass transfer between liquid and gas [42].

Helical tubular bioreactors are a combination of both horizontal and vertical reactors, and consist of a fluorescent lamp in the middle as a light source. External cooling and degassing units are used to operate such a reactor successfully and air is supplied from the bottom of the reactor with the help of a flow meter. Degassing is achieved at the top of the reactor to remove excess oxygen [41].

Photo-loss and inhibition are the main problems that occur in a tubular reactor of small diameter and with a less-dense culture. The array of tubes in this type of reactor provide protection from excess light, thereby increasing algal growth. Increasing the diameter can also overcome the problem due to the resulting high temperature and light, to a certain extent, but this can be problematic for nutrient mass transfer in the system. Ugwu et al. demonstrated an increase in volumetric productivity of biomass of 63% using a static mixer in a 12.5 cm diameter reactor as compared to that without any mixing unit in the same reactor [43]. The static mixer was made of glass material, helical in shape, and fitted inside the tubular reactor to improve the mixing capacity of the reactor. Different methods used to control the reactor temperature are presented in Table 13.2.

Table 13.2 Different methods followed to maintain the temperature in tubular reactor system.

Sl. no.	Methods to maintain temperature	Advantage	Disadvantage	Reference
1	Shading with dark sheets	Reduced photoinhibition	May lead to photo deficient condition	[43]
2	Sprinkling with water	No shading effect	Loss of water and energy	[44]
3	Immersion	Reflection of lights on water can improve the light supply	Moving parts are needed to operate the reactor	[45]
4	Heat exchange	Easy to handle	Energy consumption at large scale	[41]

Unlike in an airlift or bubble column reactor, the gas transfer rate is quite low in tubular reactors, which creates substrate (CO_2) limitation and oxygen accumulation. Lehr and Posten used a third-generation smart-controlled (G3) bioreactor for injecting CO_2 into multiple locations along the axis of the bioreactor for improving mass transfer [46]. A tubular reactor was integrated with an airlift reactor for effective gas stripping in which CO_2 is added into the riser column to remove dissolved O_2 and further head space prevents the accumulation of O_2 in this system. The advantages and disadvantages of various photobioreactors are provided in Table 13.3.

13.3 Control systems and their strategies in photobioreactors

The impacts of control systems and their strategies in biological processes, particularly in photobioreactor systems, is gaining importance due to the many challenges and opportunities associated with them. Control over a photobioreactor's operating parameters is highly desirable in order to achieve the maximum algal biomass production and growth rate. Microalgae cultures have been conventionally cultivated in raceway ponds/open ponds due to their easy operation and low cost [47]. Unfortunately, this type of open system is not suitable for high volumetric productivity and high value-added products due to its low efficiency of light utilization, lack of pH and temperature control, high risk of contamination, poor liquid–gas mass transfer, and low yields of algal biomass [48]. Hence closed-loop photobioreactor systems such as tubular, airlift, bubble column, and flat panel photobioreactors, can be used to achieve controlled conditions of temperature, pH, mixing, etc.

13.3.1 pH control in tubular photobioreactor

In a photobioreactor system, pH control accounts for 30% of the production cost, which can be reduced by using an automatic control system that utilizes a data acquisition system (DAQ), a pH probe, automatic pneumatic valves for CO_2 supply, and a peristaltic pump for supplying an alkaline solution. DAQ acts as an interface between the pH probe and peristaltic pump, thereby allowing continuous monitoring and control of the pH at the desired level through commercially available software for data acquisition and monitoring. Similarly, carbon losses can be reduced to less than 30% through the appropriate design and operation of a PBR.

Table 13.3 Advantages and disadvantages of different photobioreactors for algal cultivation [42,43].

Sl. No	Photobioreactor configuration	Advantages	Disadvantages
1.	Bubble column photobioreactor	1. Low capital cost 2. High surface-area-to-volume ratio 3. Lack of moving parts 4. Satisfactory heat and mass transfer 5. Relatively homogenous culture environment 6. Efficient release of O_2 and residual gas mixture	1. Considerable degree of back mixing in both liquid and gas phases, which adversely affects product conversion 2. Short gas phase residence time
2.	Airlift photobioreactor	1. Mixing is done by bubbling the gas through a sparger in the riser tube without the need for mechanical agitation 2. Provides axial mixing pattern allowing liquid culture to pass through dark and light phases giving flashing light effect to algal cells 3. High photosynthetic efficiency	1. Performance depends upon proper design and operating conditions, for example, gas flow rate and light and dark phase 2. The amount of gas that does not disengage in the disengagement zone gets trapped by liquid moving downward in the downcomer 3. Complex and difficult to scale-up
3.	Tubular photobioreactor	1. Offers large surface area for illumination 2. Suitable for outdoor cultures 3. Fairly good biomass productivities 4. Relatively cheap	1. Gradients of pH 2. Non-uniform distribution of dissolved oxygen and CO_2 along the tubes 3. Fouling problem 4. Some degree of wall growth 5. Requires large amount of land
4.	Flat-plate photobioreactor	1. Offers large surface area for illumination 2. Suitable for outdoor cultures 3. Good for immobilization of algae 4. Good light path 5. Good biomass productivities 6. Relatively cheap 7. Easy to clean up 8. Low oxygen build-up	1. Scale-up requires many compartments and support materials 2. Difficulty in controlling culture temperature 3. Some degree of wall growth 4. Possibility of hydrodynamic stress to some algal strains

It is required to design highly developed control strategies to reduce carbon losses even further by considering mixing and mass transfer in the system. The model predictive control (MPC) strategy was established to control pH, which minimizes the carbon losses to less than 5%. In this strategy, the lower layer is controlled by a proptional integral (PI) controller pulse feed forward compensator for tracking the reference. The response of pH to changes in CO_2 supply is modeled using a first-order equation with a dead-time system. These advanced control strategies are used to reduce CO_2 losses and operating costs at the large-scale level [48].

13.3.2 Control of biomass production in tubular photobioreactors

Biomass production and productivity in PBRs are controlled by nonlinear predictive control and proportional-integral derivatives (PID) with feedback linearization for operating a PBR with constant density mode. A PID controller estimates the difference between a desired set point and an observed set point and then rectifies the difference using proportional, integral, and derivatives. These control strategies are used to maintain the algal culture at an optimal biomass density to attain a large amount of biomass with a predetermined feeding profile as a reference. Experiments conducted by Abdollahi and Dubljevic in 2012 used interior point optimization, moving horizon estimation, and MPC for maximizing lipid production by a heterotrophic microalga in a fed-batch bioreactor [49]. Interior point optimization is an algorithm that solves the linear and nonlinear optimization problems; a moving horizon estimator considers the measurements obtained along with its noise to calculate the estimated value. MPC uses the data from the experiments to estimate any changes in the dependent variable.

13.3.3 Fluid dynamics in photobioreactors

CO_2 injection, mass transfer, hydrodynamics, and photon intensity and distribution are examples of some of the most important factors for designing a photobioreactor. Computational fluid dynamics (CFD) plays an important role in solving any fluid flow problems in various PBRs with the aid of computer and numerical equations. CFD models for PBRs are constructed by considering physicochemical properties for evaluating various factors to reduce laborious experiments. A proper selection of the turbulence model (flow model), bubble size, and grid resolution is very important for increasing the accuracy of prediction. Software used for CFD consists of several models based on the phase of the fluids. Simulations of tubular PBRs are simple due to the absence of moving parts as compared to baffled reactors. Two-phase and multiphase are the basic flow models for performing simulations, and the latter is the most suitable for realistic modeling [50]. Choosing a model for deciding on a flow is important and would depend on the gas hold up of the dispersed fluid; the two-phase model is used with a gas hold up of up to 30% [51]. Pfleger et al. studied the different flow regimes in a bubble column PBR and found that a turbulent regime correlates well with experimental behavior, whereas a laminar regime showed a chaotic flow pattern [52]. $K-\varepsilon$ is the most common model used in CFD software, where K is the turbulence kinetic energy (m^2/s^2) and ε is the rate of dissipation of the kinetic energy (m^2/s^3). Mixture properties and mixture velocities are used to find the properties of turbulence flow. $K-\varepsilon$ is divided into three types based on the condition of the phases (Table 13.4).

13.4 Species transport models for bubble movement

Bubbles inside PBRs are analyzed through the species transport model in multiphase, which is used to solve conservation equations elaborating diffusion, convection, and reaction sources for individual bubbles inside the system [50]. There are several models available for species modeling, which are discussed further here.

13.4.1 Eulerian–Eulerian model

This model is used for gas–liquid and liquid–liquid systems, and considers the mean of the Navier–Stokes equations over the volume by considering arbitrary particle and continuous phase. This solves continuity, mass,

Table 13.4 Three different $K-\varepsilon$ models along with their conditions of use.

Sl. no	Model	Condition to use
1	Dispersed $K-\varepsilon$	Secondary phase is dilute with continuous primary phase
2	Two-phase $K-\varepsilon$	Dominant turbulence
3	Mixture $K-\varepsilon$	i. Stratified multiphase flow with phases separate ii. When ratio of the density between phases is close to 1

and momentum for solid and liquid phases and volume fractions. Liquid, gas, and particles are considered as continuum and their equations are solved in the same manner. Pressure is considered constant in continuous and dispersed phases. This model includes drag force and virtual mass effect when there is a relation between the movement of the primary and secondary phases and the involvement of a drag force. This model treats bubble mass as a continuous medium with a fluid regime [50].

13.4.2 Lagrangian–Eulerian model

This model considers each and every particle (bubbles) as well as the movement of the particles due to turbulent dispersion in column reactors [53,54]. The trajectories of each and every particle are calculated based on the external force acting on them. The Lagrangian–Eulerian model includes two steps, namely (1) the flow field is estimated from a balance equation of quasi homogenous gas–liquid dispersion in which the average density changes based on the volume of gas in the liquid (gas hold up) and (2) local gas hold up is estimated by considering all the particles in the flow field. Both these steps are repeated to attain close values. This model helps in determining bubble trajectory as well as the interaction effects of bubble–bubble and bubble–liquid and includes provisions for adding in the mass transfer effect due to physical and chemical parameters, bubble–bubble interaction to form new bubbles, and repeated dispersion [55]. This model incorporates bubble trace, which provides spatiotemporal details of the bubble, and it can be estimated within specific boundary, thereby enabling this model free from numerical diffusion for dispersed phase element [54]. Since bubbles are considered as sphere in this model, its application is restricted to bubble column reactors with minuscule bubbles and a homogenous regime [55]. The simulation of bubble breakage and coalescence is possible in heterogeneous regimes by including large deformable bubbles.

13.4.3 Volume of fluid model

Bubble–bubble interaction and high turbulence are essential for modeling a continuous swarm of bubbles, which cannot be achieved through single particle studies [55]. The volume of fluid (VOF) is the most suitable parameter for bubble column reactors, which also helps in finding the position of the interface between two immiscible layers. This is more suitable for volume tracking with complicated boundary conditions because it uses only a set of transport equations with jump boundary conditions [56]. VOF uses Navier–stokes equations for bubble movement simulation; continuity and momentum equations are solved in domain and then utilized in all phases. Volume fraction is a marker for interface location, which is obtained by finding the solution to a continuity equation. A momentum equation for the mixture fluid can be solved using the properties of the fluid materials, which tend to peak across the interface. Mixed fluids turbulence and energy equations are considered, while properties like surface tension, wall adhesion effect, and mixture of species are used in this model by considering phases as compressible. This model is suitable for analyzing the rise velocity of a bubble as well as the internal interaction of spherical cap bubbles, and the inclusion of bubble breakup and coalescence has made VOF highly effective. The coalescence of particles happens when the distance between two bubbles is smaller than computational cells. Modeling for large equipment with a dispersed multiphase flow can be a memory consuming process.

13.4.4 Bubble size estimation

Bubbles formed through sparging move from the bottom to the top of a reactor, thereby pushing the liquid, and turbulence is created because of the movement of the bubbles and liquid in opposite directions. A uniform regime of bubbles is obtained at a low gas velocity, whereas turbulent and annular regimes can be observed at a high flow rate. The size of a bubble is influenced by the geometrical and physical properties of the reactor and the fluid. Eqs. (13.1) and (13.2) were used to estimate bubble diameter by Wilkinson and Pohorecki et al. [57,58].

$$B_d = 3g^{-0.44}\sigma^{0.34}\mu_l^{0.22}\rho_l^{-0.45}\rho_g^{-0.11}v_{sg}^{-0.02} \tag{13.1}$$

$$B_d = 0.289\sigma^{0.442}\mu_l^{-0.048}\rho_l^{-0.552}\rho_l^{-0.552}v_{sg}^{-0.124} \tag{13.2}$$

where g (m/s^2) equals gravity, σ (N/m) equals the surface tension of the liquid, ρ_l and ρ_g (kg/m^3) represent the density of the liquid and gas, and v_{sg} (m/s) represents the gas superficial velocity. Eqs. (13.1) and (13.2) under-estimate the gas hold up and axial liquid velocity when the bubble size exceeds 8 mm. This is due to a high velocity of bubbles, which tends to reduce the gas hold up volume in liquid and lowers axial liquid velocity.

13.4.5 Computational fluid dynamics-based design for an enclosed horizontal bioreactor for algae cultivation

An enclosed horizontal bioreactor (HBR) is a type of raceway pond for cultivating algae on a large scale consisting of a closed polythene sheet with various process control units to achieve higher productivity as compared to PBRs. Pirasaci et al. designed a closed HBR coupled with paddle wheels to reduce the operational cost with high productivity using CFD [59]. Mixing can increase the productivity as compared to that in a static system. The formation of dead zones and power consumption are reduced by optimizing the aspect ratio, wheel diameter, spatial alignments, and depth of the reactor. Dead zones are defined as areas in a reactor in which the liquid velocity is less than 0.2 m/s. Modeling aspects were evaluated in a small-scale reactor of 150 L volume and 3 m^2 surface area, which was scaled-up to a pilot plant of 40 m^2 surface area. Pirasaci et al. reported that designing an HBR with baffles and increasing the size and number of the paddle wheels can reduce the occurrence of dead zones; on the other hand, increasing the aspect ratio increased the number of dead zones [59]. Increasing the reactor height can reduce power consumption and it can also enhance the CO_2 fixing capacity, whereas an HBR with a closed cover can show an enhanced efficiency of up to 95%. Raceway ponds with baffles can reduce the number of dead zones by enhancing the flow rate and by saving one tenth of the power consumption as compared to unbaffled reactors. Polyethylene microbeads with different drag coefficients were used to measure liquid velocity in order to determine the various constants to perform the CFD analysis. Materials with the same weight and density will have different drag forces depending on the shape of the material, which can influence aerodynamics and hydrodynamics. Pirasaci et al. found that an HBR length and breadth ratio of 2:1 reduced the number of dead zones, but power consumption was increased. In order to keep the power consumption minimal in an HBR the recommended L:B ratio is 8:1 [59].

13.5 Light intensity and distribution in photobioreactors

Microalgae absorb light energy in the 600–700 nm wavelength for photosynthesis. Outdoor systems can accumulate a lot of heat, which tends to reduce biomass production during sunny days. As controlling and monitoring natural light is difficult, indoor reactors with solar and artificial lights are preferred. Incandescent, halogen, fluorescent, compact fluorescent bulbs, and LED are the different sources of artificial light. LED bulbs can save 80%–90% on energy use as compared to incandescent bulbs, and they do not contain toxic metals such as mercury and lead, unlike fluorescent bulbs. When electron transition happens in the conduction band, light is emitted through a band gap in the semiconductor. A high concentration of microalgae can create a self-shading effect, whereas a high light intensity can inhibit microalgal activity. Optimization of light intensity is needed to achieve a high algal specific growth rate. Flat-plate, raceway pond, and horizontal tubular reactors can effectively utilize light energy. Reactors with a large surface area can efficiently harvest light energy. Light intensity diminishes from the outer layer to the center of reactors in high dense conditions. Tubular PBRs are constructed using glass, plastic, acrylic, polyethylene, and perspex materials, which are cheap and easy to maintain for photosynthetic organisms. Constructing large-scale reactors using glass material can increase the reactor cost. The surface area of tubular reactors is higher than that of any other reactor, which tends to yield a large amount of algal biomass.

The success rate of a PBR design depends on maximizing its efficiency to utilize light energy, process control for attaining maximum yield, and minimizing operational energy use to reduce the production cost. Cellular respiration due to the day–night cycle can reduce the total biomass produced in daytime to 58% in nighttime [59]. Light fraction and cycle time are two important factors that can influence day–night cycle. A high light fraction value and a low cycle time value can influence biomass production positively, while negative effects can be observed in the opposite conditions [32]. Light fraction is defined as the ratio of light exposure period-to-total day–night cycle, and one cycle time is equivalent to the time required to complete a cycle in a PBR.

13.5.1 Modeling of photosynthetic and biomass rate in photobioreactors

Biomass concentration is an important parameter in autotrophic cultivation, which can be estimated by subtracting compounds such as NADH, ATP, and other organics produced during photosynthesis and endogenous respiration [60]. Photosynthetic rate is influenced by various parameters such as light intensity, hydrogen and hydronium ion concentration, nutrient content, temperature, dissolved oxygen, and CO_2. Phototrophic cells need 100 μs to convert absorbed photons into NADPH and ATP, and during this time, cells cannot absorb light; hence based on this effect, light can be flashed to provide optimal energy [61,62]. Light gradient occurs in high cell density systems, but can be avoided using a short light cycle. Different models, as represented in Table 13.5, were evaluated for photosynthetic rate [63,64] among which hyperbolic tangent and light inhibition models were found to be the best. The Poisson model is based on the strike rate of photons on the photosynthetic unit (PSU), Eq. (13.6). This model considers light as a substrate for algal photosynthesis for predicting the rate of photosynthesis. Several models have been proposed by modifying the Monod equation, Eq. (13.3), to achieve an extremely high prediction accuracy. It is important to consider photoinhibition and optimal light intensity conditions under outdoor conditions due to dynamic daylight changes. Predictions are precise at low cell concentrations, whereas accuracy is reduced under high cell density conditions. Increases in cell density tend to create light gradient, and uniform mixing

Table 13.5 Different models for photosynthetic rate modeling based on light intensity.

Sl. no	Model equations	Parameters (units)
1.	Monod-based model $\phi = \frac{\phi_m I}{I_c + I}$ (13.3)	ϕ and ϕ_m—specific and maximum specific rate of photosynthesis $\left(\frac{\text{mol } O_2}{h \cdot (\text{mol Chla})}\right)$ I—light intensity reaching the cell $\left(\frac{E}{M^2 h}\right)$ I_c—constant (NA)
2.	Hyperbolic tangent model $\phi = \phi_m \tan h(\beta I)$ (13.4)	β— constant $\left(\frac{m^2 - h}{E}\right)$
3.	Light inhibition model $\mu = \frac{\varphi_{av}}{C_1 + C_2 I^2{}_{av}}$ (13.5)	C_1, C_2—constants $\left(\frac{h - W}{M^2}, \frac{h - M^2}{W}\right)$
4.	Poisson model $\varphi = \phi_m(1 - \exp(\alpha_d / \phi_m))$ (13.6)	α_d—constant based on the strike rate, $\left(\alpha = \frac{\text{gC} - \text{m}^2\text{s}}{\text{g} \cdot \text{Chlh} - \mu\text{mol}}\right)$

is needed to ensure that all the cells are exposed to a uniform light intensity. A reactor with a high-density culture has a dark region in the center and a light region near the reactor wall, and the time spent by the cells in each of these regions is called the cycle time. Reducing the cycle time will be helpful in attaining a uniform light exposure, which is important from a modeling perspective.

13.5.2 Models used for assessing biomass production rate in photobioreactors

Type-one models are based on the incident light intensity on the top layer of a reactor to predict biomass production or photosynthesis rate, and are represented in Eq. 13.7 [65]. This model is convenient as it needs only light intensity, but parameters such as X_m and μ_m depend on cell concentration, which can vary depending on the reactor system and operating conditions used, thus limiting its application to outdoor conditions. Considering the effect of cell density and reactor geometry, a subclass of the type-one model (Eq. 13.8) was developed as a function of average light intensity for predicting photosynthesis rate. The subclass type-one model considers different parts of a reactor with different light intensities as different systems and then it averages the biomass production rate. Obtaining the average light intensity value is helpful in predicting biomass production or photosynthesis rate in outdoor systems within a specific range of light intensity, cell density, and geometry of the system used.

Type-two models use several photosynthesis rates from small areas of a reactor based on the light intensity experienced by algal cells and then adds up the results from all the areas in the reactor for getting an accurate result. The type-two model can be precise in prediction with different ranges of parameters such as operational conditions, light intensities, cell concentrations, and reactor geometries [66]. Eq. (13.9) is basic for type-two models and can be modified based on different parameters.

Type-three models analyze the amount of light utilized by each and every microalga during their movement inside a reactor then sums up their photosynthetic rate for individual cells with their cycle time. This model differentiates the reactor into several parts; for example, in an airlift reactor the central riser is differentiated as the dark zone and the downcomer is defined as the light zone [67]. Cell flow trajectories are important to determine the cycle time, and can be determined using CFD analysis by considering the different parameters. This model focuses specifically on the PSU instead of individual cells. The rate of photosynthesis is proportional to the turnover number of individual units that are moving from an excited state to a ground state. PSU exists in three different states during the photosynthesis process, namely the ground state in which PSU utilizes photons to achieve an excited state, the excited state in which electrons are delivered to NADPH and ATP, and the recovery state in which denatured PSU proteins are stabilized to restart this cyclic process. The type-three model is represented in Eq. (13.10). This model includes short light cycles, which make it more suitable for predicting photosynthesis rates in outdoor cultivation, whereas the exclusion of light distribution, cell trajectories, and biological response can create deviations when predicting photosynthesis or biomass production rates. This model includes the flashlight effect, which helps in reducing operational costs by optimizing mixing and light energy through uniform light distribution (Table 13.6).

13.5.3 Modeling light distribution in a photobioreactor

In an isotropic reactor without a scattering effect, light distribution is determined using the Beer–Lambert law, Eq. (13.11), which considers the exponential reduction of light from the outer wall to the center of the reactor. Acién Fernández et al. used a radiative transfer equation (RTE) in a cylindrical photobioreactor and then determined that the Beer–Lambert law overpredicts by twofold as compared to RTE [68]. Eq. (13.12) represents radiative transfer, which is based on incident, absorbed, transmitted, and scattered radiation (Table 13.7).

13.5.4 Modeling based on temperature and intensity of light on photosynthesis

Uncoupled [Eq. (13.13)] and coupled [Eq. (13.14)] models are the two most prominent models used for predicting photosynthetic activity using light intensity and temperature. These models are the product of the heat (Arrhenius) and light (Monod) equations. Rate of deactivation was included in the equation for increasing precision. Dermoun et al. developed a light inhibition model based on the coupled model [69]. The limiting step for photosynthesis is dependent on temperature for which the values for a large number of unknown parameters need to be determined for accurate predictions.

Table 13.6 Equations used for type I, II, and III models.

Sl. no.	Model equations	Parameters (units)
1.	Type I model $\mu = \mu_m\left(1 - \frac{X}{X_m}\right)$ (13.7) Where, $\mu_m = \frac{C_1 L}{C_2 T + L}\exp\left(-\frac{C_3 L}{RT}\right)$ $X_m = (K_4 + K_5 T)\exp\left(-\frac{C_6 L}{RT}\right)$ $L = \frac{I_0}{I_{max}}$ $\mu = \mu_m(P, N)\theta_\mu^{T-T_0}\frac{I_0}{C + I_0} - \frac{\lambda_m}{h}\theta_\lambda T - T_0$	$C_1(day^{-1})$, $C_2(°C^{-1})$, C_3(J/mol), C_4(cells/mL), C_5(cells/mL/°C), C_6(J/mol)—constant I_{max} —maximum incident light (lux) R—gas constant (J/mol/K) P, N—concentration of phosphorus and nitrogen (mol) h—height of the cultivation system (cm) C, θ_μ, θ_λ—constant
2.	Type I subclass $\mu = \mu_m(T)\frac{I_{av}}{K + I_{av}}$ (13.8) Where, $\mu_m(T) = \mu_{m,o}\exp(-E_a/kT)$	E_a—activation energy (KJ/mol) K—Boltzmann constant (J/K) $\mu_{m,o}$—constant

3.	Type II model $\phi = \int_V \phi(l) \cdot \mathrm{d}V \quad (13.9)$ Where, $\phi(l) = \phi_m \frac{I_0 \exp(-\sigma X l)}{I_k + I_0 \exp(-\sigma X l)}$	ϕ and ϕ_m—volumetric and maximum rate of photosynthesis $\left(\frac{mg \cdot O_2}{g \cdot biomass - h}\right)$ l—distance from the wall of the reactor (cm) I_k—saturation constant (lux) I_0—incident light (lux) σ—extinction coefficient (L/mol/cm) X—biomass concentration (mg/lit)
4.	Type three model $\phi = \phi_s \frac{I}{1 + KI + K'I^2} \quad (13.10)$	ϕ—photosynthesis rate $\left(\frac{mg \cdot O_2}{g \cdot biomass - h}\right)$ $\phi_s \left(\frac{mg \cdot O_2}{g \cdot biomass - h}\right)$, K (lux), K' (lux)—constants

Table 13.7 Equations used to model light distribution.

Sl. no	Model equations	Parameters (units)
1.	Beer–Lambert equation $I(l) = I_0 \exp(-\sigma X l)$ (13.11)	X—cell concentration (g/L) $I(l)$—local light intensity (lux) l—distance from external surface (cm) σ—excitation coefficient
2.	Radiative transfer $I(l) = I_0 \exp\left(-\frac{K_1 X l}{K_2 + X}\right)$ (13.12)	$I(l)$—local light intensity (lux) l—distance from the wall of the reactor to specific location (cm) K_1 (m^{-1}) and K_2 (kg/m^3)—species dependent empirical constant X—biomass concentration (g/L)
3.	Coupled model $\mu = 2\mu_m(T)(1+\beta_1)\frac{I/I_{opt}(T)}{1 + 2\beta_1 I/I_{opt}(T) + (I/I_{opt}(T))^2}$ (13.13)	$\mu_m(T)$—maximum specific growth rate at T (day^{-1}) β_1—constant I—intensity of light (lux) $I_{opt}(T)$—maximum light intensity at T (°C) lux
4.	Uncoupled model $\mu(I) = \mu_{m,0}(I)\frac{\exp(-E_a/KT)}{1 + K\exp(-E'_a/KT)}$ (13.14)	$\mu(I)$—specific growth rate at the intensity I (day^{-1}) $\mu_{m,0}$—maximum specific growth (day^{-1}) K—Boltzmann constant (J/K) T—temperature (°C) E_a and E'_a—activation energy for photosynthesis and enzyme denaturation (J)

Table 13.8 Equations for Bodenstein number and axial dispersion coefficient for bubble column and airlift reactors.

Sl. no	Equation	Parameters (units)
1.	Bodenstein number $B_{o=}\dfrac{V_iD}{E_r}$ (13.15)	When, $B_0 < 0.1$ (mixing is perfect), $B_0 > 20$ (plug flow), B_0—dimensionless number V_i—liquid linear velocity (cm^3/s) E_r—axial dispersion coefficient, (cm^2/s) D—distance between point of injection and detection of tracer, (cm)
2.	Axial dispersion coefficient for bubble column reactor $\dfrac{\partial M_f}{\partial t} = E_r\dfrac{\partial M_f}{\partial r^2}$ (13.16) $\frac{M_i}{M_f} = 1 + \frac{2D}{\pi\gamma}\sum_{n=1}^{\infty}\left[\frac{1}{n}\sin\left(\frac{n\pi\gamma}{D}\right)\cos\left(\frac{n\pi\gamma}{D}\right)\right]\exp\left[-\left(\frac{n\pi}{L}\right)^2 E_r t\right]$	M_f—final concentration of tracer, (mg/mL) M_i—concentration of tracer at time t (mg/mL) D—distance between point of injection and probe (cm) γ—height of the liquid along with tracer (cm)
3.	Axial dispersion coefficient of airlift reactor $\dfrac{\partial M_f}{\partial t} = E_r\dfrac{\partial M_f}{\partial r^2} + V_i\dfrac{\partial M_f}{\partial r^2}$ $\sum_{s=1}^{1}(C_t)_p = (B_0/4\pi l)^{\frac{1}{2}}\sum_{p=1}^{1}\exp[-B_0(s-\theta)^2/4\theta]$ (13.17)	$\theta = \frac{t_i}{t_c}$, t_i—instant time (s) t_c—circulation time (s) $s = \frac{L_i}{L_o}$ L_i—distance covered by liquid at the specific time (cm) L_0—length of the circulation loop (cm)

13.6 Kinetics of mixing in airlift and bubble reactors

Mixing time, circulation time, axial dispersion coefficient, and Bernstein numbers are important variables in studying the mixing characteristics of fluids. Phase separation occurs if mixing is not proper, which tends to create three different solid, liquid, and gas phases in the reactor. Mixing in airlift and bubble column reactors are related through certain terms, namely axial dispersion coefficient (E_r), mixing time (t_m), circulation time (t_c), Froude number (F_r), and Bernstein number (B_0). t_m is the time required to attain a uniform concentration of substrate from the time of the addition of the substrate. Axial and radial mixing are responsible for mass transfer in these types of reactors. Axial mixing happens due to rising gas bubbles, which is indicated by E_r. The flow rate, dimensions, and properties of fluids can influence the dispersion coefficient [70,71]. The B_0 helps to determine the degree of mixing in these reactors, which is represented in Eq. (13.15) (Table 13.8). The E_r in bubble column reactors

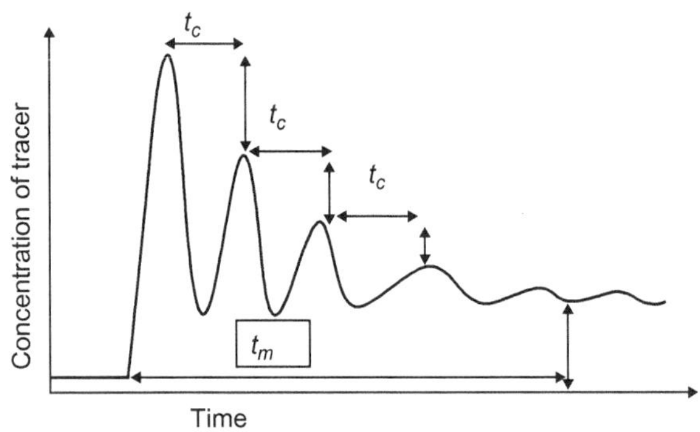

Figure 13.2 Mixing pattern in airlift and bubble column reactors.

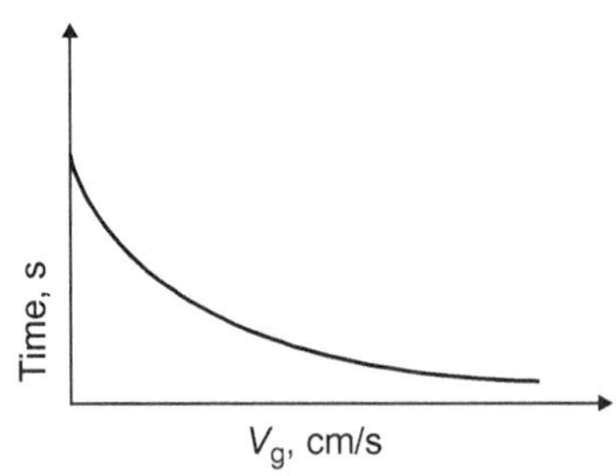

Figure 13.3 Mixing time versus gas velocity of Bando et al. model.

Table 13.9 Equations for cycle time and mixing time.

Sl. no	Model equations	Parameters
1	$V_i = \frac{L_0}{t_c}$ (13.18)	Where, V_i—liquid circulation velocity (cm/s) L_o—length of circulation loop (cm) t_c—circulation time
2	$t_m = cV_G^{-0.5}D^{1.5}\left(\frac{L_D}{D}\right)^{1.2}\left(\frac{D_d}{D}\right)^{-1.4}\left(1-\frac{D_d}{D}\right)^{-1.1}$ (13.19)	Where V_G—gas velocity (cm/s) D—diameter of the vessel (cm) M—constant (2.2 and 2.6 draft tube and annulus sparger) L_D—height of the liquid dispersed (cm) D_d—diameter of draft tube (cm)

is estimated using Eq. (13.16), whereas Eq. (13.17) is used in airlift reactors [23,35,72].

Mixing in a bubble column reactor is not uniform, but it is faster than in an airlift reactor, however, increases in gas velocity or t_c can reduce the t_m. Increasing gas velocity beyond 0.02 m/s can reduce the difference between the downcomer and riser gas holdup because of turbulence and a high number of microbubbles. The t_m can be reduced by enhancing the height of dispersion above the baffle [44,73–75]. Mixing in these reactors can be estimated using tracers such as bromocresol blue, NaCl, and NaOH. The time taken to attain a 95% homogenous concentration of tracer in the liquid from the time of the tracer was injected is called t_m. The t_c is the distance between one crest and trough of peak; the patterns of mixing and circulation are shown in Fig. 13.2 and Eq. (13.18). The t_m is predicted in an airlift reactor using Eq. (13.19) and its pattern is represented in Fig. 13.3, for certain conditions, namely 50–11 cm vessel diameter, 5–40 value for ratio of gas dispersion length-to-diameter, and a V_G value greater than 0.01 m/s (Table 13.9).

13.7 Conclusion

Microalgae are important to the achievement of a sustainable process by which simultaneous waste detoxification and resource recovery are possible. Airlift, bubble column, flat-plate, and tubular PBRs are the most prominent reactors used for resource recovery and CO_2 sequestration. Designing PBRs by considering important design aspect equations to reduce power consumption using CFD and light modeling for biomass and photosynthesis rates have improved the performance of these reactors. Developments are in progress for complete automation of PBRs with respect to climatic conditions, light intensity, and substrate concentration in wastewater using DAQ to attain next-generation PBRs for large-scale application.

References

[1] A. Xia, J. Cheng, J.D. Murphy, Innovation in biological production and upgrading of methane and hydrogen for use as gaseous transport biofuel, Biotechnol. Adv. 34 (2016) 451–472.

[2] D. Batstone, T. Hülsen, C. Mehta, J. Keller, Platforms for energy and nutrient recovery from domestic wastewater: a review, Chemosphere 140 (2015) 2–11.

[3] M.B. Johnson, Z. Wen, Production of biodiesel fuel from the microalga *Schizochytrium limacinum* by direct transesterification of algal biomass, Energy Fuels 23 (2009) 5179–5183.

[4] N. Mallick, Biotechnological potential of immobilized algae for wastewater N, P and metal removal: a review, Biometals 15 (2002) 377–390.

[5] T. Hülsen, E.M. Barry, Y. Lu, D. Puyol, J. Keller, D.J. Batstone, Domestic wastewater treatment with purple phototrophic bacteria using a novel continuous photo anaerobic membrane bioreactor, Water Res. 100 (2016) 486–495.

[6] T.M. Mata, A.A. Martins, N.S. Caetano, Microalgae for biodiesel production and other applications: a review, Renew. Sustain. Energy Rev. 14 (2010) 217–232.

[7] Y. Chisti, Biodiesel from microalgae, Biotechnol. Adv. 25 (2007) 294–306.

[8] G. Markou, E. Nerantzis, Microalgae for high-value compounds and biofuels production: a review with focus on cultivation under stress conditions, Biotechnol. Adv. 31 (2013) 1532–1542.

[9] C. Sambusiti, M. Bellucci, A. Zabaniotou, L. Beneduce, F. Monlau, Algae as promising feedstocks for fermentative biohydrogen production according to a biorefinery approach: a comprehensive review, Renew. Sustain. Energy Rev. 44 (2015) 20–36.

[10] S.E. Fanta, W.R. Hill, T.B. Smith, B.J. Roberts, Applying the light: nutrient hypothesis to stream periphyton, Freshw. Biol. 55 (2010) 931–940.

[11] B. Wang, C.Q. Lan, M. Horsman, Closed photobioreactors for production of microalgal biomasses, Biotechnol. Adv. 30 (2012) 904–912.

[12] I.T.D. Cabanelas, J. Ruiz, Z. Arbib, F.A. Chinalia, C. Garrido-Pérez, F. Rogalla, et al., Comparing the use of different domestic wastewaters for coupling microalgal production and nutrient removal, Bioresour. Technol. 131 (2013) 429–436.

[13] W. Zhou, Z. Wang, J. Xu, L. Ma, Cultivation of microalgae *Chlorella zofingiensis* on municipal wastewater and biogas slurry towards bioenergy, J. Biosci. Bioeng. 126 (2018) 644–648.

[14] M. Podevin, I.A. Fotidis, D. De Francisci, P. Møller, I. Angelidaki, Detailing the start-up and microalgal growth performance of a full-scale photobioreactor operated with bioindustrial wastewater, Algal Res. 25 (2017) 101–108.

[15] P. Praveen, J.Y.P. Heng, K.-C. Loh, Tertiary wastewater treatment in membrane photobioreactor using microalgae: comparison of forward osmosis & microfiltration, Bioresour. Technol. 222 (2016) 448–457.

[16] H.V. de Mendonça, J.P.H.B. Ometto, M.H. Otenio, I.P.R. Marques, A.J.D. dos Reis, Microalgae-mediated bioremediation and valorization of cattle wastewater previously digested in a hybrid anaerobic reactor using a photobioreactor: comparison between batch and continuous operation, Sci. Tot. Environ. 633 (2018) 1–11.

[17] L. Zhu, Z. Wang, Q. Shu, J. Takala, E. Hiltunen, P. Feng, et al., Nutrient removal and biodiesel production by integration of freshwater algae cultivation with piggery wastewater treatment, Water Res. 47 (2013) 4294–4302.

[18] J.-C. Lee, K. Baek, H.-W. Kim, Semi-continuous operation and fouling characteristics of submerged membrane photobioreactor (SMPBR) for tertiary treatment of livestock wastewater, J. Clean. Prod. 180 (2018) 244–251.

[19] J. Shi, B. Podola, M. Melkonian, Application of a prototype-scale twin-layer photobioreactor for effective N and P removal from different process stages of municipal wastewater by immobilized microalgae, Bioresour. Technol. 154 (2014) 260–266.

[20] A. Sheng, M. Bilad, N. Osman, N. Arahman, Sequencing batch membrane photobioreactor for real secondary effluent polishing using native microalgae: process performance and full-scale projection, J. Clean. Prod. 168 (2017) 708–715.

[21] Z. Arbib, J. Ruiz, P. Álvarez-Díaz, C. Garrido-Pérez, J. Barragan, J.A. Perales, Long term outdoor operation of a tubular airlift pilot photobioreactor and a high rate algal pond as tertiary treatment of urban wastewater, Ecol. Eng. 52 (2013) 143–153.

[22] X. Tan, H. Chu, Y. Zhang, L. Yang, F. Zhao, X. Zhou, *Chlorella pyrenoidosa* cultivation using anaerobic digested starch processing wastewater in an airlift circulation photobioreactor, Bioresour. Technol. 170 (2014) 538–548.
[23] X. Gao, B. Kong, R.D. Vigil, Multiphysics simulation of algal growth in an airlift photobioreactor: effects of fluid mixing and shear stress, Bioresour. Technol. 251 (2018) 75–83.
[24] F. Di Caprio, P. Altimari, F. Pagnanelli, Integrated microalgae biomass production and olive mill wastewater biodegradation: optimization of the wastewater supply strategy, Chem. Eng. J. 349 (2018) 539–546.
[25] Q.-H. Shen, Y.-P. Gong, W.-Z. Fang, Z.-C. Bi, L.-H. Cheng, X.-H. Xu, et al., Saline wastewater treatment by *Chlorella vulgaris* with simultaneous algal lipid accumulation triggered by nitrate deficiency, Bioresour. Technol. 193 (2015) 68–75.
[26] F. Gao, Z.-H. Yang, C. Li, Y.-J. Wang, W.-H. Jin, Y.-B. Deng, Concentrated microalgae cultivation in treated sewage by membrane photobioreactor operated in batch flow mode, Bioresour. Technol. 167 (2014) 441–446.
[27] F. Gao, Z.-H. Yang, C. Li, G.-M. Zeng, D.-H. Ma, L. Zhou, A novel algal biofilm membrane photobioreactor for attached microalgae growth and nutrients removal from secondary effluent, Bioresour. Technol. 179 (2015) 8–12.
[28] A. Solovchenko, S. Pogosyan, O. Chivkunova, I. Selyakh, L. Semenova, E. Voronova, et al., Phycoremediation of alcohol distillery wastewater with a novel *Chlorella sorokiniana* strain cultivated in a photobioreactor monitored on-line via chlorophyll fluorescence, Algal Res. 6 (2014) 234–241.
[29] L. Xu, P.J. Weathers, X.R. Xiong, C.Z. Liu, Microalgal bioreactors: challenges and opportunities, Eng. Life Sci. 9 (2009) 178–189.
[30] H. Qiang, A. Richmond, Optimizing the population density in *Isochrysis galbana* grown outdoors in a glass column photobioreactor, J. Appl. Phycol. 6 (1994) 391–396.
[31] H. Blenke, Loop reactors, Advances in Biochemical Engineering, vol. 13, Springer, 1979, pp. 121–214.
[32] M.J. Barbosa, M. Janssen, N. Ham, J. Tramper, R.H. Wijffels, Microalgae cultivation in air-lift reactors: modeling biomass yield and growth rate as a function of mixing frequency, Biotechnol. Bioeng. 82 (2003) 170–179.
[33] A. Contreras, F. García, E. Molina, J. Merchuk, Interaction between CO_2-mass transfer, light availability, and hydrodynamic stress in the growth of *Phaeodactylum tricornutum* in a concentric tube airlift photobioreactor, Biotechnol. Bioeng. 60 (1998) 317–325.
[34] K. Loubière, E. Olivo, G. Bougaran, J. Pruvost, R. Robert, J. Legrand, A new photobioreactor for continuous microalgal production in hatcheries based on external-loop airlift and swirling flow, Biotechnol. Bioeng. 102 (2009) 132–147.
[35] F. Gao, Y.-Y. Peng, C. Li, W. Cui, Z.-H. Yang, G.-M. Zeng, Coupled nutrient removal from secondary effluent and algal biomass production in membrane photobioreactor (MPBR): effect of HRT and long-term operation, Chem. Eng. J. 335 (2018) 169–175.
[36] K. Issarapayup, S. Powtongsook, P. Pavasant, Flat panel airlift photobioreactors for cultivation of vegetative cells of microalga *Haematococcus pluvialis*, J. Biotechnol. 142 (2009) 227–232.
[37] M. Tredici, R. Materassi, From open ponds to vertical alveolar panels: the Italian experience in the development of reactors for the mass cultivation of phototrophic microorganisms, J. Appl. Phycol. 4 (1992) 221–231.
[38] Q. Hu, H. Guterman, A. Richmond, A flat inclined modular photobioreactor for outdoor mass cultivation of photoautotrophs, Biotechnol. Bioeng. 51 (1996) 51–60.

[39] N. Zou, A. Richmond, Effect of light-path length in outdoor flat plate reactors on output rate of cell mass and of EPA in *Nannochloropsis* sp, Prog. Ind. Microbiol., Elsevier, 1999, pp. 351–356.
[40] Y. Bando, H. Hayakawa, M. Nishimura, Effects of equipment dimensions on liquid mixing time of bubble column with draft tube, J. Chem. Eng. Jpn. 31 (1998) 765–770.
[41] Y. Watanabe, D.O. Hall, Photosynthetic CO_2 fixation technologies using a helical tubular bioreactor incorporating the filamentous cyanobacterium *Spirulina platensis*, Energy Convers. Manag. 36 (1995) 721–724.
[42] C. Ugwu, H. Aoyagi, H. Uchiyama, Influence of irradiance, dissolved oxygen concentration, and temperature on the growth of *Chlorella sorokiniana*, Photosynthetica 45 (2007) 309–311.
[43] C. Ugwu, H. Aoyagi, H. Uchiyama, Photobioreactors for mass cultivation of algae, Bioresour. Technol. 99 (2008) 4021–4028.
[44] K.J. George, A.K. Jhawar, A. Prakash, Investigations of flow structure and liquid mixing in bubble column equipped with selected internals, Chem. Eng. Sci. 170 (2017) 297–305.
[45] P. Carlozzi, B. Pushparaj, A. Degl'Innocenti, A. Capperucci, Growth characteristics of *Rhodopseudomonas palustris* cultured outdoors, in an underwater tubular photobioreactor, and investigation on photosynthetic efficiency, Appl. Microbiol. Biotechnol. 73 (2006) 789–795.
[46] F. Lehr, C. Posten, Closed photo-bioreactors as tools for biofuel production, Curr. Opin. Biotechnol. 20 (2009) 280–285.
[47] O. Jorquera, A. Kiperstok, E.A. Sales, M. Embirucu, M.L. Ghirardi, Comparative energy life-cycle analyses of microalgal biomass production in open ponds and photobioreactors, Bioresour. Technol. 101 (2010) 1406–1413.
[48] E. Molina, J. Fernández, F. Acién, Y. Chisti, Tubular photobioreactor design for algal cultures, J. Biotechnol. 92 (2001) 113–131.
[49] E. Sforza, D. Simionato, G.M. Giacometti, A. Bertucco, T. Morosinotto, Adjusted light and dark cycles can optimize photosynthetic efficiency in algae growing in photobioreactors, PLOS One 7 (2012) e38975.
[50] J.P. Bitog, I.-B. Lee, C.-G. Lee, K.-S. Kim, H.-S. Hwang, S.-W. Hong, et al., Application of computational fluid dynamics for modeling and designing photobioreactors for microalgae production: a review, Comput. Electron. Agric. 76 (2011) 131–147.
[51] J. Gimbun, C.D. Rielly, Z.K. Nagy, Modelling of mass transfer in gas–liquid stirred tanks agitated by Rushton turbine and CD-6 impeller: a scale-up study, Chem. Eng. Res. Des. 87 (2009) 437–451.
[52] D. Pfleger, S. Gomes, N. Gilbert, H.-G. Wagner, Hydrodynamic simulations of laboratory scale bubble columns fundamental studies of the Eulerian–Eulerian modelling approach, Chem. Eng. Sci. 54 (1999) 5091–5099.
[53] E. Delnoij, F. Lammers, J. Kuipers, W.P.M. van Swaaij, Dynamic simulation of dispersed gas–liquid two-phase flow using a discrete bubble model, Chem. Eng. Sci. 52 (1997) 1429–1458.
[54] J. Joshi, Computational flow modelling and design of bubble column reactors, Chem. Eng. Sci. 56 (2001) 5893–5933.
[55] A. Akhtar, V. Pareek, M. Tadé, CFD simulations for continuous flow of bubbles through gas–liquid columns: application of VOF method, Chem. Product. Process. Model. 2 (2007).
[56] R. Krishna, J. Van Baten, Simulating the motion of gas bubbles in a liquid, Nature 398 (1999) 208.

[57] P.M. Wilkinson, Physical Aspects and Scale-Up of High Pressure Bubble Columns, Rijksuniversiteit Groningen, 1991.
[58] P. Sobieszuk, P. Cygański, R. Pohorecki, Bubble lengths in the gas–liquid Taylor flow in microchannels, Chem. Eng. Res. Des. 88 (2010) 263–269.
[59] T. Pirasaci, C. Wickramaratne, F. Moloney, D. Yogi Goswami, E. Stefanakos, Dynamics of phase change in a vertical PCM capsule in the presence of radiation at high temperatures, Appl. Energy 206 (2017) 498–506.
[60] F.C. Rubio, F.G. Camacho, J.M.F. Sevilla, Y. Chisti, E.M. Grima, A mechanistic model of photosynthesis in microalgae, Biotechnol. Bioeng. 81 (2003) 459–473.
[61] J.U. Grobbelaar, The influence of light/dark cycles in mixed algal cultures on their productivity, Bioresour. Technol. 38 (1991) 189–194.
[62] H.-P. Luo, M.H. Al-Dahhan, Analyzing and modeling of photobioreactors by combining first principles of physiology and hydrodynamics, Biotechnol. Bioeng. 85 (2004) 382–393.
[63] A.D. Jassby, T. Platt, Mathematical formulation of the relationship between photosynthesis and light for phytoplankton, Limnol. Oceanogr. 21 (1976) 540–547.
[64] H.Y. Lee, L.E. Erickson, S.S. Yang, Kinetics and bioenergetics of light-limited photoautotrophic growth of *Spirulina platensis*, Biotechnol. Bioeng. 29 (1987) 832–843.
[65] J. Doucha, F. Straka, K. Lívanský, Utilization of flue gas for cultivation of microalgae *Chlorella* sp. in an outdoor open thin-layer photobioreactor, J. Appl. Phycol. 17 (2005) 403–412.
[66] Y.-S. Yun, J.M. Park, Kinetic modeling of the light-dependent photosynthetic activity of the green microalga *Chlorella vulgaris*, Biotechnol. Bioeng. 83 (2003) 303–311.
[67] X. Wu, J.C. Merchuk, Simulation of algae growth in a bench scale internal loop airlift reactor, Chem. Eng. Sci. 59 (2004) 2899–2912.
[68] F.G.A. Fernández, F.G. Camacho, J.A.S. Pérez, J.M.F. Sevilla, E.M. Grima, A model for light distribution and average solar irradiance inside outdoor tubular photobioreactors for the microalgal mass culture, Biotechnol. Bioeng. 55 (1997) 701–714.
[69] D. Dermoun, D. Chaumont, J.-M. Thebault, A. Dauta, Modelling of growth of *Porphyridium cruentum* in connection with two interdependent factors: light and temperature, Bioresour. Technol. 42 (1992) 113–117.
[70] W.D. Deckwer, R. Burckhart, G. Zoll, Mixing and mass transfer in tall bubble columns, Chem. Eng. Sci. 29 (1974) 2177–2188.
[71] Y.T. Shah, B.G. Kelkar, S.P. Godbole, W.-D. Deckwer, Design parameters estimations for bubble column reactors, AIChE J. 28 (1982) 353–379.
[72] A. Sánchez Mirón, M.C. Cerón García, F. García Camacho, E. Molina Grima, Y. Chisti, Mixing in bubble column and airlift reactors, Chem. Eng. Res. Des. 82 (2004) 1367–1374.
[73] A.S. Mirón, M.-C.C. García, F.G. Camacho, E.M. Grima, Y. Chisti, Mixing in bubble column and airlift reactors, Chem. Eng. Res. Des. 82 (2004) 1367–1374.
[74] Y. Chisti, M. Moo-Young, On the calculation of shear rate and apparent viscosity in airlift and bubble column bioreactors, Biotechnol. Bioeng. 34 (1989) 1391–1392.
[75] M. Gavrilescu, R.Z. Tudose, Residence time distribution of the liquid phase in a concentric-tube airlift reactor, Chem. Eng. Process. Process. Intensif. 38 (1999) 225–238.

CHAPTER 14

An overview of bioreactor configurations and operational strategies for dark fermentative biohydrogen production

Arindam Sinharoy, Manoj Kumar and Kannan Pakshirajan
Department of Biosciences and Bioengineering, Indian Institute of Technology Guwahati, Guwahati, India

14.1 Introduction

About 80% of the energy consumed in the world is dependent on fossil fuel sources and this is increasing due to rapid industrialization and urbanization. This has not only led to the inevitable depletion of limited fossil fuels [1], but it is also a major concern to the world due to its effect on global climate change, environmental degradation, and health problems [2]. As a result of fossil fuel combustion, pollutants such as CO_x, NO_x, SO_x, C_xH_x, ash, droplets of tar, soot, and other organic compounds are released into the atmosphere [3]. Thus, increasing focus is being placed on clean energy supply and environmental protection for overall sustainable development. Environment-friendly energy sources are currently the most highlighted theme in the energy and environmental sectors.

Renewable energy, in the form of solar, wind, hydroelectric, geothermal, biofuels, and hydrogen, reduces dependency on fossil fuels for energy for various applications. Because most forms of renewable energy are derived either directly or indirectly from natural processes, for example, from the sun, obtaining a constant supply is not a problem, unlike in the case of fossil fuels. The use of renewable energy also provides environmental, economic, and political benefits [4].

Hydrogen (H_2), an environment-friendly, renewable, clean energy, yields 2.75 times more energy than that of any of the hydrocarbon fuels upon combustion. It also serves to mitigate global warming and ever growing pollution problems, since it produces only water as the sole by-product with zero emissions of air pollutants such as NO_x, SO_x, and

Bioreactors
DOI: https://doi.org/10.1016/B978-0-12-821264-6.00014-0

greenhouse gases [5]. Apart from bioenergy applications, it has a wide range of other potential functions. It is a potent electron donor in various reductive processes, both in the chemical and biotechnological industries. Besides, H_2 is frequently used in the ammonia production process and in the hydrogenation of edible oil, petroleum, coal, and shale oil [2]. Also, electricity can be produced directly from H_2 using fuel cell technologies. Currently, scientists and engineers are focusing their interests in developing new and efficient technologies for H_2 production through the electrolysis of water or the catalytic reduction of methane (CH_4) and carbohydrates [6].

The conventional process of H_2 production from fossils and water involves enormous energy consumption and greenhouse gas emissions [7]. On the contrary, biological H_2 production from renewable biomass is highly advantageous, mainly because it can be produced by converting biomass to photo-energy using H_2-producing bacteria or from a variety of renewable resources or even from wastewater via a low energy demanding continuous process [4,8]. Fig. 14.1 shows various thermochemical, catalytic, and biological processes for H_2 production and their applications.

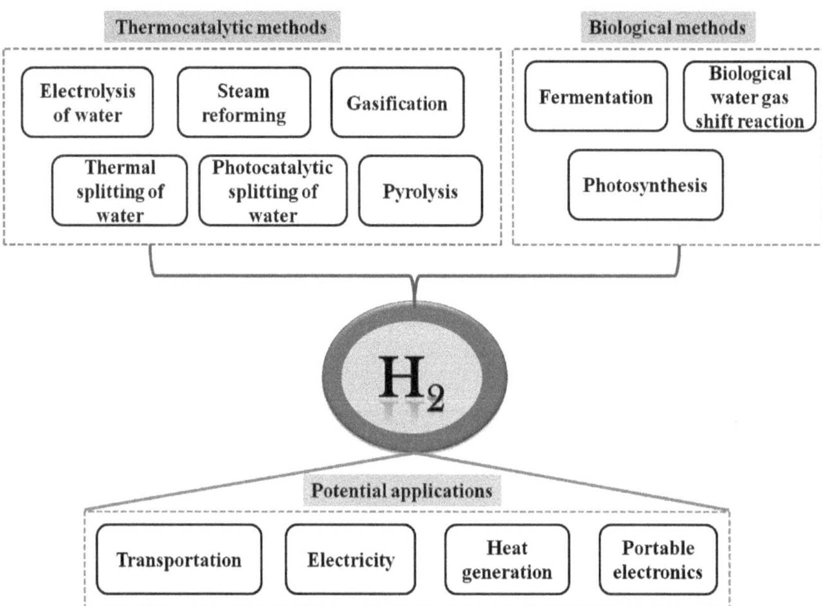

Figure 14.1 Different thermochemical, catalytic, and biological processes for H_2 production and their applications.

Biological H_2 production processes can be broadly divided into two main classes, namely (1) light-dependent autotrophic process, and (2) light-dependent or independent heterotrophic process. In autotrophic conversion, microalgae and photosynthetic bacteria produce H_2 by directly utilizing solar energy. On the other hand, in heterotrophic processes, photosynthetic bacteria or anaerobic bacteria produce H_2 from organic compounds under light or dark conditions, respectively [5,8]. In another process, known as water gas shift reaction, carboxydotrophic bacteria produce H_2 from carbon monoxide (CO) [7,9]. Fig. 14.2 provides an overview of the different biohydrogen production processes known till date.

In the photolysis process, algae or cyanobacteria produce H_2 directly through water-splitting during photosynthesis. Light is absorbed by the organisms and due to the transfer of electrons to hydrogenase and/or nitrogenase enzymes H_2 is produced [10]. In another method, certain algae are shown to produce H_2 photosynthetically under sulfur deficient anaerobic

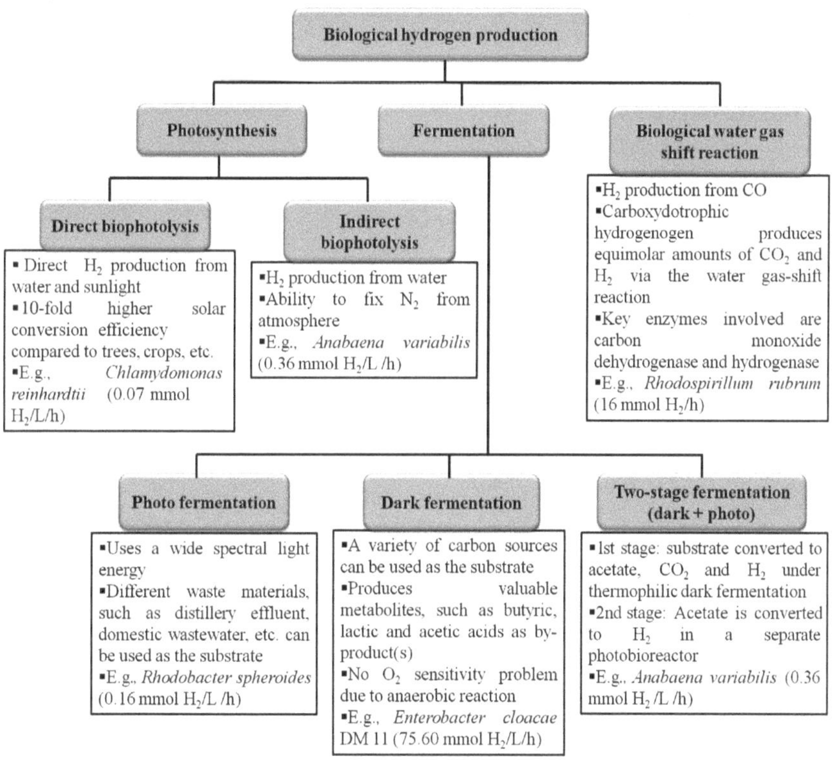

Figure 14.2 Overview of different biohydrogen production processes [5,8,10,11].

conditions [12]. The energy for the process is derived from light using the hydrogenase enzyme system. Since the main substrate is water, and is plentiful, direct photolysis seems to be more attractive than the other methods for biohydrogen production. However, there are a number of challenges that need to be taken care of before the process can be commercialized successfully. Few such drawbacks include the need for a large land area to capture maximum light and its low production rate, and so forth. [11].

In the photofermentation process, purple, nonsulfur, photosynthetic bacteria produce H_2 utilizing organic substrates under nitrogen deficient conditions using light energy. Unlike green algae or cyanobacteria, these bacteria are unable to split water and use organic acids derived from organic compounds as electron donors. Nitrogenase is a key enzyme in this process. Major advantages of this process include the significantly high yield of H_2 and noninhibition due to the presence of O_2. The limitations with this process include the requirement of organic acids, reduced enzyme activity, and high energy requirement [13].

Most popular among the various biological H_2 production routes is dark fermentation, wherein H_2 can be produced from carbon-rich substrate by anaerobic bacteria without any light requirement [14]. Carbohydrates, primarily glucose, are the carbon source for dark fermentation using anaerobic bacteria, and this primarily leads to the production of acetic and butyric acids along with H_2 gas. In contrast to direct and indirect photolysis, pure H_2 is not produced during dark fermentation, but carbon dioxide (CO_2) and occasionally CH_4 and hydrogen sulfide (H_2S) are formed [15]. This process seems advantageous as it leads to high energy production, requires no light, and utilizes a variety of organic compounds including different organic wastes and wastewaters for H_2 production [16]. However, H_2 production by this process is highly dependent on the process conditions such as pH, hydraulic retention time (HRT), gas partial pressure, among others [10].

A large number of studies on biohydrogen production are based on small-scale batch systems and process optimization; however, biohydrogen production using continuous bioreactor systems is essential for the successful scale-up of this technology.

14.2 Bioreactors for hydrogen fermentation

Bioreactor design and process development are key areas of research for the successful scale-up of biohydrogen production [10]. Different modes

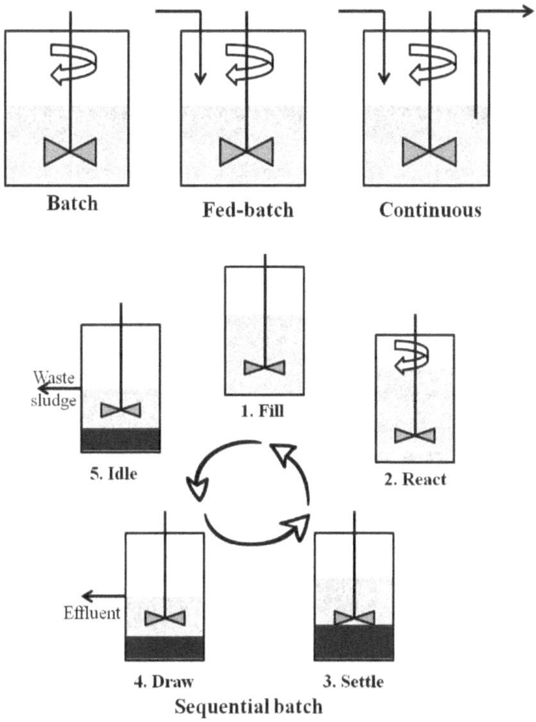

Figure 14.3 Various bioreactor operational modes for biohydrogen production.

of operation, mainly batch, fed batch, sequential batch, and continuous, have been studied for H_2 production from a variety of substrates (Fig. 14.3).

Under a batch operation mode, a bioreactor is operated as a closed system with the addition of substrate and inoculum at the start of the batch process. In fed batch mode, substrate is added into the reactor at a high concentration toward the depletion of the added substrate, but without increasing the working reactor volume beyond its capacity. During sequential batch mode operation, the reactor is operated for a fixed time with provisions for filling, fermenting, settling, and withdrawing the reactor contents at predetermined time intervals. Whereas in continuous operation mode, the substrate is continuously supplied to the reactor with continuous product recovery. In the case of H_2 production, sequential batch and continuous modes of operation are successfully followed. During reactor operation, both liquid and gaseous samples are withdrawn periodically at a set time interval and analyzed for substrate conversion,

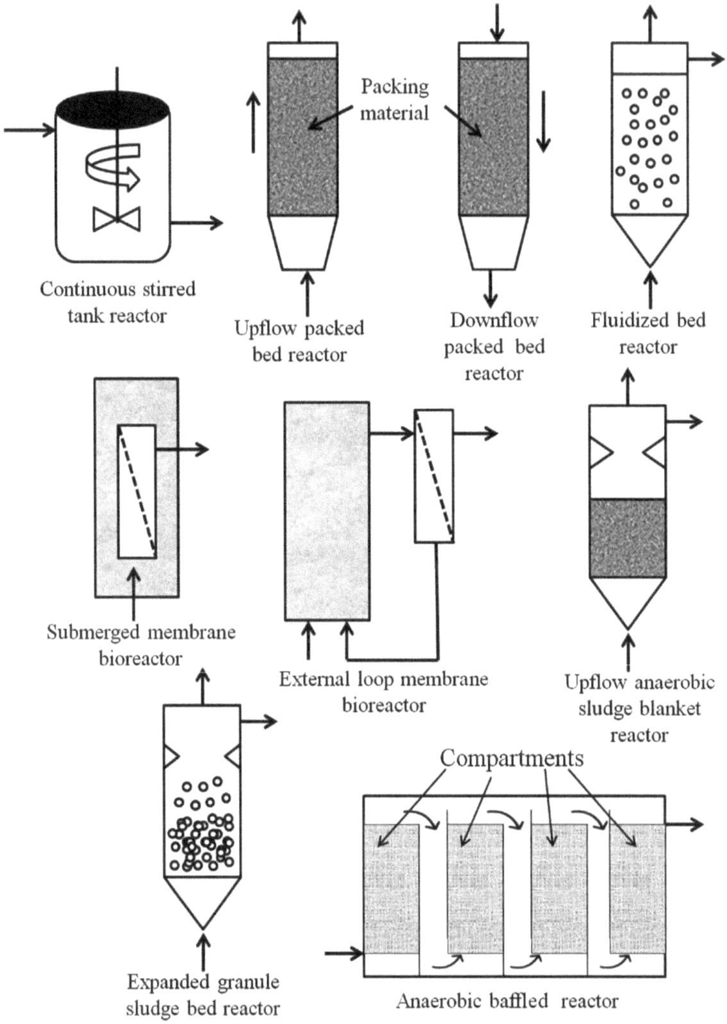

Figure 14.4 Various bioreactor systems used for fermentative biohydrogen production [2,17–23].

biomass concentration, headspace biogas composition, H_2 production, and volatile fatty acids production, among others. The continuous stirred tank reactor (CSTR) is the most commonly used bioreactor system, not only for H_2 production, but also for various other biochemical processes; although other types of reactors such as packed bed bioreactor (PBR) or fixed bed reactor, membrane bioreactor (MBR), fluidized bed reactor (FBR), upflow anaerobic sludge blanket reactor (UASB), expanded granular sludge bed (EGSB), and anaerobic baffled reactor (ABR), among others, have been examined under continuous operation mode. Fig. 14.4 is a schematic of these different types of bioreactors. Table 14.1 provides details about biohydrogen production in different bioreactors and Table 14.2 compares the commonly used bioreactors for H_2 production along with their advantages and drawbacks. A detailed discussion on these different bioreactor configurations for H_2 production is provided here.

14.2.1 Continuous stirred tank reactor

The simplest type of bioreactor used for H_2 production is the stirred tank reactor or CSTR, which is operated with continuous feeding and withdrawal of its contents. The advantages of using CSTR for biohydrogen production are good mixing and mass transfer and a relatively low start-up period (Table 14.2) [62]. However, as in the case of any suspended growth systems, the main concerns are cell washout and low biomass retention at high dilution rates, and, hence, it requires rigorous monitoring and control [17]. A number of studies have reported H_2 production from a wide range of substrates using CSTR (Table 14.1) [66], among which the salient studies are discussed further.

Palomo-Briones et al. studied biohydrogen production using CSTR with heat-treated anaerobic sludge biomass and investigated the effect of an HRT in the range 6–24 h [24]. Lactose at a 20 g/L concentration was used as the substrate. The highest volumetric H_2 production rate of 2000 ± 149 mL/L/day with a H_2 yield of 0.86 mol H_2/mol lactose was reported for a 6 h HRT. Detailed analysis of the microbial community and metabolites formed in the system showed the predominance of *Clostridiaceae, Lachnospiraceae, Enterobacteriaceae* spp. for an enhanced production of H_2 at low HRT values (6 h). Whereas at a long HRT, microbial species belonging to *Sporolactobacillaceae–Streptococcaceae* genera outnumbered H_2-producing bacteria, which led to higher levels of lactate than H_2 in the reactor. These findings clearly indicate HRT is an important operational parameter that controls the microbial community and metabolic flux in such continuous bioreactor systems to produce the desired product.

Luo et al. [67] examined anaerobic H_2 production from cassava stillage using a CSTR and showed that heat pretreatment was ineffective in preventing methanogenesis compared to H_2 production under mesophilic temperature conditions. However, under thermophilic conditions, the H_2

Table 14.1 Biohydrogen production in different bioreactors.

Carbon source	Biomass type	Operational condition			Hydrogen production		References
		HRT (h)	T (°C)	pH	Rate	Yield	
Continuous stirred tank reactor							
Lactose	Anaerobic sludge	6	37	5.9	2.0 L/L/day	0.86 mol H_2/mol lactose	[24]
Glucose	Mixed culture	8	23	5.5	NA	1.77 mol H_2/mol glucose	[25]
Tofu processing waste	Mixed culture	8	60	7.6	8.17 L/L/day	1.2 mol H_2/mol hexose	[26]
Rice straw hydrolysate	Mixed culture	4	37	5.5	10 L/L/day	0.069 mol H_2/mol T-sugar	[27]
Rice straw	Mixed culture	4	37	5.5	16.32 L/L/day	1.02 mol H_2/mol hexose	[18]
Sugarcane syrup	*Clostridium butyricum*	3	37	NA	17.5 L/L/day	NA	[28]
Cheese whey	Mixed culture	24	35	5.2	2.9 L/L/day	0.78 mol H_2/mol glucose	[29]
Food waste hydrolysate	Mixed culture	6	55	˃4	8.49 L/L/day	85.6 mL/g food waste	[30]
Galactose	Anaerobic granular sludge	8	37	5.5	11.9 L/L/day	2.14 mol H_2/mol galactose	[31]
Expanded granular sludge bed reactor							
Glucose	Anaerobic granular sludge	10	25	5.5	4.1 L/L/day	1.6 mol H_2/mol glucose	[32]
Molasses	Mixed culture	6–1	35	4.2–4.4	0.71 L/L/ h	3.47 mol/mol sucrose	[17]

Starch-containing wastewater	Mixed culture	8	30	3.95	1.64 L/L/day	0.11 L/g COD	[17]
Glucose	Mixed culture	10	30	5.5	4.23 L/L/day	0.92 mol H_2/mol hexose	[33]
Anaerobic fluidized bed reactor							
Synthetic wastewater containing glucose	Mixed culture	8	23	5	NA	2.11 mol H_2/mol glucose	[19]
Glucose	Granular sludge	1	37	4	7.6 L/L/h	1.7 mol H_2/mol glucose	[34]
Glucose	Methanogenic granular sludge	10	30	5.5	7.0 L/L/day	3.5 mol H_2/mol hexose	[33]
Glucose	Activated sludge and digested sludge	1	37	4	2.36 L/L/h	NA	[35]
Glucose	Mixed culture	2	30	6.4	0.6 L/L/h	2.49 mol H_2/mol glucose	[36]
Glucose	Mixed culture	1	30	3.8	1.28 L/L/h	2.29 mol H_2/mol glucose	[37]
Glucose	Mixed culture	1	30	5.5	1.21 L/L/h	2.59 mol H_2/mol glucose	[38]
Glucose	Mixed culture	1	30	6–7	0.76 L/L/h	2.45 mol H_2/mol glucose	[39]
Sucrose	Methanogenic sludge	24	35	4.5–5	NA	5.1 mmol H_2/gas	[40]
Sucrose	Anaerobic sludge	2	40	NA	1.80 L/L/h	4.26 mol H_2/mol sucrose	[41]

(*Continued*)

Table 14.1 (Continued)

Carbon source	Biomass type	Operational condition			Hydrogen production		References
		HRT (h)	T (°C)	pH	Rate	Yield	
Membrane bioreactor							
Glucose	Mixed culture	9	35	5.5	5.8 L/L/day	1.19 mol H_2/mol glucose	[42]
Fructose	Mixed culture	1	35	6.7	2.75 L/L/h	1.3 mol H_2/mol hexose	[35]
Glucose	Anaerobic mixed microflora	9	35	5.5	NA	1.62 mol H_2/mol glucose	[42]
Food waste	Mixed culture	14	55	5.5	10.7 L/L/day	111.1 mL H_2/g VS	[20]
Glucose	Mixed mesophilic microflora	NA	35	6	0.2–0.25 L/day /g substrate	1.1 mol H_2/mol glucose	[23]
Packed bed reactor							
Oat straw hydrolysate	Mixed culture	24–6	28	5.5	3.3 mmol/L/h	2.9 mol H_2/mol hexose	[43]
Glucose	Mixed culture	10	27	5.5	6.01 L/L/day	1.29 mol H_2/mol glucose	[44]
Sucrose	Mixed culture	2–4	25	6	8.9 L/L/day	2.0 mol H_2/mol sucrose	[22]
Cane molasses	Mixed culture	10–1	60	6.5	1.7 L/L/h	16.82 mmol/ g COD	[3]
Glucose	Mixed culture	0.5–2	30	NA	NA	2.48 mol H_2/mol glucose	[45]
Soft-drink wastewater	mixed culture	0.5	25	6.5	0.4 L/L/h	3.5 mol H_2/mol sucrose	[46]

Sucrose	Mixed culture	2	25	6.5	128.13 mL/L/h	4.22 mol H_2/mol sucrose	[40]
Cheese whey	Mixed culture	24	30	5	1.0 L/L/day	0.668 mol H_2/mol lactose	[47]
Sucrose	Mixed culture	2	25	6.5	195.6 mL/L/h	6.9 mol H_2/mol sucrose	[48]
Sucrose	Mixed culture	2	25	NA	4.83 L/L/day	3.22 mol H_2/mol sucrose	[49]
Sugarcane vinasse	Mixed culture	12	55	6.5	1117.2 mL/L/day	2.4 mol H_2/mol carbohydrates	[50]
Galactose	Mixed culture	2	37	5.5–6.0	65.5 L/L/day	2.60 mol/mol hexose	[2]
Glucose	Mixed culture	8	37	7	23.4 L/L/day	1.0 mol H_2/mol glucose	[50]
Glucose	Mixed culture	2–4	37	6	2.32 L/L/day	1.34 mol H_2/mol glucose	[51]
Sugarcane vinasse	Mixed culture	24	25	6.5	509.5 mL/L/day	3.2 mol H_2/mol carbohydrates	[52]
Upflow anaerobic sludge blanket reactor							
Galactose	Mixed culture	2	37	5.5–6.2	56.8 L/L/day	2.25 mol/mol galactose	[2]
Starch	Mixed culture	24	55	5.0	5.0 L/L/day	1.7 mol H_2/mol glucose	[53]
Cheese whey	Mixed culture	24	30	5	122 mL/L/day	NA	[54]
Cheese whey powder	Mixed culture	18	22–25	5.9	0.38 L/L/day	1.31 mol H_2/mol hexose	[47]

(Continued)

Table 14.1 (Continued)

Carbon source	Biomass type	Operational condition			Hydrogen production		References
		HRT (h)	T (°C)	pH	Rate	Yield	
Cassava wastewater	Mixed culture	NA	37	5.5	0.39 L/L/day	39.83 l H_2/kg COD	[55]
Glycerol	Mixed culture	NA	37	5.5	6.0 mmol/L/h	579.8 mmol H_2/mol glycerol	[56]
Galactose	Mixed culture	6	37	NA	12.5 L/L/day	1.5 mol/mol galactose	[57]
Cheese whey powder	Heat-treated mixed culture	6	NA	5.3	1.67 L/L/day	NA	[58]
Anaerobic baffled reactor							
Sucrose	Sewage sludge	8	35	5.5	10.9 L/L/day	1.7 mol H_2/mol Sucrose	[21]
Monoethylene glycol	Mixed culture bacteria	36	30	7.09	438.07 mL/L/day	122.23 mL H_2/g COD	[59]
Tapioca wastewater	Mixed culture	6	32	9	883.19 mL/L/day	13.4 mL H_2/g COD	[60]
Municipal food waste	Mixed culture	26	26	6.5	NA	4.9 mol H_2/mol hexose	[61]

Table 14.2 Comparison of different bioreactors used for biohydrogen production.

Reactor type	Advantages	Disadvantages	References
CSTR	• Continuous operation • Good temperature control • Well suited for two-phase system • Good operational control • Simple construction • Low operating cost • Easy to clean	• Lowest conversion per unit volume • Channeling due to poor agitation	[10,51,62]
UASB	• High conversion rate of feed to hydrogen • Withstand high organic loading rates • Low energy requirement • Low nutrients requirement • Good operational control	• Requires expert design and construction supervision • Difficult to maintain proper hydraulic conditions (upflow and settling rate must be balanced)	[10,28,56,63]
EGSB	• High biomass retention • High upflow velocity • High mass transfer • Withstand high organic loading rates • Low energy requirement • Low footprint	• Kinetics dependent on the temperature • Slow start-up • Prone to inhibition by toxic compounds	[17,64]
PBR	• High conversion per unit mass of catalyst • Low operating cost • Continuous operation	• Undesired thermal gradients may exist • Poor temperature control • Channeling may occur • Difficult to maintain and clean	[35,43,50,51]

(Continued)

Table 14.2 (Continued)

Reactor type	Advantages	Disadvantages	References
FBR	• Uniform particle mixing • Uniform temperature gradient • Capable of continuous operation • Catalyst can be continuously regenerated with the use of an auxiliary loop	• Bed-fluid mechanics not well understood • Severe agitation can result in biomass destruction and dust formation • Scale-up is difficult	[10,19,34,51]
MBR	• High biomass retention • High conversion efficiency • High volumetric H_2 production rates	• Membrane fouling is a potential menace • High capital cost	[23,62]
ABR	• High resistance to shock loadings • Good biomass retention • Less sludge production • Compartmentalized structure enables different microbial metabolisms within the same system • Withstand sudden changes in pH and temperature	• Needs expert design • Long start-up phase • Pretreatment required to prevent clogging due to large size particles	[59,65]

ABR, Anaerobic baffled reactor; *CSTR*, continuous stirred tank reactor; *EGSB*, expanded granular sludge bed; *FBR*, fluidized bed reactor; *MBR*, membrane bioreactor; *PBR*, packed bed bioreactor; *UASB*, upflow anaerobic sludge blanket reactor.

yield increased fivefold to 69.6 mL H_2/g VSS and volatile fatty acid production decreased due to the inhibitory effect of the temperature on homoacetogens. Thus operational temperature rather than pretreatment of biomass plays an important role in biohydrogen production. Further, it is reported that under a high organic loading rate (OLR), that is, at >10 g VSS/L/day, the two-phase thermophilic CSTR showed a stable performance with 56.6 mL H_2/g VSS and 249 mL CH_4/g VSS, compared with that of a single-stage system. In the single-stage CSTR, low pH due to the accumulation of acetate and propionate resulted in its poor performance.

Shen et al. examined the effect of five different OLRs (4.0–30 g COD/L/day) on fermentative H_2 production [25]. The results showed that a maximum H_2 yield of 1.78 mol H_2/mol glucose was achieved at a moderate loading rate of 22 g COD/L/day. Further increase in the organic loading created excess burden on the system and reduced the H_2 yield and H_2 production rate due to substrate inhibition. Kim et al. carried out continuous fermentative H_2 production from tofu (soybean curd) processing waste (TPW) using anaerobic mixed microflora under thermophilic (60°C) conditions in a CSTR [68]. In order to increase the available soluble carbon for the bacteria, the TPW was diluted with tap water and hydrolyzed using 0.5% hydrogen chloride (HCl). The ratio of soluble chemical oxygen demand (COD) to total COD increased several-fold from 14% to 60%, and the soluble carbohydrate concentration was increased threefold, from 2.4 g/L to 7.2 g/L. This pretreatment of the substrate also increased the H_2 production by 2.8-fold. In continuous operation mode and at shift, a stable volumetric H_2 production rate of 8.17 ± 0.32 L of H_2/L/day and a H_2 yield of 1.20 ± 0.05 mol H_2/mol hexose were achieved using the pretreated TPW as the substrate.

Liu et al. used acid pretreated rice straw hydrolysate (RSH) as a substrate for H_2 production using a CSTR [27]. The rice straw was completely solubilized using concentrated sulfuric acid (H_2SO_4) at 40°C. The effect of initial pH and substrate concentration on biohydrogen production was assessed at 37°C under a batch mode of operation. The maximum H_2 yield and titer were 0.44 mol H_2/mol total sugar and 97.30–0.17 mL, respectively, with a substrate utilization of 81.55%. Further, a mixture of food industry wastewater with RSH was used for biohydrogen production using CSTR under continuous operational mode. The results showed 1.5 times higher H_2 production rate (10 ± 1.17 L/L/day) compared with that under batch mode, with an H_2

yield of 0.69 mol H_2/mol total sugar. A denaturing gradient gel electrophoresis analysis of the biomass confirmed the presence of *Clostridium pasteurianum*, reported to produce H_2 from acetic acid and butyric acid as the intermediate.

A few modifications to the conventional CSTR have been suggested for achieving a high H_2 production and yield using different substrates. Liu et al. studied biohydrogen production from RSH using a continuously external circulating bioreactor (CECBR), in which the original hydrolysate contained 40−50 g total sugar/L, that was adjusted to 20 g total sugar/L with tap water for effective feeding [18]. The working volume of the CECBR was 300 mL with a volumetric feed circulating rate of 9.6 L/min. The average H_2 production rates of 5.52 L/L/day and 16.32 L/L/day were achieved for 4 h and 8 h HRT, respectively. The value of H_2 production rate at HRT 4 h was three times more than that at HRT 8 h. The H_2 yields were 0.72 and 1.02 mol H_2/mol hexose respectively for these two HRTs. Biomass washout was observed at HRT less than 2 h in the CECBR, which resulted in a low H_2 production as well as ineffective utilization of RSH. Hence a minimum HRT of 4 h is found to be effective for H_2 production from RSH in a CECBR.

Han et al. examined a continuous mixed immobilized sludge reactor with activated carbon as the biosupport material and enzyme hydrolyzed food waste as the substrate for H_2 production [50]. The effects of immobilized sludge packing ratio (10%−20%, v/v) and OLR (8−40 kg/m^3/day) on biohydrogen production were examined. The best H_2 production rate of 353.9 mL/h/L was obtained at 15% packing ratio and 40 kg/m^3/day OLR. Kumar et al. investigated the performance of a galactose fed continuous reactor producing H_2 [2]. The results showed that process disturbances reduced H_2 production as well as large variations in the microbial diversity. High H_2 yields of 2.01 ± 0.05 and 2.14 ± 0.03 mol/mol galactose were obtained at 12 and 8 h HRT, respectively. The peak energy generation rate and energy yield were 134 kJ/L/day and 612 kJ/mol galactose consumed, respectively.

In some studies, a combination of two or more reactors have been shown to be effective for improving the biohydrogen production efficiency. For example, Venetsaneas et al. used cheese whey as a substrate for the production of H_2 and CH_4 in a two-stage continuous system [29]. In the first stage, mesophilic fermentative H_2 production from undiluted cheese whey was investigated at a fixed pH value of 5.2 and 24 h HRT. The highest H_2 production rate of 2.9 ± 0.2 L/L/day with a yield of

0.78 ± 0.05 mol H_2/mol glucose consumed was achieved in the reactor with pH control. Effluent from the hydrogenogenic reactor was further processed for biogas production in a continuous mesophilic anaerobic bioreactor. The methanogenic reactor was operated at an HRT of 20 days and produced approximately 1 L CH_4/day, corresponding to a yield of 6.7 L CH_4/L of the influent. A COD removal efficiency of 95.3% was accomplished using this two-stage process, indicating the superior performance of the system compared to other reactor systems. In another study, Nualsri et al. combined a CSTR (1 L) with a UASB reactor (24 L) for H_2 and CH_4 production [28]. The CSTR was fed with 25 g COD/L sugarcane syrup to produce H_2. Effluent from the CSTR was directly fed to the UASB reactor for CH_4 production. A maximum H_2 production rate of 17.5 L/L/day and CH_4 production rate of 2.25 L/L/day were achieved at optimum HRT values of 3 h in the CSTR and 3 d in the UASB reactor with a total energy production rate of 270 kJ/L/day. The two-stage reactor system showed stable performance over 200 days with a COD removal of 97.5%. The volatile fatty acids to alkalinity ratio in the UASB reactor was below the critical value of 0.4 at all HRTs, indicating its stable performance for a long-term CH_4 production process with direct feeding of raw hydrogenogenic effluent.

14.2.2 Packed bed reactor

PBR is operated with biosupport materials packed inside for the H_2-producing bacteria to grow and form biofilm. A wide range of materials including glass beads, expanded clay (EC), perlite, activated carbon, ceramic, coconut coir, synthetic polymers, and plastic materials have been used as biosupport material [35,43,45,69]. Support materials for superior biofilm formation should be inert and have a high specific surface area, rough surface, high porosity, and so forth. Owing to its ability to retain high biomass concentrations in the reactor, a PBR system can achieve high conversion rates. However, the mixing regime in a PBR is poor compared with that in a CSTR, resulting in a low mass transfer and low yield of product to substrate (Table 14.2). In order to improve mixing, a recirculation loop is often introduced in PBR-type systems. Based on the flow regime inside the reactor, PBRs can be classified as either upflow or downflow (Fig. 14.3).

Roy et al. studied H_2 production using cane molasses in a PBR with coconut coir as the support material under a continuous mode of operation [58]. A high H_2 production rate of 1.7 L/L/h was achieved at a dilution rate and recycle ratio of 0.8 h and 0.6, respectively. The substrate utilization kinetics revealed that at a flow rate of 245 mL/h, the external film mass transfer coefficient and first order substrate utilization constant were 55.4% and 44.6%, respectively. The viable biomass concentration in the reactor was directly proportional to the recycle ratio with the highest value obtained at a recycle ratio of 0.6, which enhanced the H_2 production rates by 9%. Viable cell count was also directly proportional to the recycle ratio within the range 0.1–0.6. Experiments performed with a PBR employing the Taguchi design showed the most significant effects of pH followed by dilution rate and recycle ratio on H_2 production. Lima and Zaiat [40] optimized biohydrogen production in an anaerobic PBR by varying the degree of back mixing. The experimental results showed that at an optimum recycle ratio of 0.6 the maximum yield was 4.22 mol H_2/mol of sucrose. It was hypothesized that under this optimum condition the reactor performed as eight CSTRs in series.

Leite et al. explored the potential of using an anaerobic PBR to treat low contents of organic matter to generate H_2 and organic acids [45]. The bioreactor was fed with glucose-based synthetic wastewater and operated at an HRT ranging from 0.5 to 2 h. A stable H_2-producing microbial biofilm was developed over EC beads (4.8–6.3 mm size) as the support material. The main parameter affecting H_2 production was found to be alkalinity, and the system could be operated without any addition of a pH buffer. The average H_2 yield was 2.48, 2.15, and 1.81 mol H_2/mol of glucose for influent $NaHCO_3$ concentrations of 0, 1000, and 2000 mg/L, respectively.

Zhang et al. investigated the effect of packing material on biohydrogen production in two upflow reactors fed with sucrose (10 g/L) containing synthetic wastewater [35]. The results showed that the reactor with packing was more effective than that without any packing as a high biomass concentration can be achieved using biosupport materials. The H_2 production rate increased from 0.56 to 6.17 L/L/day when the HRT was decreased from 24 to 2 h. The maximum H_2 yield of 1.22 mol/mol sucrose was obtained at 14 h HRT, and granular sludge aggregates were formed in both the reactors. Junior et al. studied four different support materials for biomass growth, namely EC, charcoal (CH), porous ceramic (PC), and low-density polyethylene (LDP) for H_2 production from sugarcane vinasse using an anaerobic upflow PBR [69]. The reactors (each with 2.3 L volume) with different packing materials were operated simultaneously at 24 h HRT, 36.2 kg $COD/m^3/day$ OLR and 25°C. Extremely high volumetric H_2 production rates of 509.5, 404, 81.4, and 10.3 mL/L/day and maximum H_2 yields of 3.2, 2.6, 0.4, and 0.05 mol H_2/mol of total carbohydrates for LDP, EC, CH, and PC, respectively, were obtained in the study. These results, thus, indicate the strong influence of support material for biomass growth on H_2 production in PBR systems. Genetic analysis of the biomass from the LDP-filled reactor revealed the presence of H_2-producing species such as *Clostridium* and *Pectinatus* spp., lactic acid bacteria, and other nonfermentative organisms.

In another similar study, Kumar et al. explored thin, low-porous polyethylene and thick, high-PC as support materials for biofilm formation in a PBR-type reactor for H_2 production [31]. The results showed that the thick and porous biofilm developed on the ceramic carrier favored more biomass growth than the polyethylene material. However, it also favored more propionic acid production than H_2 production. On the other hand, the polyethylene material that supported the formation of a thin biofilm allowed for a high H_2 production. The glucose utilization was more than 97% using both the carrier materials. Maximum H_2 productivity was achieved in the case of the reactor with the polyethylene material as the biosupport.

Arraiga et al. studied H_2 production using an oat straw acid hydrolysate−fed biotrickling filter (BF) packed with perlite [43]. The effect of influent COD (1.2−35 g/L) and HRT (24 h and 6 h) on H_2 production were examined. A maximum H_2 production rate of 81.4 mL/L/h and H_2 yield of 2.9 mol H_2/mol hexose were obtained for influent COD of 0.05 g COD/L/h (HRT 24 h) and 2.9 g COD/L/h (HRT 12 h), respectively. The maximum H_2 content in the biogas produced was 45.4% (v/v) with the remaining percentage being CO_2. However, this study suggests that special care needs to be taken to avoid reactor clogging for the successful scale-up of the system. Anzola-Rojas et al. explored a novel anaerobic downflow fixed bed reactor for long-term H_2 production, in which the OLR was varied from 12 to 96 g COD/L/day to find the optimum conditions [33]. The volumetric H_2 production rate and yield obtained indicated that the specific organic load (SOL) rather than the OLR is important. A stable H_2 production rate of 8.9 L/L/day and yield of 2.0 mol H_2/mol sucrose converted was obtained at a SOL value between 3.8 and 6.2 g sucrose/g VSS/day.

Peixoto et al. studied fermentative H_2 production from soft-drink wastewater using two upflow anaerobic PBRs, one with a nutrient supplement addition and the other without any nutrient adjuvant [46]. Experimental results showed the reactor with the nutrient supplement added to it performed better in terms of H_2 yield (3.5 mol H_2/mol of sucrose), H_2 production rate (0.4 L/L/h), and H_2 content in the biogas (15.8%). The difference in performance between the two reactors was attributed to changes in the metabolic pathway for H_2 production due to the nutrient supplement addition. Genetic analysis of the biomass samples revealed the presence of *Clostridium* sp., *Enterobacter* sp., and *Klebsiella* sp. in both the reactors.

Castello et al. studied H_2 production from cheese whey using an upflow anaerobic PBR [47]. The effect of three different OLRs (22, 33, and 37 g COD/L/day) on reactor performance was evaluated with a fixed HRT of 24 h. The results showed the positive effects of increasing the OLR from 22 to 33 g COD/L/day and adjusting the pH to a value greater than 5 on H_2 production (H_2 of 1 L/L/day). The maximum H_2 yield obtained was 1.1 mol H_2/mol lactose. This study demonstrated a stable performance of the PBR throughout the operating period without any problems such as clogging, methanogenesis, and solvent production. Microbiological analysis of the biomass from the reactor showed the presence of a mixed population dominated by *Clostridium* and *Klebsiella* spp. Junior et al. investigated the effect of different OLRs on H_2 production using an upflow anaerobic PBR fed with sugarcane vinasse [69]. A maximum H_2 production rate of 1117.2 mL/L/day and H_2 yield of 2.4 mol H_2/mol total carbohydrates at an optimum OLR of 84.2 kg COD/m^3/day was reported. Microbial analysis of the reactor biomass showed the predominant presence of *Thermoanaerobacterium* and *Thermosaccarolyticum* spp.

Sivagurunathan et al. studied H_2 fermentation from galactose using a fixed bed reactor (FBR) packed with Lantec HD-Q-PAC material [63]. The effect of an HRT in the range of 1.5–12 h on stable H_2 production over 79 days of continuous bioreactor operation was evaluated. A maximum H_2 production rate and a H_2 yield of 65.5 L/L/day and 2.60 mol/mol hexose were achieved at 2 h HRT, respectively. The major VFAs produced were butyric and acetic acids. Microbial community analysis by quantitative real-time polymerase chain reaction revealed that population changes greatly affected the H_2 production performance. For an optimum H_2-producing condition of 2 h HRT, *Clostridium butyricum* was predominant, whereas at 1.5 h HRT increases in *Lactobacillus* sp. population led to enhanced lactic acid production, which deteriorated the bioreactor performance.

Ito et al. studied H_2 and ethanol production using *Enterobacter aerogenes* HU-101 in a PBR using glycerol-containing wastes discharged from a biodiesel production process [70]. The glycerol-containing wastes were diluted with synthetic media and supplemented with yeast extract and tryptone to enhance the production of H_2 and ethanol. The yields of H_2 and ethanol decreased with an increase in the concentration of glycerol in the wastes. The study found that H_2 and ethanol production from biodiesel wastes were much lower than that of pure glycerol due to the high salt content in the biodiesel waste. Among the two methods used for cell immobilization, the self-immobilization method yields better results than the method where biomass is immobilized on PCs. The H_2 production rate with self-immobilized biomass was 80 mmol/L/h, whereas it was 63 mmol/L/h with biomass immobilized over a PC support. The ethanol yields in both the systems were 0.8 and 0.83 mol/mol of glycerol, respectively.

Barca et al. investigated dark fermentative H_2 production by an anaerobic consortium including *Clostridium acetobutylicum* and *Desulfovibrio vulgaris* in a PBR system [51]. The effect of different process parameters, namely void hydraulic retention time (HRTv), pH, and alkalinity on H_2 production was investigated, which revealed that the H_2 production rate was increased with a decrease in HRTv from 4 to 2 h; however, the H_2 yield remained unaffected by this change. This result strongly suggests that HRTv does not affect microbial metabolism. Buitron et al. [71] used vinasse from the tequila industry as a substrate for biohydrogen production using a fixed bed bioreactor, and reported a maximum H_2 yield and H_2 production rate of 1.3 mol H_2 mol/mol glucose and 72 ± 9 mL/L/h, respectively. The H_2 content in the biogas produced was 64% (v/v) with the remaining percentage as CO_2. The main soluble metabolites in the liquid phase were acetic, butyric, and iso-butyric acids.

14.2.3 Anaerobic fluidized bed reactor

An AFBR is another type of attached growth system in which carrier materials and a microbial biofilm are in suspension due to a high upward fluid velocity. This "bed fluidization" causes efficient mixing, high mass transfer, and increases the conversion efficiency (Table 14.2) [10]. Several studies have reported biohydrogen production using AFBR with a variety of carrier materials (Table 14.1).

Barros et al. studied H_2 production using three different AFBRs with different support materials, namely polystyrene, ground tire, and polyethylene terephthalate [19]. The total volume of the individual reactors was 4192 cm^3 and each of these was fed with 4000 mg/L of glucose-containing media. The best performance in terms of H_2 production was achieved by the reactor with ground tire as the support material with a yield of 2.11 mol H_2/mol of glucose and 60% (v/v) H_2 content in the biogas. The main soluble metabolites were acetic acid, butyric acid, lactic acid, and ethanol, along with small amounts of propionic acid.

Zhang et al. examined biofilm sludge and granular sludge in two AFBRs operated at a 5.5 pH and 37°C for H_2 production [38]. The effect of HRT (0.125−3 h) and glucose concentration (5−120 g/L) on H_2 production was studied by keeping the OLR constant at 40 g glucose/L/h. The H_2 yield ranged between 0.4 and 1.7 mol/mol glucose in both the reactors with a maximum yield obtained at 0.25 h HRT and 10 g/L glucose concentration in the feed. Under the same optimum conditions, maximum H_2 production rates of 7.6 and 6.6 L/L/h were achieved using the biofilm reactor and granular reactor, respectively. The study found that compared with the carrier-based biofilm reactor, the granule-based reactor was more advantageous in terms of biomass retention without any washout of the support carriers; hence, its performance in terms of substrate conversion and H_2 production was better than the other reactor systems.

Zhang et al. used a mixed culture of hydrogen-producing bacteria immobilized on granular activated carbon (GAC) in an AFBR [35]. The reactor was maintained at a constant 4.0 pH and temperature of 37°C. The effect of HRT was analyzed by varying it from 4 h to 0.5 h and keeping the influent glucose concentration fixed at 10 g/L; the effect of the glucose concentration was further studied in the range of 10−30 g/L at a fixed HRT of 1 h. The biogas produced in the system was free from CH_4 and contained 61%−57% (v/v) H_2. The soluble products in the effluent stream were mainly acetate and butyrate with relatively low quantities of other volatile fatty acids and alcohols. The H_2 production rate and specific H_2 production rate were 2.36 L/L/h and 4.34 mmol H_2/g VSS/h, respectively, which linearly correlated with the effective OLR and glucose conversion rate. The attached biomass concentration was also very high (~21.5 g/L). Hence it was concluded that due to this high biomass retention, the reactor could be successfully operated even at a high OLR.

Amorim et al. evaluated H_2 production using an AFBR containing EC particles (2.8−3.35 mm size) as a biosupport material and fed with glucose-based synthetic wastewater [37]. The study found that the H_2

yield increased from 1.41 to 2.49 mol H_2/mol glucose with a decrease in HRT from 8 to 2 h. However, there was a slight decrease in yield to 2.41 mol H_2/mol glucose at 1 h HRT. The H_2 content in the biogas also increased from 8% to 35% (v/v) with a decrease in HRT; the biogas produced was totally devoid of CH_4. Similar to other studies reported in the literature, the major soluble metabolites formed during H_2 fermentation were found to be acetic acid and butyric acid.

In another study, the potential of EC as a support material for H_2 production in an AFBR was demonstrated [37]. The reactor was inoculated with thermal pretreated anaerobic sludge and operated at different HRTs (8−1 h) at a controlled temperature of 30°C and 3.8 pH. The overall substrate conversion rate in the reactor was 92−98% of glucose for an influent concentration of 2000 mg/L. The H_2 production rate increased with a decrease in the HRT, reaching a maximum of 1.28 L/L/h at 1 h HRT. Whereas a maximum H_2 yield of 2.29 mol H_2/mol glucose and a H_2 content of 37% (v/v) in the biogas were obtained at 2 h HRT. This study suggests that the thermal pretreatment of biomass, the selection of suitable biosupport material, and optimum operating conditions are key to ensuring a stable performance of AFBRs over a long period.

Barros et al. compared polystyrene and EC as biosupport materials for biohydrogen production using two AFBRs fed with 4000 mg/L of glucose [38]. The results showed that the reactor containing the EC as the biosupport material performed better than the reactor with polystyrene as the biosupport material, which yielded 2.59 mol H_2/mol glucose. The attached biomass contents (as total volatile solids or TVS) were 0.805 mg per g of polystyrene and 1.1 mg per g EC, indicating the superiority of EC particles as a biosupport material. The H_2 content in the biogas obtained was in the range of 16%−47% for the reactor containing the polystyrene and 22%−51% for the reactor with EC.

Shida et al. assessed the effect of OLR (19.0−140.6 kg COD/m^3/day) and pH buffer (sodium bicarbonate) addition on H_2 production in two simultaneously operated glucose-fed AFBRs [39]. EC particles were used as the support material for biofilm formation in both the AFBRs. The reactor without sodium bicarbonate added showed the highest H_2 yield of 2.45 mol H_2/mol glucose at 84.3 kg COD/m^3/day OLR. Furthermore, sodium bicarbonate favored *Clostridium* spp. proliferation in the reactor, whereas in the other reactor with no pH buffer, other species such as *Enterobacter*, *Klebsiella*, *Veillonellaceae*, *Chryseobacterium*, *Sporolactobacillus*, *Burkholderiaceae*, and others, along with *Clostridium* spp. proliferated.

Munoz-Paez et al. compared the H_2 production at different operational temperatures and OLRs in a lab-scale AFBR at a fixed HRT of 1 day [72]. The results suggest that an increase in OLR had a positive effect on the bioreactor performance as the H_2 concentration in the biogas increased 1.8-fold and H_2 productivity 2.2-fold. Lin et al. used an ethylene–vinyl acetate (EVA) copolymer to immobilize sewage sludge in a three-phase fluidized bed reactor for H_2 production [41]. The study reports that an immobilized cell packing ratio of 10% (v/v) and a liquid recycle rate of 5 L/min (23% bed expansion) were found to be optimum for H_2 production. The performance of the reactor fed with sucrose-based synthetic medium was examined under different sucrose concentrations and HRTs. A high volumetric H_2 production rate of 1.80 ± 0.02 L/L/h was achieved at 40 g influent COD/L and at 2 h HRT, whereas, a maximum H_2 yield of 4.26 ± 0.04 mol H_2/mol sucrose was obtained at 20 g influent COD/L and at 6 h HRT. The H_2 content in the biogas obtained was stable at 40% (v/v) or above throughout the study period.

14.2.4 Membrane bioreactor

Anaerobic MBRs have long been studied for wastewater treatment even at pilot scale. However, their application in fermentative H_2 production is recent (Table 14.1). Based on the arrangement of the membrane in a bioreactor the system can be classified into two broad types, namely external loop MBR, where the membrane is outside and connected to the reactor with a link, and submerged MBR, where the membrane module is immersed in the bioreactor (Fig. 14.3). MBR-type reactors provide several advantages over other reactor types such as high biomass retention, high conversion rates of organic substrate, high volumetric H_2 production rate, and so on [62]. The use of ultra- and microfiltration membrane systems following fermentation produces clean effluent that does not require any further cleanup [68]. However, membrane fouling and the cost of membrane installation and replacements are major drawbacks of this type of reactor system (Table 14.2) [73].

Lee et al. investigated the influence of solids retention time (SRT) on continuous H_2 production in a submerged MBR fed with glucose-based media using a mixed mesophilic microbial consortium [53]. The bioreactor in this study was continuously operated at the four different SRTs of 2, 4, 12.5, and 90 days and at a fixed HRT of 9 h. Stable biogas production with an H_2 content of 50.8%–60% (v/v) was obtained at an SRT in the range 2–12.5 days, without any CH_4 production. The H_2 production was increased from 17.62 to 26.1 L/day when the SRT was increased from 2 to 12.5 days, but it decreased to 9.1 L/day at 90 days SRT. The best H_2 yield of 1.19 mol H_2/mol of glucose was observed at a 2 days SRT and the highest H_2 production rate of 5.8 L/L/day was obtained at a 12.5 days SRT. According to the authors, the decrease in H_2 yield at a prolonged SRT is due to the low concentration of volatile suspended solids/total suspended solids (VSS/TSS) and high concentration of extracellular polymeric substances (EPSs).

Zhang et al. used an MBR fabricated by integrating a hollow-fiber microfiltration membrane module with a CSTR to boost H_2 production through high dilution rate operations [35]. Three different carbon substrates, namely glucose, sucrose, and fructose, were used in the reactor. The results showed that the MBR operated better in handling cell washout compared with the CSTR, leading to a high H_2 production rate and biomass concentration. Due to the high biomass retention inside the reactor, the system performed efficiently even at a low HRT of 1 h with H_2 production rates of 1.48, 2.07, and 2.75 L/L/h, with glucose, sucrose, and fructose as the sole substrate, respectively. The yields of H_2 for glucose, sucrose, and fructose were 1.27, 1.39, and 1.36 mol H_2/mol hexose, respectively, which revealed that fructose is the most preferable substrate for H_2 production. Butyrate and acetate were the major soluble metabolites in this study, accounting for 70%–85% of the total soluble product formed. The estimated values for H_2 production rate and yield from stoichiometric correlation were in conformity with the experimental results.

Lee et al. studied H_2 fermentation using an MBR fed with food waste and operated under thermophilic conditions [20]. The effect of three different OLRs (70.2, 89.4, and 125.4 kg COD/m^3/day) on H_2 production was investigated. The total biogas production rates were found to be 22.4, 32.8, and 62.5 L/day for the three respective OLR values. The maximum H_2 yield and production rate in the study were 111.1 mL/g VSS added and 10.7 L/L/day at an OLR of 125.4 Kg COD/m^3/day. The results found a high total carbohydrate utilization efficiency of more than 96% throughout the study period of 90 days. Genetic analysis of the microbial community present in the reactor system revealed the predominance of *Clostridium* sp. strain Z6. The H_2 production was considerably improved by keeping the HRT low and by increasing the OLR. Also, the system showed a high cell retention capability due to it being able to process a high organic load.

Lee et al. compared the performance of an MBR and a CSTR for H_2 production by a mixed anaerobic consortium [20]. The experimental results showed a stable H_2 content of 51% (v/v) for the MBR and 58% (v/v) for the CSTR over a period of 35 days. No CH_4 gas was produced even at a long SRT of 90 days in this study. The H_2 production rate was 2.43–2.56 L/L/day in the MBR, which was 2.6-times higher than that in the CSTR (0.95–0.97 L/L/day). In another study, the effect of $FeSO_4$ addition on H_2 production in an MBR using anaerobic mixed microflora under mesophilic condition was investigated by Lee et al. (2008b). A maximum H_2 production of 41.6 L/day was obtained with the addition of 10.9 mg $FeSO_4$/L. The hydrogenase activity was six-times higher than without any added $FeSO_4$ in the media.

Lee at al. reported membrane fouling in an H_2-producing submerged MBR by investigating the EPSs produced by bacteria during H_2 fermentation [35]. The experimental setup in this study consisted of a 1.4 L submerged membrane filtration tank and a 3 L H_2 fermenter. Intermittent suction was provided to maintain a stable filtration performance in the system. The results showed that the EPS accumulation in the MBR was due to the adsorption of the EPS onto the polymeric membrane surface during its continuous operation. Characterization of the EPS revealed that it consisted of 179 mg/L of proteins, 58 mg/L of carbohydrates and small amounts of Cu^{2+}, Mg^{2+}, Zn^{2+} (1.6–3.3 mg/L). The EPS produced in this process had a high chelation potential that led to the formation of ligand complexes with metals or cations, which also resulted in the clogging of the membrane.

Shen et al. investigated the effect of different OLRs ranging from 4.0 to 30 g COD/L/day on fermentative H_2 production using a lab-scale continuously operated MBR [25]. The study found that with an increase in OLR, the biomass content, colloidal hydrophobicity, and mean particle diameter of the biomass increased, causing more rapid membrane fouling. The bound and soluble EPS contents of the biomass in the MBR were found to be higher than those in an H_2-producing CSTR system and in an aerobic activated sludge system. From this study, membrane fouling in an MBR occurs in two distinct phases, an initial rapid fouling followed by slow fouling. The rate of fouling in both these stages can greatly increase with increases in biomass concentration.

Shen et al. compared the performance of an MBR with that of a CSTR for H_2 production at five different OLRs ranging from 4.0 to

30 g COD/L/day, and found that the H_2 yield in the CSTR was significantly higher than that in the MBR at a low OLR (1.25 vs 0.97 mol H_2/mol glucose, respectively) [25]. But at a high OLR of 30 g COD/L/day both the reactors performed poorly with H_2 yields of 1.77 and 1.49 mol H_2/mol glucose, respectively, for CSTR and MBR. At an intermediate OLR of 22 g COD/L/day both the CSTR and MBR showed the same value for H_2 yield of 1.78 mol H_2/mol glucose. However, the H_2 production rate at this OLR was higher by 50% in the MBR than in the CSTR. Hence it was concluded that CSTRs are easy to operate and provide superior H_2 yields at low OLRs, while MBRs are suited for operating at moderate to optimum OLRs. In a similar work, Kim et al. compared the results of H_2 production from TPW using a mixed anaerobic consortium in an MBR and a CSTR both operating under thermophilic (60°C) condition [26]. A stable volumetric H_2 production rate of 8.17–0.32 L/L/day and a H_2 yield of 1.20–0.05 mol H_2/mol hexose at 8 h HRT were achieved in the CSTR. These results were low when compared with those obtained using the MBR at the same operating condition. Moreover, maximum H_2 yields of 1.87 mol H_2/mol hexose and 1.00 mol H_2/mol hexose were achieved at 8 h and 2 h HRT, respectively, using the MBR. The highest H_2 production rate of 19.86 L/L/day was obtained in the study at 4 h HRT. Thus the main advantages of MBR compared with other bioreactor systems are the increase in retention time of solid substrate and the high microbial concentration, which enhances the effective substrate utilization for H_2 production.

14.2.5 Upflow anaerobic sludge blanket

The UASB reactor is popular for biohydrogen production due to its high biomass retention capability, high treatment efficiency, and capacity to withstand high OLRs and HRTs (Table 14.2) [8]. Sivagurunathan et al. evaluated H_2 fermentation from galactose using a UASB reactor [63]. A maximum H_2 production rate and a maximum H_2 yield of 56.8 L/L/day and 2.25 mol/mol galactose, respectively, were achieved at a 2 h HRT. A further reduction in the HRT to 1.5 h lowered the H_2 production rate and yield to 48.3 L/L/day and 1.44 mol/mol galactose respectively. A microbial community analysis in the study showed that *Proteobacetria*, particularly *Clostridium* spp., were present to the extent of 72% during the reactor operation at a 2 h HRT. The reactor pH was maintained in the range 5.5–6.2 by adding a carbonate buffer into the nutrient medium, and the main soluble metabolites produced during the H_2 fermentation were acetate and butyrate along with small quantities of lactate and propionate. Thus HRT is reported as the key parameter for successful H_2 fermentation using a UASB reactor.

Akutsu et al. investigated the effect of the HRT, pH, and substrate concentration on thermophilic H_2 production from starch using an UASB reactor [1]. A maximum H_2 yield of 1.7 mol H_2/mol glucose was obtained at a 48 h HRT, corresponding to 8 kg COD/m^3/day. Thermophilic H_2-producing granules of 0.5–4.0 mm diameter each with heat-treated methanogenic granules as the nuclei were successfully formed during the study. Drastic changes in the metabolic pathway and metabolites formed were observed in the UASB due to changes in the operating conditions.

Castelló et al. examined H_2 production from unsterilized cheese whey using a UASB reactor [54]. The results showed that H_2 production was low (122 mL/L/day) in the bioreactor even for the highest OLR of 20 g COD/L/day, which revealed that the carbon flux was mainly toward methanogenesis. Microbiological studies showed the presence of fermentative organisms belonging to the genera *Megasphaera*, *Anaerotruncus*, *Pectinatus*, and *Lactobacillus*, which were responsible for H_2 production.

Carrillo-Reyes et al. studied the effect of different operational strategies and biomass types (i.e., either granular or disaggregated granules) for H_2 production during the start-up phase of four UASB reactors [47]. Volumetric H_2 production rates of 0.38 and 0.36 L/L/h were obtained in the reactors operated at a constant OLR that contained either of the two biomass types. On the other hand, methanogenesis was mainly observed in the reactors operated at a high OLR. Biomass in the form of disaggregated granules was found to be more active than the granular biomass in terms of high specific H_2-producing activity. However, CH_4 production and changes in the operating conditions are seen to be major drawbacks for achieving stable H_2 production.

Intanoo et al. investigated H_2 production using a two-stage UASB reactor fed with cassava wastewater [55]. For the first stage, the UASB system was operated at 37°C and pH 5.5 at different COD loading rates. A high specific H_2 production rate of 0.39 L/L/day along with a 39.83 L H_2/kg COD of H_2 yield were achieved for a COD loading rate of 25 kg/m^3/day. The biogas contained was 36.4% (v/v) H_2 and 63.6% CO_2 with no CH_4. In the second CH_4 production stage, the UASB was fed with the effluent produced from the previous H_2 production stage.

A maximum specific CH_4 production rate of 0.91 L/L/day and CH_4 yield of 115.23 L/kg COD removed were obtained for a COD loading rate of 8 kg/m^3/day in the second stage.

Recently, Sittijunda, and Reungsang studied the effect of different OLRs (25, 37.5, 50, and 62.5 g/L/day) on the production of biohydrogen, 1,3-propanediol (1,3-PD), and ethanol from glycerol using a UASB reactor [56]. The optimum OLR for H_2 and ethanol production was found to be 50 g/L/day, whereas for 1,3-PD production, it was slightly higher (62.5 g/L/day). The values for H_2 production, H_2 yield, and ethanol and 1,3-PD production were 134.2 mmol/L, 579.8 mmol H_2/mol glycerol, and 78.9 mmol/L and 60.9 mmol/L, respectively. These values were, however, low when compared with those obtained using pure glycerol as the substrate. However, when crude glycerol was used all the values were reduced. Microbial analysis showed that *Clostridium* sp., *Enterobacter* sp., *Firmicutes* bacterium, *Actinobacterium* sp., and *Klebsiella* sp., were predominantly present in the UASB reactor.

Sivagurunathan et al. used anaerobic sludge that was pretreated by repeated heating for H_2 production from galactose using a UASB reactor operated at 6 h HRT and 37°C temperature [31]. The H_2 production rate and H_2 yield gradually increased up to 9.1 L/L/day and 1.1 mol/mol galactose, respectively. The H_2 production performance was improved by 37% when the biomass was heat pretreated at 80°C for 30 min. This study demonstrates that repeated heat treatment of biomass weeds out strains that are not H_2 producers from the mixture. Genetic analysis of the biomass revealed significantly low levels of *Klebsiella*, *Prevotella*, and *Lactobacillus* species, and high levels of *Clostridium*, *Eneterococcus*, and *Citrobacter* species, responsible for efficient utilization of the substrate for H_2 production.

Carrillo-Reyes et al. investigated various factors such as pH, HRT, organic shock loading, and repeated heat treatment of biomass for enhancing H_2 production in a UASB reactor fed with cheese whey [58]. From the results obtained in the study, it was found that repeated heat treatment of granular sludge biomass was the best strategy to completely inhibit methanogens, which led to a high H_2 production rate of 1.67 L/L/day. However, the main problem with this heat treatment strategy was the intermittent stopping of the reactor operation. Moreover, hydrogenotrophic acetogenesis could not be completely inhibited by this strategy. Organic shock loading is another strategy that was found to be effective for reducing the CH_4 production in this study, which also increased the H_2 production rate by 172% without intermittent stopping of the reactor operation.

14.2.6 Expanded granular sludge bed

EGSB reactors represent a variation of the UASB reactor, distinguished by a high upflow velocity, which is often achieved by effluent recycling. The advantages of the EGSB reactor for hydrogenogenic conversion of organic wastewater include a high biomass retention capacity and less mass transfer limitation (Table 14.2) [64]. Moreover, it can operate at a high OLR (20–40 kg/m^3/day) and serves to filter low chain fatty acids containing wastewater by adsorbing the fats onto sludge biomass granules present, followed by their slow degradation in the reactor [70]. However, critical aspects such as the process conditions to obtain hydrogenate granules have been scarcely investigated in EGSB reactors (Table 14.1).

Barcenas-Ruiz et al. investigated the performance of an EGSB reactor for biohydrogen production [32]. The effect of different biomass pretreatment methods and upward liquid velocities (ULV) on granules formation, their structure, and hydrogenogenic activity was investigated. Heat pretreated biomass produced more hardy and resilient granules compared with those produced with other pretreatment methods. Also, an increase in ULV from 2.5 m/h to 4.5 m/h resulted in the formation of larger granules with high protein-to-carbohydrate ratios. Furthermore, the increase in ULV caused the selective washout of propionate-producing bacteria, which was the main soluble metabolite, and due to which the H_2 production rate and H_2 yield increased to 600 mL H_2/g VSS/day and 1.5 mol of H_2/mol of glucose, respectively. These results demonstrate the importance of inoculum pretreatment and optimum operating conditions for achieving stable H_2 production using granular biomass in EGSB reactors.

Guo et al. used an EGSB reactor with a GAC immobilized mixed microbial consortium for H_2 production from molasses-containing wastewater [17]. The reactor was operated at 35°C without any pH control. The maximum H_2 production rate in the reactor was 0.71 L/L/h with a H_2 yield of 3.47 mol/mol sucrose and specific H_2 production rate of 3.16 mmol H_2/g VSS/h. The H_2 content in the biogas was estimated to be 30–53% (v/v) without any CH_4 produced throughout the operating period. Acetate and ethanol were the main soluble products along with small quantities of propionate, butyrate, and valerate. The average attached biofilm concentration was estimated to be 17.1 g/L, which facilitated H_2 production efficiently. At high OLRs, this bioreactor system with its high biomass retention capacity is promising for simultaneous H_2 production and high strength wastewater treatment.

In another study, H_2 production from starch-containing wastewater was examined using an EGSB reactor system with GAC as the biosupport material [24]. High values of H_2 production and COD removal efficiency capability over a period of 400 days were achieved. The maximum H_2 production rate reported in this study was 1.64 L/L/day under a 1.0 g starch/L/day OLR, pH 4.42, and 4 h HRT conditions. The maximum H_2 yield was found to be 0.11 L/g COD at a 0.5 g starch/L/day OLR, pH 3.95, and 8 h HRT. The H_2 content in the biogas was 35%–65% (v/v), which was slightly better than that obtained with molasses-containing wastewater. However, the COD removal rates were low (31.1%) even at 24 h HRT, and the major soluble products were ethanol, acetate, and butyrate. The average attached biofilm concentration was about 8.26 g/L. The capability of the EGSB system to operate at a low pH reduced the cost of alkali addition to maintain the pH inside the bioreactor, which seems to be economically viable for scale-up. The experimental data was fitted to a back propagation neural network and linear regression model that could predict the H_2 production and COD removal in the system.

14.2.7 Anaerobic baffled reactor

The ABR design was developed during the 1980s primarily for wastewater treatment. It provides numerous advantages over other more conventional anaerobic reactors including its high resilience to shock loadings, long biomass retention time, low sludge production, and so on. This reactor is also capable of partially separating different phases of anaerobic catabolism, which provides good protection to microbial populations inside the reactor from recalcitrant compounds and sudden changes in operational parameters such as pH and temperature (Table 14.2). The structure of the ABR provides flexibility for reactor modification such as the insertion or deletion of certain parts to improve the process efficiency [65].

Ran et al. [74] studied biohydrogen production using a novel ABR system with 3.46 L of total volume. The reactor consisted of four equal chambers, out of which the first was used for H_2 production and the other three as continuous microbial electrolysis cells for CH_4 production. The system was operated at 35 ± 1°C temperature, a 24 h HRT, and influent COD of 3500–4000 mg/L. The results showed that the H_2 content in the first chamber was 20.7% (v/v) and the CH_4 content in the last three compartments were 98.0%, 93.6%, and 70.1% (v/v), respectively. An overall 98% COD removal was achieved using this bioreactor system.

Jurgensen et al. [75] evaluated the effect of a low pH condition on H_2 production using a 200 L bench scale ABR fed with wheat starch. A total biogas production of 230 L/day with 42% (v/v) H_2 and 11% (v/v) CH_4 at a mean residence time of 29 h and influent COD of 4 g/L were achieved in this study. Gas produced in different compartments of the reactor varied largely in terms of volume and composition, with a maximum H_2 concentration of 60% (v/v) observed in the first compartment.

Tawfik et al. investigated a two-stage ABR for H_2 production from municipal food waste [61]. Average yields of 250 mL H_2/g VS at a 26 h HRT and 58 kg COD/m^3/day OLR and 370 mL H_2/g VS at a 26 h HRT and 35 kg COD/m^3/day OLR were obtained in the first and second stages, respectively. The total H_2 yield in the two-stage process was estimated to be 4.9 mol H_2/mol hexose. The study also found that in the first stage the conversion of mainly particulate matter to H_2 takes place, and in the second stage the soluble substrate is utilized. In another study, Tawfik and El-Qelish investigated the effect of different OLRs (29, 36, and 47 g COD total/L/day) on H_2 production from municipal food waste and kitchen wastewater using a mesophilic ABR [59]. The HRT was kept constant at 1.6 days. The results illustrated that H_2 production dropped from 6.0 ± 0.5 to 5.4 ± 1.04 L/day with an increase in the OLR from 29 to 36 g COD/L/day. A further increase in the OLR to 47 g COD/L/day did not affect the H_2 production much, which was almost constant at 5.3 ± 1.04 L/day. An overall 57% COD removal and 81% carbohydrate utilization was achieved in this study.

Lay et al. studied biohydrogen production using a novel anaerobic baffled stacking reactor [21]. The reactor consisted of three chambers with provision for recirculation from the second or third chamber to the first or second chamber, which enhanced the mass transfer between the substrate and microbes, leading to a high H_2 production rate in the reactor. The effect of different recirculation modes and substrate concentrations (10–30 g COD/L) on H_2 production at a fixed HRT of 8 h, pH 5.5, and temperature of 35°C was investigated in this study. A maximum H_2 production rate of 10.9 ± 1.5 L/L/day, yield of 1.7 ± 0.2 mol H_2/mol sucrose, and H_2 content of 40 ± 2.4% (v/v) were achieved due to higher recirculation from the third to the first or second chamber. Whereas recirculation from the second to the first chamber, decreased the mixing efficiency, substrate utilization, and H_2 production.

Elreedy et al. investigated the simultaneous production of H_2, ethanol, and CH_4 from monoethylene glycol (MEG)-containing petrochemical wastewater using a four chamber anaerobic packed bed baffled reactor packed with polyurethane sheets [76]. The reactor was operated at a constant HRT of 36 h and different OLRs of 0.67, 1, 2, and 4 g COD/L/day. The results showed high volumetric H_2 and CH_4 production rates of 438.07 ± 43.02 and 237.80 ± 21.67 mL/L/day, respectively, at 4 g COD/L/day OLR. The ethanol in the liquid effluent stream increased from 57.15 ± 2.31 to 240.19 ± 34.69 mg/L when the OLR was increased from 0.67 to 4 g COD/L/day. A maximum MEG biodegradation of 98% was achieved at a low OLR of 0.67 g COD/L/day. H_2 and ethanol production were maximal in the first chamber, whereas CH_4 production was maximal in the last two compartments. Genetic analysis of the microbial population in the reactor revealed the dominance of *Proteobacteria* (36.62%), *Firmicutes* (20.85%), and *Bacteroidetes* (3.44%) species.

Thanwised et al. evaluated the effect of HRT (24, 18, 12, 6, and 3 h) on H_2 production from tapioca wastewater by a mixed anaerobic consortium using an ABR [60]. The results showed that the H_2 production rate increased with a decrease in HRT, that is, from 164.45 ± 4.14 mL/L/day at a 24 h HRT to 883.19 ± 7.89 mL/L/day at a 6 h HRT, while a further decrease in the HRT to 3 h lowered the H_2 production rate to 748.54 ± 13.84 mL/L/day. A maximum COD removal of 29.3% was achieved at 6 h HRT.

Lu et al., studied the effect of substrate concentration on photofermentative H_2 production using a 4 L pilot-scale baffled photofermentative reactor [77]. A high H_2 production rate of 202.64 ± 8.83 mol/m^3/day was achieved in the third chamber of the reactor for a substrate concentration of 20 g/L. The H_2 content was in the range of 42.19 ± 0.94% (v/v)-49.71 ± 0.27% (v/v). The H_2 production rate increased with an increase in OLR from 3.3 to 20 g/L/day, but further increases in the OLR up to 25 g/L/day reduced the H_2 production rate.

Bioreactor configuration is one of the key parameters affecting the yield and volumetric productivity of biohydrogen. Besides, the start-up time and steady state of performance of a system may vary depending upon the reactor design, thus affecting its overall performance at a large scale [62]. Furthermore, OLR, HRT, SRT, pH, temperature, and biomass pretreatment are important process parameters that affect biohydrogen production, and, therefore, need to be optimized for successful scale-up of a given system [10]. Table 14.3 summarizes the different optimum conditions

Table 14.3 Different optimum conditions for operating various types of bioreactors for biohydrogen production.

Bioreactor type	Operational strategies	References
CSTR	• Start-up period: Not required • OLR: 22–40 g COD/L/day • HRT: 3–8 h • Biosupport material: Not essential • Effluent recycling: Not required • Gas solids separation: Not required	[25,30,31]
PBR	• Start-up period: 10–20 days • OLR: 1–40 g COD/L/day • HRT: 0.5–4 h • Effluent recycling: Not required • Biosupport material: Essential • Gas solids separation: Beneficial	[3,43,45,51]
FBR	• Start-up period: 3–4 days • OLR: 20–40 g COD/L/day • HRT: 0.25–4 h • Biosupport material: Essential • Effluent recycling: Required • Gas solids separation: Beneficial	[19,37,39]
UASB	• Start-up period: 4–16 days • OLR: 25–50 g COD/L/day • HRT: 2–6 h • Biosupport material: Not essential • Effluent recycling: Not required • Gas solids separation: Essential	[55–57,78]
ABR	• Start-up period: 4–16 days • OLR: 10–30 g COD/L/day • HRT: 6–8 h • Biosupport material: Not essential • Effluent recycling: Not required • Gas solids separation: Essential	[61,78]
MBR	• Start-up period: 3–4 days • Gas solids separation: Beneficial • Effluent recycling: Not required • Biosupport material: Not required • OLR: 4–22 g COD/L/day • HRT: 1–9 h	[42,62,78]
EGSB	• Start-up period: 4–16 days • Gas solids separation: Required • Effluent recycling: Required • Biosupport material: Not essential • OLR: 20–50 g COD/L/day • HRT: 4–10 h	[17,33,64,78]

ABR, Anaerobic baffled reactor; *CSTR*, continuous stirred tank reactor; *EGSB*, expanded granular sludge bed; *FBR*, fluidized bed reactor; *MBR*, membrane bioreactor; *PBR*, packed bed bioreactor; *UASB*, upflow anaerobic sludge blanket reactor.

required to operate various types of bioreactors for biohydrogen production.

14.3 Conclusion

The depletion of fossil fuel reserves has resulted in an emergent need for alternative renewable energy sources mainly in the form of biofuels. Biohydrogen production from various substrates, particularly utilizing waste resources, is highly promising as it not only produces clean energy but also valorizes organic waste. In the past few decades, extensive research has been carried out to significantly improve H_2 yields and volumetric production. However, the technological development of laboratory-based research to large-scale installation is the need of the hour. Therefore research should be focused on bioreactor design and operating strategies for improving H_2 production on a large scale. Furthermore, high H_2 production rates and yields should be achieved so that biohydrogen production processes can out compete conventional thermochemical catalytic production processes.

References

[1] S. Kumari, D. Das, Improvement of biohydrogen production using acidogenic culture, Int. J. Hydrogen Energy 42 (2017) 4083–4094.
[2] P. Sivagurunathan, P. Anburajan, G. Kumar, S.-H. Kim, Effect of hydraulic retention time (HRT) on biohydrogen production from galactose in an up-flow anaerobic sludge blanket reactor, Int. J. Hydrogen Energy 41 (2016) 21670–21677.
[3] S. Roy, M. Vishnuvardhan, D. Das, Continuous thermophilic biohydrogen production in packed bed reactor, Appl. Energy 136 (2014) 51–58.
[4] B.S.F. Boodhun, A. Mudhoo, G. Kumar, S.-H. Kim, C.-Y. Lin, Research perspectives on constraints, prospects and opportunities in biohydrogen production, Int. J. Hydrogen Energy 42 (2017) 27471–27481.
[5] P.C. Hallenbeck, M. Abo-Hashesh, D. Ghosh, Strategies for improving biological hydrogen production, Bioresour. Technol. 110 (2012) 1–9.
[6] C.M. Kalamaras, A.M. Efstathiou, Hydrogen production technologies: current state and future developments, Conf. Pap. Energy 2013 (2013) 9.
[7] A. Sinharoy, N.A. Manikandan, K. Pakshirajan, A novel biological sulfate reduction method using hydrogenogenic carboxydotrophic mesophilic bacteria, Bioresour. Technol. 192 (2015) 494–500.
[8] I.Z. Boboescu, V.D. Gherman, G. Lakatos, B. Pap, T. Bíró, G. Maróti, Surpassing the current limitations of biohydrogen production systems: the case for a novel hybrid approach, Bioresour. Technol. 204 (2016) 192–201.
[9] K. Pakshirajan, J. Mal, Biohydrogen production using native carbon monoxide converting anaerobic microbial consortium predominantly *Petrobacter* sp., Int. J. Hydrogen Energy 38 (2013) 16020–16028.

[10] K.-Y. Show, D.-J. Lee, J.-S. Chang, Bioreactor and process design for biohydrogen production, Bioresour. Technol. 102 (2011) 8524–8533.
[11] J. Turner, G. Sverdrup, M.K. Mann, P.-C. Maness, B. Kroposki, M. Ghirardi, et al., Renewable hydrogen production, Int. J. Energy Res. 32 (2008) 379–407.
[12] T.V. Laurinavichene, S.N. Kosourov, M.L. Ghirardi, M. Seibert, A.A. Tsygankov, Prolongation of H2 photoproduction by immobilized, sulfur-limited *Chlamydomonas reinhardtii* cultures, J. Biotechnol. 134 (2008) 275–277.
[13] L. Singh, Z.A. Wahid, Methods for enhancing bio-hydrogen production from biological process: a review, J. Ind. Eng. Chem. 21 (2015) 70–80.
[14] P.C. Hallenbeck, J.R. Benemann, Biological hydrogen production; fundamentals and limiting processes, Int. J. Hydrogen Energy 27 (2002) 1185–1193.
[15] D.B. Levin, L. Pitt, M. Love, Biohydrogen production: prospects and limitations to practical application, Int. J. Hydrogen Energy 29 (2004) 173–185.
[16] I.K. Kapdan, F. Kargi, Bio-hydrogen production from waste materials, Enzyme Microb. Technol. 38 (2006) 569–582.
[17] W.-Q. Guo, N.-Q. Ren, X.-J. Wang, W.-S. Xiang, Z.-H. Meng, J. Ding, et al., Biohydrogen production from ethanol-type fermentation of molasses in an expanded granular sludge bed (EGSB) reactor, Int. J. Hydrogen Energy 33 (2008) 4981–4988.
[18] C.-M. Liu, S.-Y. Wu, C.-Y. Chu, Y.-P. Chou, Biohydrogen production from rice straw hydrolyzate in a continuously external circulating bioreactor, Int. J. Hydrogen Energy 39 (2014) 19317–19322.
[19] A.R. Barros, E.L. Silva, Hydrogen and ethanol production in anaerobic fluidized bed reactors: performance evaluation for three support materials under different operating conditions, Biochem. Eng. J. 61 (2012) 59–65.
[20] D.-Y. Lee, K.-Q. Xu, T. Kobayashi, Y.-Y. Li, Y. Inamori, Effect of organic loading rate on continuous hydrogen production from food waste in submerged anaerobic membrane bioreactor, Int. J. Hydrogen Energy 39 (2014) 16863–16871.
[21] C.-H. Lay, C.-Y. Huang, C.-C. Chen, C.-Y. Lin, Biohydrogen production in an anaerobic baffled stacking reactor: recirculation strategy and substrate concentration effects, Biochem. Eng. J. 109 (2016) 59–64.
[22] M. del Pilar Anzola-Rojas, M. Zaiat, H. De Wever, Improvement of hydrogen production via ethanol-type fermentation in an anaerobic down-flow structured bed reactor, Bioresour. Technol. 202 (2016) 42–49.
[23] Z. Trad, J. Akimbomi, C. Vial, C. Larroche, M.J. Taherzadeh, J.-P. Fontaine, Development of a submerged anaerobic membrane bioreactor for concurrent extraction of volatile fatty acids and biohydrogen production, Bioresour. Technol. 196 (2015) 290–300.
[24] R. Palomo-Briones, E. Razo-Flores, N. Bernet, E. Trably, Dark-fermentative biohydrogen pathways and microbial networks in continuous stirred tank reactors: novel insights on their control, Appl. Energy 198 (2017) 77–87.
[25] L. Singh, Z.A. Wahid, M.F. Siddiqui, A. Ahmad, M.H.A. Rahim, M. Sakinah, Biohydrogen production from palm oil mill effluent using immobilized *Clostridium butyricum* EB6 in polyethylene glycol, Process. Biochem. 48 (2013) 294–298.
[26] M.-S. Kim, D.-Y. Lee, D.-H. Kim, Continuous hydrogen production from tofu processing waste using anaerobic mixed microflora under thermophilic conditions, Int. J. Hydrogen Energy 36 (2011) 8712–8718.
[27] C.-M. Liu, C.-Y. Chu, W.-Y. Lee, Y.-C. Li, S.-Y. Wu, Y.-P. Chou, Biohydrogen production evaluation from rice straw hydrolysate by concentrated acid pre-treatment in both batch and continuous systems, Int. J. Hydrogen Energy 38 (2013) 15823–15829.
[28] C. Nualsri, P. Kongjan, A. Reungsang, Direct integration of CSTR-UASB reactors for two-stage hydrogen and methane production from sugarcane syrup, Int. J. Hydrogen Energy 41 (2016) 17884–17895.

[29] N. Venetsaneas, G. Antonopoulou, K. Stamatelatou, M. Kornaros, G. Lyberatos, Using cheese whey for hydrogen and methane generation in a two-stage continuous process with alternative pH controlling approaches, Bioresour. Technol. 100 (2009) 3713–3717.

[30] W. Han, D.N. Liu, Y.W. Shi, J.H. Tang, Y.F. Li, N.Q. Ren, Biohydrogen production from food waste hydrolysate using continuous mixed immobilized sludge reactors, Bioresour. Technol. 180 (2015) 54–58.

[31] G. Kumar, J.-H. Park, P. Sivagurunathan, S.-H. Lee, H.-D. Park, S.-H. Kim, Microbial responses to various process disturbances in a continuous hydrogen reactor fed with galactose, J. Biosci. Bioeng. 123 (2017) 216–222.

[32] C.D. Bárcenas-Ruiz, J. Carrillo-Reyes, L. Arellano-García, L.B. Celis, F. Alatriste-Mondragón, E. Razo-Flores, Pretreatment and upward liquid velocity effects over granulation in hydrogen producing EGSB reactors, Biochem. Eng. J. 107 (2016) 75–84.

[33] C. Cisneros-Pérez, J. Carrillo-Reyes, L.B. Celis, F. Alatriste-Mondragón, C. Etchebehere, E. Razo-Flores, Inoculum pretreatment promotes differences in hydrogen production performance in EGSB reactors, Int. J. Hydrogen Energy 40 (2015) 6329–6339.

[34] Z.-P. Zhang, J.-H. Tay, K.-Y. Show, R. Yan, D. Tee Liang, D.-J. Lee, et al., Biohydrogen production in a granular activated carbon anaerobic fluidized bed reactor, Int. J. Hydrogen Energy 32 (2007) 185–191.

[35] Z.-P. Zhang, K.-Y. Show, J.-H. Tay, D.T. Liang, D.-J. Lee, W.-J. Jiang, Effect of hydraulic retention time on biohydrogen production and anaerobic microbial community, Process. Biochem. 41 (2006) 2118–2123.

[36] E.L.C. De Amorim, A.R. Barros, M.H.R.Z. Damianovic, E.L. Silva, Anaerobic fluidized bed reactor with expanded clay as support for hydrogen production through dark fermentation of glucose, Int. J. Hydrogen Energy 34 (2009) 783–790.

[37] G.M. Shida, A.R. Barros, C.Md Reis, E.L.Cd Amorim, M.H.R.Z. Damianovic, E. L. Silva, Long-term stability of hydrogen and organic acids production in an anaerobic fluidized-bed reactor using heat treated anaerobic sludge inoculum, Int. J. Hydrogen Energy 34 (2009) 3679–3688.

[38] A.R. Barros, E.L. Cavalcante de Amorim, C.M. Reis, G.M. Shida, E.L. Silva, Biohydrogen production in anaerobic fluidized bed reactors: effect of support material and hydraulic retention time, Int. J. Hydrogen Energy 35 (2010) 3379–3388.

[39] G.M. Shida, L.T. Sader, E.L. Cavalcante de Amorim, I.K. Sakamoto, S.I. Maintinguer, N.K. Saavedra, et al., Performance and composition of bacterial communities in anaerobic fluidized bed reactors for hydrogen production: effects of organic loading rate and alkalinity, Int. J. Hydrogen Energy 37 (2012) 16925–16934.

[40] D.M.F. Lima, M. Zaiat, The influence of the degree of back-mixing on hydrogen production in an anaerobic fixed-bed reactor, Int. J. Hydrogen Energy 37 (2012) 9630–9635.

[41] C.-N. Lin, S.-Y. Wu, J.-S. Chang, J.-S. Chang, Biohydrogen production in a three-phase fluidized bed bioreactor using sewage sludge immobilized by ethylene–vinyl acetate copolymer, Bioresour. Technol. 100 (2009) 3298–3301.

[42] D.-Y. Lee, Y.-Y. Li, Y.-K. Oh, M.-S. Kim, T. Noike, Effect of iron concentration on continuous H2 production using membrane bioreactor, Int. J. Hydrogen Energy 34 (2009) 1244–1252.

[43] S. Arriaga, I. Rosas, F. Alatriste-Mondragón, E. Razo-Flores, Continuous production of hydrogen from oat straw hydrolysate in a biotrickling filter, Int. J. Hydrogen Energy 36 (2011) 3442–3449.

[44] J. Carrillo-Reyes, M.A. Cortés-Carmona, C.D. Bárcenas-Ruiz, E. Razo-Flores, Cell wash-out enrichment increases the stability and performance of biohydrogen

producing packed-bed reactors and the community transition along the operation time, Renew. Energy 97 (2016) 266–273.
[45] J.A.C. Leite, B.S. Fernandes, E. Pozzi, M. Barboza, M. Zaiat, Application of an anaerobic packed-bed bioreactor for the production of hydrogen and organic acids, Int. J. Hydrogen Energy 33 (2008) 579–586.
[46] G. Peixoto, N.K. Saavedra, M.B.A. Varesche, M. Zaiat, Hydrogen production from soft-drink wastewater in an upflow anaerobic packed-bed reactor, Int. J. Hydrogen Energy 36 (2011) 8953–8966.
[47] J. Carrillo-Reyes, L.B. Celis, F. Alatriste-Mondragón, E. Razo-Flores, Different start-up strategies to enhance biohydrogen production from cheese whey in UASB reactors, Int. J. Hydrogen Energy 37 (2012) 5591–5601.
[48] E.D. Penteado, C.Z. Lazaro, I.K. Sakamoto, M. Zaiat, Influence of seed sludge and pretreatment method on hydrogen production in packed-bed anaerobic reactors, Int. J. Hydrogen Energy 38 (2013) 6137–6145.
[49] D.M.F. Lima, W.K. Moreira, M. Zaiat, Comparison of the use of sucrose and glucose as a substrate for hydrogen production in an upflow anaerobic fixed-bed reactor, Int. J. Hydrogen Energy 38 (2013) 15074–15083.
[50] A.D.N.F. Júnior, J. Wenzel, C. Etchebehere, M. Zaiat, Effect of organic loading rate on hydrogen production from sugarcane vinasse in thermophilic acidogenic packed bed reactors, Int. J. Hydrogen Energy 39 (2014) 16852–16862.
[51] C. Barca, D. Ranava, M. Bauzan, J.-H. Ferrasse, M.-T. Giudici-Orticoni, A. Soric, Fermentative hydrogen production in an up-flow anaerobic biofilm reactor inoculated with a co-culture of *Clostridium acetobutylicum* and *Desulfovibrio vulgaris*, Bioresour. Technol. 221 (2016) 526–533.
[52] A.D.N.F. Júnior, C. Etchebehere, M. Zaiat, Mesophilic hydrogen production in acidogenic packed-bed reactors (APBR) using raw sugarcane vinasse as substrate: influence of support materials, Anaerobe 34 (2015) 94–105.
[53] Y. Akutsu, D.-Y. Lee, Y.-Z. Chi, Y.-Y. Li, H. Harada, H.-Q. Yu, Thermophilic fermentative hydrogen production from starch-wastewater with bio-granules, Int. J. Hydrogen Energy 34 (2009) 5061–5071.
[54] E. Castelló, C. García y Santos, T. Iglesias, G. Paolino, J. Wenzel, L. Borzacconi, et al., Feasibility of biohydrogen production from cheese whey using a UASB reactor: links between microbial community and reactor performance, Int. J. Hydrogen Energy 34 (2009) 5674–5682.
[55] P. Intanoo, P. Chaimongkol, S. Chavadej, Hydrogen and methane production from cassava wastewater using two-stage upflow anaerobic sludge blanket reactors (UASB) with an emphasis on maximum hydrogen production, Int. J. Hydrogen Energy 41 (2016) 6107–6114.
[56] S. Sittijunda, A. Reungsang, Fermentation of hydrogen, 1,3-propanediol and ethanol from glycerol as affected by organic loading rate using up-flow anaerobic sludge blanket (UASB) reactor, Int. J. Hydrogen Energy 42 (2017) 27558–27569.
[57] P. Sivagurunathan, P. Anburajan, G. Kumar, J.-H. Park, S.-H. Kim, Recovering hydrogen production performance of upflow anaerobic sludge blanket reactor (UASBR) fed with galactose via repeated heat treatment strategy, Bioresour. Technol. 240 (2017) 207–213.
[58] J. Carrillo-Reyes, L.B. Celis, F. Alatriste-Mondragón, E. Razo-Flores, Decreasing methane production in hydrogenogenic UASB reactors fed with cheese whey, Biomass Bioenergy 63 (2014) 101–108.
[59] A. Elreedy, A. Tawfik, A. Enitan, S. Kumari, F. Bux, Pathways of 3-biofules (hydrogen, ethanol and methane) production from petrochemical industry wastewater via anaerobic packed bed baffled reactor inoculated with mixed culture bacteria, Energy Convers. Manag. 122 (2016) 119–130.

[60] P. Thanwised, W. Wirojanagud, A. Reungsang, Effect of hydraulic retention time on hydrogen production and chemical oxygen demand removal from tapioca wastewater using anaerobic mixed cultures in anaerobic baffled reactor (ABR), Int. J. Hydrogen Energy 37 (2012) 15503−15510.
[61] A. Tawfik, A. Salem, M. El-Qelish, Two stage anaerobic baffled reactors for biohydrogen production from municipal food waste, Bioresour. Technol. 102 (2011) 8723−8726.
[62] P. Bakonyi, N. Nemestóthy, V. Simon, K. Bélafi-Bakó, Fermentative hydrogen production in anaerobic membrane bioreactors: a review, Bioresour. Technol. 156 (2014) 357−363.
[63] P. Sivagurunathan, G. Kumar, P. Bakonyi, S.-H. Kim, T. Kobayashi, K.Q. Xu, et al., A critical review on issues and overcoming strategies for the enhancement of dark fermentative hydrogen production in continuous systems, Int. J. Hydrogen Energy 41 (2016) 3820−3836.
[64] L. Singh, Z.A. Wahid, Enhancement of hydrogen production from palm oil mill effluent via cell immobilisation technique, Int. J. Energy Res. 39 (2015) 215−222.
[65] W.P. Barber, D.C. Stuckey, The use of the anaerobic baffled reactor (ABR) for wastewater treatment: a review, Water Res. 33 (1999) 1559−1578.
[66] Y. Ueno, M. Tatara, H. Fukui, T. Makiuchi, M. Goto, K. Sode, Production of hydrogen and methane from organic solid wastes by phase-separation of anaerobic process, Bioresour. Technol. 98 (2007) 1861−1865.
[67] G. Luo, L. Xie, Z. Zou, W. Wang, Q. Zhou, Evaluation of pretreatment methods on mixed inoculum for both batch and continuous thermophilic biohydrogen production from cassava stillage, Bioresour. Technol. 101 (2010) 959−964.
[68] E. Jeong, H.-W. Kim, J.-Y. Nam, H.-S. Shin, Enhancement of bioenergy production and effluent quality by integrating optimized acidification with submerged anaerobic membrane bioreactor, Bioresour. Technol. 101 (2010) S7−S12.
[69] A.D.N.F. Júnior, C. Etchebehere, M. Zaiat, Mesophilic hydrogen production in acidogenic packed-bed reactors (APBR) using raw sugarcane vinasse as substrate: influence of support materials, Anaerobe 34 (2015) 94−105.
[70] G. Collins, C. Foy, S. McHugh, V. O' Flaherty, Anaerobic treatment of 2,4,6-trichlorophenol in an expanded granular sludge bed−anaerobic filter (EGSB−AF) bioreactor at 15°C, FEMS Microbiol. Ecol. 53 (2005) 167−178.
[71] G. Buitrón, D. Prato-Garcia, A. Zhang, Biohydrogen production from tequila vinasses using a fixed bed reactor, Water Sci. Technol. 70 (2014) 1919−1925.
[72] D.M. Fontes Lima, M. Zaiat, The influence of the degree of back-mixing on hydrogen production in an anaerobic fixed-bed reactor, Int. J. Hydrogen Energy 37 (2012) 9630−9635.
[73] F. Gallucci, E. Fernandez, P. Corengia, M. van S. Annaland, Recent advances on membranes and membrane reactors for hydrogen production, Chem. Eng. Sci. 92 (2013) 40−66.
[74] Z. Ran, Z. Gefu, J.A. Kumar, L. Chaoxiang, H. Xu, L. Lin, Hydrogen and methane production in a bio-electrochemical system assisted anaerobic baffled reactor, Int. J. Hydrog. Energy. 39 (2014) 13498−13504.
[75] L. Jürgensen, E.A. Ehimen, J. Born, J.B. Holm-Nielsen, Hydrogen production using an anaerobic baffled reactor−mass balances for pathway analysis and gas composition profiles, Int. J. Hydrog. Energy. 40 (2015) 12154−12161.
[76] S. Krishnan, L. Singh, M. Sakinah, S. Thakur, Z.A. Wahid, M. Alkasrawi, Process enhancement of hydrogen and methane production from palm oil mill effluent using two-stage thermophilic and mesophilic fermentation, Int. J. Hydrogen Energy 41 (2016) 12888−12898.

[77] M. Kumar, A. Sinharoy, K. Pakshirajan, Process integration for biological sulfate reduction in a carbon monoxide fed packed bed reactor, J. Environ. Manage. 219 (2018) 294–303.
[78] K. Rajeshwari, M. Balakrishnan, A. Kansal, K. Lata, V. Kishore, State-of-the-art of anaerobic digestion technology for industrial wastewater treatment, Renew. Sustain. Energy Rev. 4 (2000) 135–156.

CHAPTER 15

Bioreactor for algae cultivation and biodiesel production

Rashmi Chandra[1], Garima Vishal[2], Carlos Eduardo Gámez Sánchez[1] and Janet Alejandra Gutiérrez Uribe[3]
[1]Tecnologico de Monterrey, School of Engineering and Science, Toluca, Mexico
[2]Department of Chemical Engineering, Indian Institute of Technology, New Delhi, India
[3]Tecnologico de Monterrey, School of Engineering and Science, Puebla, Mexico

15.1 Introduction

Microalgae and cyanobacteria biotechnology have gained a huge amount of attention in the past two decades. Algae can be microbial factories producing various compounds other than lipids for biodiesel [1]. Being composed of lipids (7%–23%), carbohydrates (5%–23%), and proteins (6%–52%) microalgae have a range of components of value [2]. Microalgae produce commercially important value-added products used in food, nutraceutical, cosmetic, and pharmaceutical active compounds that have great importance outside of fuels [3]. Algal biodiesel has substantial potential to be used as a substitute to petrochemical diesel because of its technical advantage of requiring no change in present engine design and infrastructure. Algal biodiesel is biodegradable, nontoxic with a favorable combustion emission profile, and produces much less carbon monoxide, sulfur dioxide, and unburned hydrocarbons than petroleum-based diesel fuel. Algal biodiesel contributes toward carbon neutrality [4]. Algae sequester CO_2 from the atmosphere and convert it to oil and this burns similarly to fossil fuels in regard to greenhouse gases. However, the transformation of laboratory discoveries into industrial manufacture remains a bottleneck. The rates of bioprocess development to viable commercial venture are slow due to the difficulties associated with scale-up. Therefore bioreactor design for scale-up and operation have received much attention.

A bioreactor is a device or a vessel used to carry out biochemical reactions to convert a starting material to a given product at conditions that maximize the process. The product of conversion may be cells, biomass, or chemicals of various kinds. In this chapter, CO_2, water, and photosynthetic microbes for biomass growth and their conversion to biodiesel will

Bioreactors
DOI: https://doi.org/10.1016/B978-0-12-821264-6.00015-2

be considered. Many different kinds of bioreactors are available and sometimes a given type may be operated in different ways to obtain different results. This chapter reflects on the current state of photosynthetic biomass cultivation toward biodiesel production and carbon footprint reduction.

15.2 Algal product and chemistry of biosynthesis

The biosynthesis of lipid starts in the thylakoid and stroma of chloroplast. Algae and cyanobacteria fix CO_2 via photophosphorylation and the Calvin cycle to carbohydrates. These carbohydrates are converted to triacylglycerides via fatty acid synthases (FAS) and acetyl-CoA carboxylase (ACCase). This pathway occurs in four major steps.

1. Carbohydrates accumulation
2. Acetyl-CoA formation
3. Palmitic acid synthesis
4. Chain elongation

Photosynthates like glyceraldehyde 3-phosphate (G3P) provide an endogenous source of acetyl-CoA, which is activated acetyl-CoA synthetase in the stroma or from the cytosolic conversion of glucose to pyruvate during glycolysis [4]. This acetyl-CoA is preferentially transported from the cytosol to the plastid, where it is converted to fatty acid and subsequently to triacylglycerol (TAG) and again released into the cytosol.

The first reaction of the fatty acid pathway includes the formation of malonyl-CoA from acetyl-CoA and the catalyzation of CO_2 by the enzyme ACCase. The conversion of acetyl-CoA to malonyl-CoA is an energy intensive process and requires adenosine triphosphate (ATP). After the formation of malonyl-CoA, a four-step repeating cycle, that is, condensation, reduction, dehydration, and reduction, takes place for seven cycles and forms the principal product of FAS, namely palmitic acid (C16:0) [2]. All the reactions in this process are catalyzed by a multienzyme complex, namely FAS. When the chain length reaches 16 carbons, the product (palmitate) leaves the cycle. With each course of the cycle, the fatty acyl chain is extended by two carbons resulting in chain elongation. Palmitate is the precursor to stearate and higher chain saturated fatty acids, which are formed by the action of fatty acid elongation systems present in the smooth endoplasmic reticulum.

Algae are environment-friendly, nontoxic, do not contain sulfur, and are biodegradable [5,6]. Lipid saturation index is a significant property that determines biodiesel stability and performance properties in order to produce high quality biodiesel. Low saturated fatty acid levels (such as C16:0 and C18:0) are required for good winter operability, the highest possible monounsaturated fatty acid levels (such as C18:1) are required for good stability and winter operability, and the lowest possible polyunsaturated fatty acid levels (such as C18:3) are required to increase oxidation stability.

The principal drawback in biodiesel production is the culture concentration step since dry biomass represents only 0.1%—1% of culture weight, and besides this, unit operation is costly. These points provoked have many research programs with focus on such approaches with the aim to decrease the costs. Before economic barriers can be meaningfully addressed, many technical and engineering challenges must be tackled. Nevertheless, these economic analyses provide an indicator that the development of low-cost photobioreactors and the optimization of photosynthetic efficiency are major research and development challenges.

15.3 Cultivation bioreactors systems

15.3.1 Open pond bioreactors

The mode of reactor operation influences biomass and lipid productivity both. Open pond bioreactors mimic the natural method of microalgae cultivation. These can be categorized as natural water bodies and artificial ponds or containers. The most commonly used systems are shallow raceway ponds as shown in Fig. 15.1. These consist of shallow ponds with PVC lining or of cement or concrete that are divided by a series of baffles in order to promote the mixing of nutrients and algal biomass. These ponds are usually constructed in shallow dimensions as algae need to be exposed to sunlight, and sunlight can only penetrate water up to a certain limited depth, which is usually 1.5 ft. These kinds of ponds usually operate in a continuous mode with constant nutrient supply. CO_2 usually gets sequestered from the atmosphere. Open pond bioreactors represent the most commonly used reactor in algal biofuel technology. They are also used for cultivating algae for single cell protein production, pigments, beta carotene, etc., at industrial scale.

Open raceway ponds are low cost in terms of operation and maintenance, which results in low production costs. They can be constructed on nonagricultural land. Open pond cultivation has a few inherent demerits like poor light and CO_2 diffusion, evaporation loss, influence of external environment, contamination by predators and other fast-growing heterotrophs, which restrict the commercial production of algae in open pond/culture systems.

Figure 15.1 Raceway pond algal bioreactor.

15.3.2 Photobioreactors

Photobioreactors (PBRs) are closed systems that are kept under controlled conditions to support the growth of microalgae or cyanobacteria. They provide more favorable conditions than those found in open raceway pond bioreactors because these systems are closed and important factors like carbon, water, light, nutrients, temperature, and pH can be controlled within the system. There are various kinds of PBRs reported on that facilitate better control of culture environment systems such as CO_2 supply, water supply, temperature, light exposure, density of culture, pH, gaseous exchange, and so forth. High mass transfer is achievable in these kinds of bioreactors, especially with CO_2 sequestration. Mixing in PBRs can be provided by mechanical impellers or nonmechanically by aeration [7,8]. Nonmechanical agitation systems can be observed in airlift, bubble column, tubular, and flat-panel reactor operations, whereas under the mechanical category, continuous stirred tank reactors and fermenter models are equipped with Rushton blade impellers with adjustable agitation speed.

PBRs precisely designed for CO_2 sequestration have the option to use pure CO_2, air, or a mixture of both. This provides gaseous exchange as well as nutrient and biomass mixing. They can be operated in batch mode, continuous mode, or semicontinuous mode. A higher biomass of microalgae productivity is obtained in closed cultivation systems where contamination can be prevented under controlled conditions. Fully closed PBRs provide opportunities for monocultures of a greater variety of algae than those found in open cultivation systems. Different types of closed PBRs are discussed here.

15.3.3 Tubular or vertical photobioreactor

Tubular PBRs are made of transparent vertical tubing for proper light penetration as shown in Fig. 15.2. Certain PBRs contain gas spargers to provide tiny gas bubbles for gas exchange. This allows for proper mixing of nutrient and biomass along with CO_2 exchange. These can be classified as bubble column and airlift reactors [9–11].

Bubble column PBRs represent the most widely used PBR system. In this type of reactor, CO_2 mass transfer and mixing are carried out through the use spargers. Biomass production in a bubble column PBR depends on the photosynthetic efficiency, gas flow, and light/dark cycle (as the liquid is circulated through a central dark zone).

Airlift PBRs consist of two interconnected zones called the riser and the downcomer in an annular setup. Airlift PBRs exist in two forms, namely internal loop and external loop. In an internal loop reactor,

Figure 15.2 Tubular photobioreactor.

regions are separated either by a draft tube or a split cylinder, whereas in an external loop reactor, the riser and downcomer are separated physically by two different tubes. Mixing in these reactors is done by gas bubbling with spargers without mechanical agitation. Sparged gas moves upward and decreases the density of the riser making the liquid move upward [12,13]. The gas present in a PBR significantly influences the fluid dynamics of an airlift reactor and forces the liquid downwards. Airlift reactors have the characteristic advantage of creating circular mixing patterns in which liquid culture passes continuously through dark and light regions, giving a flashing light effect to algal cells. The biomass growth pattern of *Nannochloropsis oculata* and *Scenedesmus quadricauda* was studied inside two vertical airlift PBRs suitable for indoor operation with both saltwater and freshwater under different lighting systems. The results depicted that the biomass productivity of the cultures was found to depend on the light regimes used and the duration of operation.

15.3.4 Flat-panel photobioreactor

Flat-panel PBRs are highly photosynthetically efficient and they are widely used in scaling-up processes and mass cultivation. Dissolved oxygen accumulation is relatively low compared to airlift PBRs [12,14]. However, the lack of temperature control and gas zones are some of the inherent disadvantages observed in these kinds of PBRs.

15.3.5 Helical photobioreactor

Helical PBRs are transparent and flexible tubes of small diameters separated or attached by degassing units. A pump is used to drive the culture through a long tube to degassing units. CO_2 or air gets fed into a helical tube from the bottom, giving a better photosynthesis efficiency. This kind of reactor facilitates better gas exchange and efficient CO_2 absorption. The energy requirements in these reactors are associated with shear stress and limit their commercial use. Fouling on the inside of the reactor is another disadvantage of this system.

15.3.6 Stirred tank photobioreactor

This kind of PBR is conventionally used in laboratory studies, where agitation is provided by impellers or baffles, light is provided externally, and CO_2 is bubbled in to provide carbon for algal or cyanobacterial growth [15,16]. *Protoceratium reticulatum* growth was studied in 2 and 15 L stirred

PBRs equipped with internal spin filters showing average biomass cell productivity 3.7-times higher than that of static cultures. Low surface-area-to-volume ratio, which in turn decreases light harvesting efficiency, and high shear stress imposed due to mechanical agitation are the inherent disadvantages. Column, flat-plate, and tubular PBRs are the most commonly used systems for microalgae cultivation, but these increase the cost of operation over open raceway pond PBRs. Both systems have their own strengths and limitations. The application of either system would depend upon the product of interest.

15.4 Methods for microalgae biodiesel extraction

15.4.1 Pretreatment

Algae biomass can be treated by different processes in order to extract different compounds. There are several conversion processes, some of the most important are mechanical-based treatments [17]. After microalgae cultures reach the stationary phase of growth, the biomass is concentrated, and the products of interest can be extracted using the dry or wet biomass. In the first case, a dewatering process can be performed by centrifugation followed by a cell disruption process with the objective of breaking the cellular walls in order to release of microalgae components that are not secreted outside the cell. Generally, the methodologies used involve disruption, breakage, or disintegration [18]. On the other hand, to get a dried biomass, after the process of dewatering the biomass undergoes a thermal drying process that usually produces a paste-like biomass with a dry weight above 85% [19].

15.4.2 Solvent extraction

Most solvent-based extraction techniques used for the extraction of lipids from microalgae are based on the traditional methods used for plant oil extraction like organic solvent extraction, the Folch method, Soxhlet extraction (SE), and others. Organic solvents are absorbed within the cell wall where they cause swelling and rupture of the microalgae cell, making the cell contents available to be separated in a following step [20]. The main parameters to consider in the choosing of a solvent for the extraction of lipids from microalgae are polarity or extractability, lipid solubility, water miscibility (ability to for two-phase systems), and low toxicity [21].

15.4.3 Folch method

Lipid extraction by the Folch method relies on the use of a 2:1 chloroform/methanol mixture for the extraction of intracellular lipids, and is the basis of many solvent extraction methods used nowadays. First, a cell homogenate is equilibrated with a 25% volume saline solution and stirred. This mixture is left to stand until biphasic separation, so that the lipids settle on the upper layer [22]. Since this method was originally designed for animal cells and tissues, a preceding step involving the disruption of microalgae cell walls has to be included [20].

15.4.4 Soxhlet extraction

SE is a process in which partially soluble components of a solid sample are transferred to a liquid phase (solvent) by means of a Soxhlet extractor. This technique employs nonpolar solvents like hexane to obtain neutral lipids. The extraction consists of the placement of the solid sample in a filter paper thimble into the main chamber of the Soxhlet apparatus. Then the solvent is heated to reflux and travels into the main chamber, so that the less soluble compounds are recovered by the solvent [23]. As the extraction solvent polarity increases, a higher extraction yield from microalgae can be achieved due to the recovery of complex lipids and pigments [24]. This is an important consideration since total lipid extracts with polar solvents are complex and other metabolites different to lipids are present. SE parameters include choice of solvent, sample particle size, and extraction time [25]. SE is usually carried out at laboratory scale as it requires high solvent consumption and long extraction time.

15.4.5 Bligh and Dyer method

The Bligh and Dyer method consists of simultaneous lipid extraction and partitioning with protein precipitation in the interface between two liquid phases. It is similar to the Folch method, however, the solvent mixture composition and ratios are different. First, the lipids from a cell homogenate are extracted with a 1:2 chloroform/methanol mixture, and the chloroform phase (lipid-rich) is recovered. Microalgae lipids are extracted and measured by gravimetry. This procedure is employed in pilot- and large-scale operations [22].

An improvement to this method is the addition of 1 M NaCl, instead of water, in order to avoid binding the acidic lipids to denatured lipids. Shorter separation times have been achieved by the addition of 0.2 M phosphoric acid and HCl. An increased recovery of acidic phospholipids has been achieved by the addition of 0.5% acetic acid (v/v) [22]. Organic solvent extraction remains one of the main strategies for the recovery of valuable products from microalgae. Solvents should be chosen based on the polarity of the target compounds. For instance, TAGs are nonpolar molecules and the main lipid target for biodiesel production, hence, a nonpolar solvent is a suitable choice for extraction.

15.4.6 Mechanical methods

Cell disruption methods include solid shear, cavitation and collapse, pulsed electric fields, chemical hydrolysis, enzymatic digestion, subcritical water extraction, high-pressure homogenization, and bead milling.

15.4.7 Milling

Bead milling consists of disrupting the cell walls of microalgae by grinding and agitating the cells on a solid surface of glass beads [26]. The size of the beads for an effective disruption is in the range of 0.3–0.5 mm. Generally, the beads can be made up of zirconia–silica or zirconium oxide. The process efficiency is determined by the biomass concentration, flow rate, agitator movement type and speed, and temperature.

The process of milling can be done in shaking vessels or through agitated beads. In the shaking vessel method, the culture vessel is shaken using a vibrating platform, allowing the beads to move the microalgae cells and forcing them to collide with each other. The highest recovery of lipids through this method was performed by Ryckebosch et al., recovering 40% of lipids from a culture of *Phaeodactylum tricronutum* [27]. On the other hand, Zheng et al. extracted 11% of lipids from a culture of *Chlorella vulgaris* using a bead milling vessel [28]. In the case of the agitated beads method, the beads and the culture are agitated by a rotatory agitator inside the culture vessel, which simultaneously provides heat thereby helping in the disruption process as reported by Lee et al. The authors used this methodology and obtained an oil yield inside the range of 7.9–8.1 g/L using cultures of *Botryococcus* sp., *C. vulgaris*, and *Scenedesmus* sp. [29].

15.4.8 Pressing

The use of presses is one of the classical methods used to perform the extraction of value-added products from many sources. This method is based on the mechanical crushing of materials with a low content of moisture. First of all, dried biomass is submitted to high mechanical pressure to crush and break the cells and then squeeze the oil out of the biomass. The extraction efficiency can be improved by varying the pressure force, algal strain, and configuration of the press and pistons used. In the gel-press alternative, algae are first washed before carbohydrate extraction using diluted alkali. Residues are separated by centrifugation followed by filtration through porous silica and then concentration by evaporation. The material recovered is extruded through spinnerets into a cold solution of potassium chloride and the gelled threads are then dewatered by pressure [30].

Shear-based devices such as the French press and the Hughes press use high pressures to force a biomass solution through a small aperture. Usually the oil recovery is in the range of 70%–75%. Sometimes for enhanced oil recovery, mechanical crushing is used in addition to chemical methods. The principal drawbacks of this method are the requirement of high-cost maintenance and less efficiency compared to other methods [22].

Different products produced by microalgae including lipids, proteins, and pigments have been extracted by means of mechanical extraction. Table 15.1 shows the relation of mechanical extraction used and the yields obtained in different studies.

15.4.9 Freeze–thaw method

The freeze–thaw method favors lipid extraction from microalgae biomass since it decreases to a minimum the loss of volatile lipids due to evaporation. This method consists of the crystallization of intracellular water by freezing wet biomass at a temperature near −80°C. Afterward the samples are thawed so that the frozen cells are lysed by the expansion of ice crystals. This method is usually employed in combination with another method such as ultrasonication, microwave-assisted extraction (MAE), or

Table 15.1 Yield of value-added products extracted by mechanical extraction.

Method	Microorganism	Product	Yield (%)	References
Milling	*Chlorella vulgaris*	Lipids	11	[28]
	Chlorella protothecoides	Lipids	18.8	[31]
	Botrycoccus sp.	Lipids	28	[32]
	Phaeodactylum triconutum	Lipids	40	[27]
Pressing	*Chlorococcum infusarium*	Lipids	96.2	[33]

bead milling with the purpose of increasing the yield efficiency [34]. However, freeze–thawing cycles must be carefully managed. A study of the metabolic profile of marine microalgae after freeze–thawing under standard freeze-storage temperatures (−20°C and −78°C) for one and two cycles of 7 days each reported that unfrozen samples showed a decrease of 10% in reproducibility after one cycle and a further 7% decrease after the second cycle [35].

15.4.10 Enzymatic methods

In enzymatic extraction processes, a combination of enzymes is employed to breakdown the algal cell wall, expel lipid bodies outside the cell, and separate the lipid fraction from the lipid/protein matrix [36]. Enzymatic lysis is an alternative to mechanical cell disruption. Lytic enzymes have to be specific to the microalgae species used; the most common being cellulase and lipase due to the presence of polysaccharides like cellulose and hemicellulose in algal cell walls and lipids contained in a sac surrounded by phospholipids [34].

Aqueous enzymatic assisted extraction is a cell disruption technique for the extraction of lipids from microalgae. It most remarkable features include high selectivity, mild reaction conditions (neutral pH and incubation from 25°C to 37°C), and the absence of intensive drying steps [29]. A report by Chen et al. presents an improved method for enzymatic lysis coupled with thermal treatment for the extraction of lipids from *N. oceanica* in which the optimal extraction parameters at 37°C, pH 5.0, 1.3% of cellulase, liquid/solid ratio of 15 mL/g, and 5 h duration were found. These conditions yielded up to 28.8% of lipids [30].

The main steps for enzymatic extraction of lipids from microalgae include biomass harvesting, conditioning and the addition of enzymes, stirred incubation for the disruption of algal cell walls, the addition of solvent (if needed), centrifugation, and lipid fraction recovery [29]. Furthermore, enzymatic digestion can be used as a means for saccharification of carbohydrate biomass for bioethanol production after the removal of lipids [34]. The main microalgae products of interest obtained by enzymatic methods are lipids, carbohydrates, and proteins, and are presented in Table 15.2.

15.4.11 Supercritical fluid extraction

Supercritical fluid extraction (SFE) employs the solvating properties of a supercritical fluid by applying pressure and temperature above the critical

Table 15.2 Yield of value-added products extracted by enzymatic extraction.

Microalgae	Enzyme	Product	Yield (%)	References
Chlorella vulgaris	Cellulase	Lipids	25	[37]
	Lysozyme		23	
	Snailase		7	
Nannochloropsis oceanica	Cellulase	Lipids	28.8	[38]
		Protein	29.9	
Mixed culture of microalgae (Northwest Iran)	β-Glucosidase/ cellulase and α-amylase and amyloglucosidase	Reducing sugars	93.64	[39]
	α-Amylase and amyloglucosidase		61.19	

point of a compound or mixture. The tunable parameters to consider for SFE include solvent, temperature, pressure, extraction time, solvent flow rate, sample size, extraction time, use of modifier, and particle size [25].

SFE with CO_2 (SFE$-CO_2$) has been employed as an alternative green extraction technique to spare the use of toxic solvents [40]. The advantages of SFE$-CO_2$ consist of the low critical point of CO_2 at near room temperature at relatively low pressure (30.9°C and 73.9 bar), the fact that it is generally recognized as safe by the Food and Drug Administration, and it is environment-friendly [41,42]. Moreover, CO_2 becomes gaseous after depressurization, thus separated from the sample without residual traces of solvent, can be collected for recycling for subsequent extractions, which in itself brings economic and environmental benefits. Particularly useful for biodiesel extraction, supercritical CO_2 is highly selective for nonpolar lipids such as triglycerides and does not solubilize phospholipids [40]. Other solvents used in SFE include hydrocarbons (hexane, pentane, and butane), nitrous oxide, sulfur hexafluoride, and fluorinated hydrocarbons [42].

15.4.12 Microwave-assisted extraction

MAE relies on the contact of a dielectric polar substance (water, for instance) and a fast oscillating electric field produced by microwaves, which generates heat due to the friction caused by inter- and intramolecular movements. The heat induces the formation of water vapor in the cell, which eventually causes rupture and further leakage and release of intracellular components, led by an electroporation effect [26]. Thus MAE is regarded as a rapid, simple, safe, effective, and economical method for the extraction of lipids that does not require the preceding dewatering of samples [21]. Furthermore, microalgae pretreated by microwaves present multiple micro-fissures within the cell wall, which yields higher biooil recoveries [43].

Besides extraction, microwaves can be employed for the transesterification of oils into biodiesel and represent an attractive option since they require short reaction times (15–20 min), have low operational costs, and show efficient extraction of algal oils. One important drawback of this method is the high maintenance cost in commercial-scale settings [22]. For MAE the main parameters to be taken into account include extraction time, temperature, dielectric properties of the process mixture, solid/liquid ratio, and type and concentration of solvent [26].

15.4.13 Ultrasound-assisted extraction

Ultrasonic-assisted extractions (UAE) can recover oils from microalgae cells through cavitation [44]. During a low-pressure cycle, high-intensity small vacuum bubbles are created in the liquid. When the bubbles attain a certain size, they collapse violently during a high-pressure cycle. During the implosion, high pressures and high-speed liquid jets are produced locally, and the resulting shear forces break the cell structure mechanically. This effect supports the extraction of lipids from algae [45]. The high-pressure cycles of the ultrasonic waves support the diffusion of solvents such as hexane into the cell structure. As ultrasound breaks the cell wall mechanically by cavitational shear forces, it facilitates the transfer of lipids from the cell into the solvent [46].

Lipid recovery can be enhanced by increasing the exposure time and by using mixtures of polar and nonpolar solvents. Also, UAE favors the release of cell contents into the solvent through the mass transfer and penetration of the solvent within the cell. UAE can be performed at low temperatures, which is an ideal feature when dealing with the extraction of thermally sensitive molecules [26].

15.4.14 Pressurized liquid extraction

The wet lipid extraction process uses wet algae biomass using solvent proportionately [47]. This method resembles the solvent extraction process, but varies in the nature of biomass (wet) used. One advantage of the process is the elimination of the need for a drying step, while the interference of the moisture content with the extraction solvents and the lack of wide applicability to all kinds of solvents are the major limitations of this extraction procedure.

Hydrothermal liquefaction is a process in which biomass is converted in hot compressed water to a liquid biocrude [48]. Processing temperatures range from 200°C to 350°C with pressures of around 15–20 MPa, depending on the temperature, because the water has to remain in the subcritical region to avoid the latent heat of vaporization [48]. At these conditions, complex molecules are broken down and repolymerized to oily compounds [49]. This procedure is ideal for the conversion of high-moisture-content biomass such as microalgae because the use of a drying step on the feedstock is not necessary.

15.5 Osmotic pressure

Osmotic shock or osmotic stress is a sudden change in the solute concentration around a cell, causing a rapid change in the movement of water across its cell membrane [50]. This shock causes a release in the cellular contents of microalgae. The method is more applicable for strains cultivated in marine environments (e.g., *Nannochloropsis* sp.). Osmotic shock is also induced to release cellular components for biochemical analysis [31]. This method was applied on *Halorubrum* sp. isolated from saltern ponds. The results showed increased lipid productivities and variations in lipid compositions [51].

15.6 Pulsed electric field technologies

Pulsed electric field (PEF) processing is a method for processing cells by means of brief pulses from a strong electric field [52]. Algal biomass is placed between two electrodes and a PEF is applied. The electric field enlarges the pores of the cell membranes and expels its contents [53].

15.7 Photobioreactor in present scenarios

Innovations in PBR design bring sustainability into current algal biomass cultivation. This has led to an evolution in aesthetic designs with a symbiotic relationship between solar energy and urban ecology. The conceptual model of integrating microalgae into architecture as a promising source of bioenergy is currently being looked into. Vertical facades can become smart photosynthetic surfaces that respond to sunlight and generate biomass in the current scenario of global warming. Photosynthetic surfaces can be transformed into bio-facades that could become an integral part of futuristic smart cities. Bio-facades promise to purify the atmosphere in crowded cities from harmful carbon emissions as microalgae biomass sequesters CO_2 from the atmosphere and generates O_2. Nutrients can be obtained from buildings' liquid wastes in the form of gray and black water and can be processed via a built-in nutrient separation system. The symbiosis between urban infrastructure and PBRs promises to reduce costs via the multiple outputs like biofuels and biomass for heat and electricity generated by centralized biorefinery processing systems [54–57]. These PBRs can also be employed at the urban level as street art installations and urban canopies combining natural processes and urban settings. These installations can produce biofuel, emit light, provide shade, and release oxygen along with raising public awareness on the importance of biofuels.

15.8 Conclusion

Microalgae biomass is a versatile feedstock for biodiesel production as compared to plant feedstock. The algal biomass requirement for biodiesel production can be cultivated in photobioreactors and fermenters that provide high density biomass as well as oils. Open raceway ponds are attracting interest in industrial biomass production. Cost-cutting research with a multidisciplinary approach will help resolve some of the inherent limitations prior to upscaling. The extraction of lipids is a key aspect involved in biomass-to-biodiesel production; the selected method directly influences the potential lipid productivity of the process. So far, several methods have been employed for extracting the cellular contents (lipids) of microalgae. Each method has its own advantages and disadvantages for practical applicability. Among the processes described, solvent extraction is suitable for extracting lipids from mass cultures, but requires large volumes of solvent. The Soxhlet extraction method is applicable only when a single solvent is used and is not suitable for binary solvent applications. On the other hand, the recovery and reusability of the solvent are possible with this method. UAE can perform well when coupled with an enzymatic treatment, but both methods lack cost effectiveness and feasibility for large-scale applications. Supercritical carbon dioxide extraction, pulsed electric fields, osmotic shock, hydrothermal liquefaction, and wet lipid extraction require more optimization efforts for large-scale applications. A suitable method that is operable with both binary and single solvents, applicable at large scales, and yielding high lipid productivities is yet to be optimized to achieve enhanced microalgae lipid yields for biofuels and metabolite extraction. The conjunction of the algal biofuel production process with waste gas, wastewater, and water reclamation is a promising strategy to be considered for economic viability. The integration of algal fuel with the simultaneous production of valuable by-products will also have a positive impact on the overall process economics. Considerable interest in the algal-based biofuel market from industries in conjunction with intensified research efforts provide a testimony that the process of algal biofuels production will be economically viable when integrated with sustainable and closed loop pathways for enabling biofuels usage in the near future.

Acknowledgments

The authors would like to thank the Director of the School of Engineering and Science, Tecnologico de Monterrey, Campus Monterrey, and Campus Toluca. The authors also want to thank the Department of Sciences for Sustainable Development and the Department of Bioengineering School of Engineering and Sciences, Campus Toluca Tecnologico de Monterrey. The authors would also like to thank Aavesh green sustainability solutions S. De R. L. De. C. V. for their support.

References

[1] A.E.-F. Abomohra, W. Jin, R. Tu, S.-F. Han, M. Eid, H. Eladel, Microalgal biomass production as a sustainable feedstock for biodiesel: current status and perspectives, Renew. Sustain. Energy Rev 64 (2016) 596–606.

[2] R. Chandra, M. Rohit, Y. Swamy, S.V. Mohan, Regulatory function of organic carbon supplementation on biodiesel production during growth and nutrient stress phases of mixotrophic microalgae cultivation, Bioresour. Technol 165 (2014) 279–287.

[3] L. Soh, M. Montazeri, B.Z. Haznedaroglu, C. Kelly, J. Peccia, M.J. Eckelman, et al., Evaluating microalgal integrated biorefinery schemes: empirical controlled growth studies and life cycle assessment, Bioresour. Technol 151 (2014) 19–27.

[4] S.V. Mohan, M.P. Devi, G.V. Subhash, R. Chandra, Algae oils as fuels, Biofuels From Algae, Elsevier, 2014, pp. 155–187.

[5] X. Liu, B. Saydah, P. Eranki, L.M. Colosi, B.G. Mitchell, J. Rhodes, et al., Pilot-scale data provide enhanced estimates of the life cycle energy and emissions profile of algae biofuels produced via hydrothermal liquefaction, Bioresour. Technol 148 (2013) 163–171.

[6] K. Ullah, M. Ahmad, V.K. Sharma, P. Lu, A. Harvey, M. Zafar, et al., Algal biomass as a global source of transport fuels: overview and development perspectives, Prog. Nat. Sci. Mater. Int 24 (2014) 329–339.

[7] P. Das, W. Lei, S.S. Aziz, J.P. Obbard, Enhanced algae growth in both phototrophic and mixotrophic culture under blue light, Bioresour. Technol 102 (2011) 3883–3887.
[8] R.R. Sastre, Z. Csögör, I. Perner-Nochta, P. Fleck-Schneider, C. Posten, Scale-down of microalgae cultivations in tubular photo-bioreactors—a conceptual approach, J. Biotechnol 132 (2007) 127–133.
[9] B.K. Selvan, M. Revathi, P.S. Piriya, P.T. Vasan, D. Prabhu, S.J. Vennison, Biodiesel production from marine cyanobacteria cultured in plate and tubular photo-bioreactors, Indian J Exp Biol 51 (2013) 262–268.
[10] M.N. Kinyua, L.E. Rowse, S.J. Ergas, Review of small-scale tubular anaerobic digesters treating livestock waste in the developing world, Renew. Sustain. Energy Rev 58 (2016) 896–910.
[11] A.L. Stephenson, E. Kazamia, J.S. Dennis, C.J. Howe, S.A. Scott, A.G. Smith, Life-cycle assessment of potential algal biodiesel production in the United Kingdom: a comparison of raceways and air-lift tubular bioreactors, Energy Fuels 24 (2010) 4062–4077.
[12] N.-H. Norsker, M.J. Barbosa, M.H. Vermuë, R.H. Wijffels, Microalgal production—a close look at the economics, Biotechnol. Adv 29 (2011) 24–27.
[13] A. Scoma, L. Giannelli, C. Faraloni, G. Torzillo, Outdoor H2 production in a 50-L tubular photobioreactor by means of a sulfur-deprived culture of the microalga *Chlamydomonas reinhardtii*, J. Biotechnol 157 (2012) 620–627.
[14] G.M. Elrayies, Microalgae: prospects for greener future buildings, Renew. Sustain. Energy Rev 81 (2018) 1175–1191.
[15] M.M. Ismail, T.M. Essam, Y.M. Ragab, A.E.-kB. El-Sayed, F.E. Mourad, Remediation of a mixture of analgesics in a stirred-tank photobioreactor using microalgal–bacterial consortium coupled with attempt to valorise the harvested biomass, Bioresour. Technol 232 (2017) 364–371.
[16] D.G. Noseda, M. Recúpero, M. Blasco, J. Bozzo, M.Á. Galvagno, Production in stirred-tank bioreactor of recombinant bovine chymosin B by a high-level expression transformant clone of *Pichia pastoris*, Protein Expr. Purif 123 (2016) 112–121.
[17] F. Cherubini, G. Jungmeier, M. Wellisch, T. Willke, I. Skiadas, R. Van Ree, et al., Toward a common classification approach for biorefinery systems, Biofuels Bioprod. Bioref 3 (2009) 534–546.
[18] T. Dong, E.P. Knoshaug, R. Davis, L.M. Laurens, S. Van Wychen, P.T. Pienkos, et al., Combined algal processing: a novel integrated biorefinery process to produce algal biofuels and bioproducts, Algal Res 19 (2016) 316–323.
[19] L. Xu, D.W.W. Brilman, J.A. Withag, G. Brem, S. Kersten, Assessment of a dry and a wet route for the production of biofuels from microalgae: energy balance analysis, Bioresour. Technol 102 (2011) 5113–5122.
[20] E.M. Grima, F.A. Fernández, A.R. Medina, Downstream processing of cell-mass and products, Handbook of Microalgal Culture: Biotechnology and Applied Phycology, Blackwell Science Ltd., 2004, p. 215.
[21] S. Bensalem, F. Lopes, P. Bodénès, D. Pareau, O. Français, B. Le Pioufle, Understanding the mechanisms of lipid extraction from microalga *Chlamydomonas reinhardtii* after electrical field solicitations and mechanical stress within a microfluidic device, Bioresour. Technol 257 (2018) 129–136.
[22] R. Ranjith Kumar, P. Hanumantha Rao, M. Arumugam, Lipid extraction methods from microalgae: a comprehensive review, Front. Energy Res 2 (2015) 61.
[23] C. Salinas-Salazar, J.S. Garcia-Perez, R. Chandra, C. Castillo-Zacarias, H.M. Iqbal, R. Parra-Saldívar, Methods for extraction of valuable products from microalgae biomass, Microalgae Biotechnology for Development of Biofuel and Wastewater Treatment, Springer, 2019, pp. 245–263.

[24] F.J. Baumgardt, A. Zandoná Filho, M.V. Brandalize, D.C. da Costa, N.R. Antoniosi Filho, P.C. Abreu, et al., Lipid content and fatty acid profile of *Nannochloropsis oculata* before and after extraction with conventional solvents and/or compressed fluids, J. Supercrit. Fluids 108 (2016) 89–95.
[25] K. Sharif, M. Rahman, J. Azmir, A. Mohamed, M. Jahurul, F. Sahena, et al., Experimental design of supercritical fluid extraction – a review, J. Food Eng 124 (2014) 105–116.
[26] F. Ghasemi Naghdi, L.M. González González, W. Chan, P.M. Schenk, Progress on lipid extraction from wet algal biomass for biodiesel production, Microb. Biotechnol 9 (2016) 718–726.
[27] E. Ryckebosch, K. Muylaert, I. Foubert, Optimization of an analytical procedure for extraction of lipids from microalgae, J. Am. Oil Chemists' Soc 89 (2012) 189–198.
[28] G. Zheng, C. Li, L. Guo, W. Ruo, S. Wang, Purification of extracted fatty acids from the microalgae Spirulina, J. Am. Oil Chemists' Soc 89 (2012) 561–566.
[29] J.-Y. Lee, C. Yoo, S.-Y. Jun, C.-Y. Ahn, H.-M. Oh, Comparison of several methods for effective lipid extraction from microalgae, Bioresour. Technol 101 (2010) S75–S77.
[30] S. Amin, Review on biofuel oil and gas production processes from microalgae, Energy Convers. Manag 50 (2009) 1834–1840.
[31] Y. Shen, W. Yuan, Z. Pei, E. Mao, Heterotrophic culture of *Chlorella protothecoides* in various nitrogen sources for lipid production, Appl. Biochem. Biotechnol 160 (2010) 1674–1684.
[32] P. Prabakaran, A.D. Ravindran, A comparative study on effective cell disruption methods for lipid extraction from microalgae, Lett. Appl. Microbiol 53 (2011) 150–154.
[33] A. Karemore, R. Sen, Downstream processing of microalgal feedstock for lipid and carbohydrate in a biorefinery concept: a holistic approach for biofuel applications, RSC Adv 6 (2016) 29486–29496.
[34] N. Parfati, K.C. Rani, The effect of coprocessed superdisintegrants ratio (crospovidone–sodium starch glycolate) to the physicochemical characteristics of atenolol orally disintegrating tablets, Asian J. Pharm. Clin. Res 11 (2018) 318–324.
[35] H. Eilertsen, S. Huseby, M. Degerlund, G. Eriksen, R. Ingebrigtsen, E. Hansen, The effect of freeze/thaw cycles on reproducibility of metabolic profiling of marine microalgal extracts using direct infusion high-resolution mass spectrometry (HR-MS), Molecules 19 (2014) 16373–16380.
[36] L.S. Sierra, C.K. Dixon, L.R. Wilken, Enzymatic cell disruption of the microalgae *Chlamydomonas reinhardtii* for lipid and protein extraction, Algal Res 25 (2017) 149–159.
[37] H. Zheng, J. Yin, Z. Gao, H. Huang, X. Ji, C. Dou, Disruption of *Chlorella vulgaris* cells for the release of biodiesel-producing lipids: a comparison of grinding, ultrasonication, bead milling, enzymatic lysis, and microwaves, Appl. Biochem. Biotechnol 164 (2011) 1215–1224.
[38] L. Chen, R. Li, X. Ren, T. Liu, Improved aqueous extraction of microalgal lipid by combined enzymatic and thermal lysis from wet biomass of *Nannochloropsis oceanica*, Bioresour. Technol 214 (2016) 138–143.
[39] H. Shokrkar, S. Ebrahimi, M. Zamani, Bioethanol production from acidic and enzymatic hydrolysates of mixed microalgae culture, Fuel 200 (2017) 380–386.
[40] D. Hernández, M. Solana, B. Riaño, M. García-González, A. Bertucco, Biofuels from microalgae: lipid extraction and methane production from the residual biomass in a biorefinery approach, Bioresour. Technol 170 (2014) 370–378.
[41] D. Esquivel-Hernández, V. López, J. Rodríguez-Rodríguez, G. Alemán-Nava, S. Cuéllar-Bermúdez, M. Rostro-Alanis, et al., Supercritical carbon dioxide and

microwave-assisted extraction of functional lipophilic compounds from *Arthrospira platensis*, Int. J. Mol. Sci 17 (2016) 658.
[42] E. Reverchon, I. De Marco, Supercritical fluid extraction and fractionation of natural matter, J. Supercrit. Fluids 38 (2006) 146–166.
[43] M. Šoštarič, D. Klinar, M. Bricelj, J. Golob, M. Berovič, B. Likozar, Growth, lipid extraction and thermal degradation of the microalga *Chlorella vulgaris*, Nat. Biotechnol 29 (2012) 325–331.
[44] R. Harun, M.K. Danquah, G.M. Forde, Microalgal biomass as a fermentation feedstock for bioethanol production, J. Chem. Technol. Biotechnol 85 (2010) 199–203.
[45] F. Wei, G.-Z. Gao, X.-F. Wang, X.-Y. Dong, P.-P. Li, W. Hua, et al., Quantitative determination of oil content in small quantity of oilseed rape by ultrasound-assisted extraction combined with gas chromatography, Ultrason. Sonochem 15 (2008) 938–942.
[46] G. Cravotto, L. Boffa, S. Mantegna, P. Perego, M. Avogadro, P. Cintas, Improved extraction of vegetable oils under high-intensity ultrasound and/or microwaves, Ultrason. Sonochem 15 (2008) 898–902.
[47] A. Sathish, R.C. Sims, Biodiesel from mixed culture algae via a wet lipid extraction procedure, Bioresour. Technol 118 (2012) 643–647.
[48] P. Biller, A.B. Ross, S. Skill, A. Lea-Langton, B. Balasundaram, C. Hall, et al., Nutrient recycling of aqueous phase for microalgae cultivation from the hydrothermal liquefaction process, Algal Res 1 (2012) 70–76.
[49] A.A. Peterson, F. Vogel, R.P. Lachance, M. Fröling, M.J. Antal Jr, J.W. Tester, Thermochemical biofuel production in hydrothermal media: a review of sub- and supercritical water technologies, Energy Environ. Sci 1 (2008) 32–65.
[50] A.R. Fajardo, L.E. Cerdan, A.R. Medina, F.G.A. Fernandez, P.A.G. Moreno, E.M. Grima, Lipid extraction from the microalga *Phaeodactylum tricornutum*, Eur. J. Lipid Sci. Technol 109 (2007) 120–126.
[51] P. Lopalco, S. Lobasso, F. Babudri, A. Corcelli, Osmotic shock stimulates de novo synthesis of two cardiolipins in an extreme halophilic archaeon, J. Lipid Res 45 (2004) 194–201.
[52] M. Guderjan, P. Elez-Martínez, D. Knorr, Application of pulsed electric fields at oil yield and content of functional food ingredients at the production of rapeseed oil, Innov. Food Sci. Emerg. Technol 8 (2007) 55–62.
[53] M. Guderjan, S. Töpfl, A. Angersbach, D. Knorr, Impact of pulsed electric field treatment on the recovery and quality of plant oils, J. Food Eng 67 (2005) 281–287.
[54] R. Chandra, C. Castillo-Zacarias, P. Delgado, R. Parra-Saldivar, A biorefinery approach for dairy wastewater treatment and product recovery towards establishing a biorefinery complexity index, J. Clean. Prod 183 (2018) 1184–1196.
[55] R. Chandra, J.S. Sravan, M. Hemalatha, S. Kishore Butti, S. Venkata Mohan, Photosynthetic synergism for sustained power production with microalgae and photobacteria in a biophotovoltaic cell, Energy Fuels 31 (2017) 7635–7644.
[56] R. Chandra, G.V. Subhash, S.V. Mohan, Mixotrophic operation of photo-bioelectrocatalytic fuel cell under anoxygenic microenvironment enhances the light dependent bioelectrogenic activity, Bioresour. Technol 109 (2012) 46–56.
[57] S.V. Mohan, M. Rohit, P. Chiranjeevi, R. Chandra, B. Navaneeth, Heterotrophic microalgae cultivation to synergize biodiesel production with waste remediation: progress and perspectives, Bioresour. Technol 184 (2015) 169–178.

Index

Note: Page numbers followed by "*f*" and "*t*" refer to figures and tables, respectively.

C

D

E

N

Kolibri 360 GmbH
Ferdinand-Jühlke-Straße 7
99095 Erfurt, Deutschland
produktsicherheit@kolibri360.de